Tools, Techniques and Concepts of Plant Genetics

Tools, Techniques and Concepts of Plant Genetics

Edited by **Kiara Woods**

SYRAWOOD
PUBLISHING HOUSE

New York

Published by Syrawood Publishing House,
750 Third Avenue, 9th Floor,
New York, NY 10017, USA
www.syrawoodpublishinghouse.com

Tools, Techniques and Concepts of Plant Genetics
Edited by Kiara Woods

International Standard Book Number: 978-1-68286-169-1 (Hardback)

Printed in the United States of America.

Contents

Preface

Plant genetics is a prominent field which focuses on heredity, inheritance and variations in plants. From theories to research to practical applications, case studies related to all contemporary topics of relevance in plant genetics have been included in this book. It provides significant information of this discipline by focusing on genetic engineering, GM crops, development of fast and reliable ozone screening method, cultivars and related fields. The chapters included herein primarily emphasize on application of biotechnology in crop plants. Coherent flow of topics, student-friendly language and extensive use of examples make this book an invaluable source of knowledge for students, researchers and academicians.

This book has been the outcome of endless efforts put in by authors and researchers on various issues and topics within the field. The book is a comprehensive collection of significant researches that are addressed in a variety of chapters. It will surely enhance the knowledge of the field among readers across the globe.

It gives us an immense pleasure to thank our researchers and authors for their efforts to submit their piece of writing before the deadlines. Finally in the end, I would like to thank my family and colleagues who have been a great source of inspiration and support.

Editor

Epistasis and genotype-by-environment interaction of grain yield related traits in durum wheat

Bnejdi Fethi and El Gazzah Mohamed

Laboratoire de Génétique et Biométrie, Département de Biologie, Faculté des Sciences de Tunis, Université Tunis, El Manar, Tunis 2092, Tunisia.

Genetic control of the number of heads per plant, spikelets per spike and grains per spike was studied in two durum wheat (*Triticum durum* Desf.) crosses, Inrat 69/Cocorit71 and Karim/Ben Bechir, respectively. Separate analyses of gene effects were done using means of four generations (parents P_1 and P_2, F_1, F_2, and the two reciprocal BC_1) at two sites. A three-parameter model was inadequate to explain all traits except number of heads per plant in Inrat 69/Cocorit 71 at one site. In most cases a digenic epistatic model explained variation in generation means. Dominance effects and dominance × dominance epistasis (I) were more important than additive effects and other epistatic components. Considering the genotype-by-environment interaction, the interactive model was applied and found adequate in all majority of cases except spiklets per spike and grains per spike in Inrat 69/Cocorit71. The results of this study indicate that maintenance of heterozygosity is useful for exploitation of epistatic effects and adaptability to varied environmental conditions for spiklets per spike and grain per spike in the cross Karim/Ben Bechir. Estimates of narrow-sense heritability indicated that the genetic effect was larger than the environmental effect. The additive effect was the largest component of genetic effects.

Key words: Genetic effects, epitasis, genotype-by-environment interaction, heritability, *Triticum durum*.

INTRODUCTION

Durum wheat (*Triticum durum* Desf.) is the most important cereal crop in Tunisia and is used primarily for couscous, macaroni and various types of bread (Bnejdi and El Gazzah, 2008). The development of high-yielding wheat cultivars is the major objective of breeding programs. Knowledge of the nature, magnitude of gene effects and their contribution to the control of metric traits is important in formulating an efficient breeding program for durum wheat genetic improvement. The inheritance of grain yield in wheat has been the subject of intensive studies (Grafius, 1959; Singh et al., 1985; Menon and Sharma, 1995; Sharma et al., 2002; Heidari et al., 2005; Rebetzke et al., 2006). Grain yield in wheat is determined by component traits and is highly complex. Reported heritability estimates indicate that certain morphological traits that influence grain yield in wheat are more heritable than yield itself. Ehdaie and Waines (1989) reported direct positive effects of number of heads per plant and number of grains per head on grain yield. The inheritance of quantitative traits has been described as a 'moving target' since they are affected not only by the actions of multiple individual genes, but also by the interactions between genes and environmental factors (Lewis and John, 1999). Some genetic statistical models have been devised for plants and animals to estimate the parameters of genetic components (Mather and Jinks, 1982; Kearsey and Pooni, 1996; Lynch and Walsh, 1998; Chalh and El Gazzah, 2004). The method has been generally used to study quantitative trait inheritances and generation means analysis (Mather and Jinks, 1971), which allows testing the linear components of genotypic means. The present investigation of four generations (parental, F_1, F_2 and BC_1) studied gene action and heritability of four traits at two sites.

---

*Corresponding author. E-mail: fethibnejdi@yahoo.fr.

MATERIALS AND METHODS

The study was carried out at two locations; El Kef characterized by loam soil and a semi arid climate with an annual rainfall of about 400 - 500 mm and Tunis characterized by clayey soil and situated in the sub humid region with 600 - 700 mm under rain-fed conditions in 2005 - 2006. Parental lines were chosen for their different yield. Plants were grown in a randomised complete block design with two replications. From each plant the following data were obtained: number of heads per plant, spikelets per spike and number of grains per spike. The number of plants evaluated varied depending on the generation and was greater in generations with greater segregation, such as the F_2, BC_1P_1 and BC_1P_2.

Transforming the data by log, square root, arc-sine and arc-sine of square root had no effect on data distribution or in removing epistatic effects. Separate analysis of variance by population and by site using SAS Proc GLM (SAS, 1990) indicated that replication and generation × replications effects were not significant. Therefore generation means analysis was conducted without adjusting the data for replication.

Statistical analysis

Gene effects

The means of different generations were analyzed by a joint scaling test using the weighted least squares method (Mather and Jinks, 1982; Kearsey and Pooni, 1996; Lynch and Walsh, 1998). The observed generation means were used to estimate the parameters of a model consisting only of mean (m), additive and dominance genetic effects. The estimated parameters were used in turn to calculate the expected generation means. The goodness-of-fit between observed and expected was tested; a significant chi-squared value indicated a significant difference between the observed and expected generation means, which implied that a simple additive model did not explain the data. When the additive-dominance model was found to be insufficient, then additive × additive, additive × dominance and dominance × dominance digenic epistatic parameters were added. If a digenic epistatic parameter was not significant then it was omitted and the best fit model was applied. The weighted least-squares model that incorporates additive, dominance and digenic epistatic effects is (Hayman, 1958; Mather and Jinks, 1982; Kearsey and Pooni, 1996; Lynch and Walsh, 1998):

$$X = (C'WC)^{-1} (C'WY)$$

where X is the vector of mean, additive, dominance, additive × additive, additive × dominance and dominance × dominance parameters. W is the diagonal matrix of weights (that is, the reciprocals of the variance of generation means) and Y is the vector of generation means. The variances of the parameter estimates can be obtained from the diagonal elements of $(C'WC)^{-1}$. The expected means of the six generations were calculated using the parameter estimates, the goodness-of-fit of the observed generation means was tested with the chi-squared statistic. The significance of each parameter was determined by t-test.

Genotype-by-environment interaction

The weighted least squares method was also used to estimate environmental and genotype-by-environment interactions. This technique was applied to parents and F_1 only (Mather and Jinks, 1971). The analysis was done in three stages by three different models: the no-interactive model, the interactive model and the best-fit model.

The no-interactive model: this model involves four parameters mean (m), additive (d), dominance (h) and environmental (e) effects. If the chi-squared test revealed that the simpler model was

inadequate then the interactive model was applied. This model involves an interactive-parameter model and two genetic environment interactions; additive × environment interactions (ed) and dominance × environment (eh). When additive × environment interaction (ed) or dominance × environment interaction (eh) were not significant by t-test in the interactive model, then it was omitted and the best-fit model was applied.

Heritability

Additive variance is a component of the total genetic variance and cannot easily be distinguished from the dominance variance and environmental components. However, an estimate of the additive variance can be obtained using F_2 and backcross generations' data to calculate narrow-sense heritability:

$$\sigma^2_A = 2\sigma^2_{F2} - (\sigma^2_{BC1P1} + \sigma^2_{BC1P2})$$

Narrow-sense heritability (h^2) was estimated using F_2 and backcross generations' variance components as described by Warner (1952).

$$h^2 = [2\sigma^2_{F2} - (\sigma^2_{BC1P1} + \sigma^2_{BC1P2})] / \sigma^2_{F2}$$

Dominance variance was estimated as:

$$\sigma^2_D = (\sigma^2_{BC1P1} + \sigma^2_{BC1P2}) - \sigma^2_{F2} - \sigma^2_E$$

The environmental variance was estimated as

$$\sigma^2_E = (\sigma^2_{P1} + \sigma^2_{P2} + 2\sigma^2_{F1}) / 4 \text{ (Wright, 1968).}$$

RESULTS

Parental means and their variances in both environments are given in Table 1. In all cases, depending on the site, the means of the parents in each cross showed a tendency to be more extreme. The means of backcrosses BC_1P_1 and BC_1P_2 tended to be located close to those of their respective recurrent parents. These results confirmed the choice of parents for the present study. For most traits, F_1 generation means were higher than the mid-parent value. The F_1 and F_2 generations' means were not significantly different in the majority of cases for grains per spike and spikelets per spike. For number of heads per plant, the F_2 generation mean was significantly different from the F_1 generation mean in the majority of cases.

The estimates of the main and first order interactions and the test of the fitness of models are presented in Table 2. The adequacy of models and the magnitude of gene action depended on the cross and experimental site. The joint scaling test indicated that the additive-dominance model (three-parameter model) was inadequate to explain the nature of gene action for all traits, for both crosses at both sites, except for number of heads per plant in the Inrat 69/Cocorit 71 cross at Tunis, indicating that epistatic effects were involved in the inheritance of all traits studied.

For grains per spike at Tunis and spikelets per spike at El Kef, both models did not explain variation in generation

Table 1. Means and variances for all traits for different generations of Inrat 69/Cocorit 71 and Karim/Ben Bechir crosses grown at Tunis and El Kef sites from two replications in 2005 - 2006.

| Generation | Inrat 69/Cocorit 71 | | Karim/Ben Bechir | |
	El Kef	Tunis	El Kef	Tunis
Number of heads per plant				
P_1	6.45 ± 0.78 (20)[y]A	7.85 ± (20)A	9.35 ± 0.45 (20)A	10.15 ± 0.55(20)A
BC_1	6.02 ± 1.44 (50)AB	7.08 ± (50)B	7.82 ± 2.51(50)B	7.36 ± 0.92 (50)C
F_1	5.24 ± 1.10 (25)CD	6.44 ± (25)C	8.40 ± 1.58 (25)B	7.96 ± 1.20(25)B
F_2	5.69 ± 2.32 (72)BC	5.04 ± (72)DE	6.11 ± 3.42(72)C	6.11 ± 1.28(72)D
BC_2	4.86 ± 1.12 (43)D	4.76 ± (43)D	5.58 ± 1.78 (43)C	6.23 ± 0.84(43)D
P_2	4.04 ± 0.44 (21)E	4.28 ± 1.01 (21)E	5.66 ± 0.53 (21)C	5.23 ± 0.79 (21)E
Spikelets per spike				
P_1	23.00 ± 0.31(20)A	20.00 ± 0.42 (20)A	18.35 ± 0.55(20)A	20.40 ± 0.77 (20)A
BC_1	21.27 ± 1.45 (44)B	19.48 ± (35)AB	18.37 ± 0.75(40)A	19.26 ± 0.76 (50)B
F_1	20.15 ± 0.66(20)C	19.15 ± 0.97(20)B	18.30 ± 0.85 (20)A	19.30 ± 0.74 (20)B
F_2	20.51 ± 1.92(68)C	19.65 ± 2.76 (76)AB	18.28 ± 1.34 (50)A	17.90 ± 1.29 (74)C
BC_2	18.94 ± 0.91 (38)D	18.11 ± 1.91 (43)C	16.94 ± 1.31 (39)B	18.20 ± 0.86 (24)C
P_2	18.35 ± 0.97 (20)E	17.00 ± 0.42 (20)D	15.25 ± 0.82 (20)C	17.10 ± 0.72 (20)D
Grains per spike				
P_1	55.85 ± 7.71 (20)A	63.90 ± 12.83 (19)A	50.50 ± 13.21 (20)A	54.60 ± 6.88 (20)A
BC_1	51.42 ±14.90 (20)B	50.62 ± 43.09(50)A	45.70 ± 18.82 (50)B	49.76 ± 11.57(50)AB
F_1	53.72 ± 8.04(20)AB	57.00 ± 10.91 (25)CD	47.00 ± 11.58 (25)B	50.20 ± 22.08(25)AB
F_2	48.57 ± 18.24 (20)C	48.55 ± 57.64 (100)CD	42.18 ± 32.53 (100)C	47.26 ± 45.02(100)AB
BC_2	47.16 ± 39.57(20)C	46.53 ± 46.76 (52)D	45.15 ± 35.30(52)B	43.98 ± 40.68 (52)B
P_2	44.45 ± 6.36(20)D	49.10 ± 11.88(20)CD	37.10 ± 10.72(20)D	44.80 ± 7.74 (20)B

P_1 = better parent, P_2 = worse parent.
y = number of random plants for each generation in parentheses.
For each trait, means within a column with different letters (e.g. A, B, C, D or E) following them are significantly different using Duncan's multiple range test ($P < 0.05$).

means in the Karim/Ben Bechir cross. In the other cases the epistatic model adequately explained variation between generation means in the two crosses at both sites. The magnitude of dominance (h) and dominance × dominance (l) and additive × additive (i) when significant where more important than additive (d) effects.

The estimates of the effects of genetic, environment, genotype-by-environment interaction and the test of fitness of the model are given in Table 3. Due to the presence of allelic and non-allelic interactions analysis was invoked only in non-segregating generations. This study revealed that the non-interactive model was inadequate in all cases. Therefore the interactive model was tested and found adequate in four cases. For the cross Inrat 69/Cocorit 71 the interactive model failed to explain variation in generation mean for spiklets per spike and grains per spike. Significant environment (e) type effect was observed for all the traits except number of heads per plant in the Karim/Ben Bechir cross. The additive × environment (ed) effect was present especially in the cross Inrat 69/Cocorit71. The environment × dominance interaction (eh) effect was present in the tow crosses for

spiklets per spike and grains per spike (Table 3).

Estimates of variance components were used to calculate h^2 for both crosses and four traits (Table 4). For all traits the additive variances were positive; 0.78 - 2.60 for number of heads per plant, 10.93 - 45.98 for grains per spike and 0.63 - 1.47 for spikelets per spike. Dominance variance was negative in the majority of cases. Environmental variance was 0.86 - 1.03 for number of heads per plant, 7.54 - 14.69 for grains per spike, and 0.65 - 0.77 for spikelets per spike. For all traits, h^2 was dependent upon the cross and site and ranged from moderate to high.

DISCUSSION

There were significant differences among generation means for the three analyzed traits in all cases, revealing genetic diversity for these attributes in the materials, thus validating the genetic analysis of the traits following the technique of Mather and Jinks (1982). The analysis of gene effects revealed that both additive and dominance

Table 2. Estimates of gene effects for three quantitative traits for Inrat 69/Cocorit and Karim/Ben Bechir crosses at Tunis and El Kef sites from two replications in 2005–2006.

Site	Cross	m	d	h	i	l	j	x^2(df)
Number of heads per plant								
El Kef/	Inrat 69/Cocorit 71	5.32**	1.21**	0.15	-	-	-	4.48(3)
	Karim/Ben Bechir	5.22**	1.88**	0.36	2.27*	2.8*	-	1.53 (1)
Tunis/	Inrat 69/Cocorit 71	0.77	1.85**	10.28**	5.28**	-4.62**	-	0.84 (1)
	Karim/Ben Bechir	4.2**	2.46**	3.63**	3.44**	-	-2.68**	3.58 (1)
Grains per spike								
El Kef/	Inrat 69/Cocorit 71	40.08**	5.7**	12.75**	9.13**	-	-5.64*	0.69 (1)
	Karim/Ben Bechir	141.84**	6.7**	29.28**	12.98**	-13 .09*	-12.3**	-
Tunis /	Inrat 69/Cocorit 71	56.5**	7.4**	-32.23**	-	32.73**	-6.63*	0.810^{-3}(1)
	Karim/Ben Bechir	49.69**	5.02**	-10.76**	-	11.27**	-	0. 7 (2)
Spikelets per spike								
El Kef/	Inrat 69/Cocorit 71	22.29**	2.32**	-4.97*	-1.61**	2.82*	-	1.710^{-6}(1)
	Karim/Ben Bechir	19.31**	1.51**	-3.12	-2.51**	2.10*	-	0.24 (1)
Tunis/	Inrat 69/Cocorit 71	21.92**	1.48**	-6.28*	-3.42**	3.51*	-	0.13 (1)
	Karim/Ben Bechir	100.75**	1.65**	6.01**	3.31**	-2.15*	-119*	-

Mean (m), additive (d), dominance (h), additive × additive (i), additive × dominance (j) dominance × dominance (l) genetic effects for the model. y = m + d + h + i + j + l, where y is the generation mean.df: degrees of freedom, calculated as the number of generations minus the number of estimated genetics parameters.
*, ** indicates means and gene effects are statistically different from zero at $P < 0.05$ and $P < 0.01$, respectively.

Table 3. Estimates of the genetic, environmental and genotype-by-environment interaction components of generation means.

Cross	m	d	h	e	eh	ed	x^2(df)
Number of heads per plant							
Inrat 69/Cocorit71	5.67***	1.47***	0.19	0.47***		0.29**	2.01(1)
Karim/Ben Bechir	7.58***	2.15***	0.56**			0.31***	2.95(2)
Spikelets per spike							
Inrat 69/Cocorit71	20.49***	1.92***	0.06	1.08***	0.41***	-0.58***	
Karim/Ben Bechir	17.76***	1.59***	1.03***	0.97***	0.47**	-	0.27(1)
Grain per spike							
Inrat 69/Cocorit71	58.32***	6.55***	2.03***	3.17***	0.85*	-1.53*	-
Karim/Ben Bechir	47.14***	5.25**	2.15**	2.53***	-1.63*	-	2.32(1)

m: mean, d: additive effect, h: dominance effects, e : environment effects, eh: environment dominance effects interaction, ed : environment additive effects interaction for the model y= m + d + h + e + eh+ ed, where y equals the non-segregating generation mean. df: degrees of freedom, calculated as the number of generation minus the number of estimated genetic parameters.
* ** ***indicates means and gene effects are statistically different from zero at P <0.05, 0.01, 0.001 respectively.

effects were involved in the inheritance of most traits (Table 2). The dominance effects were greater in most cases than additive gene effects. This study also revealed a preponderance of dominance gene effects in the expression of all traits. The higher estimates for dominance than for additive effects, for the majority of

Table 4. Estimates of additive (σ^2_A), dominance (σ^2_D), and environmental (σ^2_E) variances, narrow-sense heritabilities (h^2) and genetic gain through selection (Gs) for three traits of Inrat 69/Cocorit 71 and Karim/Ben Bechir crosses at two sites (Tunis and El Kef) from two replications in 2005-2006.

Cross	Site	σ^2_A	σ^2_D	σ^2_E	h^2	Gs
Number of heads per plant						
Inrat 69/Cocorit 71	El Kef	2.08	−0.6	0.86	0.89	3.54
	Tunis	0.96	−0.17	0.90	0.56	1.64
Karim/Ben Bechir	El Kef	2.60	−0.21	1.03	0.76	4.42
	Tunis	0.78	−0.44	0.93	0.61	1.33
Grains per spike						
Inrat 69/Cocorit 71	El Kef	45.98	−13.96	7.54	1.16	78.18
	Tunis	25.42	20.57	11.63	0.44	43.22
Karim/Ben Bechir	El Kef	10.93	9.82	11.77	0.33	16.58
	Tunis	37.78	−7.46	14.69	0.83	64.23
Spikelets per spike						
Inrat 69/Cocorit 71	El Kef	1.47	−0.20	0.65	0.76	2.50
	Tunis	1.17	0.88	0.69	0.42	1.99
Karim/Ben Bechir	El Kef	0.63	−0.05	0.77	0.46	1.07
	Tunis	0.94	−0.4	0.75	0.73	1.61

Variance components calculated as follows: $\sigma^2_E = (\sigma^2_{P1} + \sigma^2_{P2} + 2\sigma^2_{F1})/4$; $\sigma^2_A = 2\sigma^2_{F2} - (\sigma^2_{BC1P1} + \sigma^2_{BC1P2})$; $\sigma^2_D = (\sigma^2_{BC1P1} + \sigma^2_{BC1P2}) - \sigma^2_{F2} - \sigma^2_E$.
Heritabilities calculated as follows: $h^2 = [2\sigma^2_{F2} - (\sigma^2_{BC1P1} + \sigma^2_{BC1P2})]/\sigma^2_F$.
Genetic gain calculated as follows: Gs = (1.76) (h^2) (σ^2_{F2}).

traits, indicated that the parents were in dispersion phase and that there was an accumulation of dominant parental genes in the hybrids (Dhanda and Sethi, 1996).

The additive-dominance model was accurate for the number of heads per plant in one cross (Inrat 69/Cocorit 71) at Tunis, similar to results of Kashif and Khaliq (2003) in bread wheat. For digenic interaction, the model was adequate in the Inrat 69/Cocorit 71 cross at both sites. By contrast, in the Karim/Ben Bechir cross the digenic model failed to explain variation between generations in two cases (Table 2), indicating more complex mechanisms of genetic control. The presence of digenic interaction in durum wheat has been reported for spike length (Sharma et al., 2003), grains per spike (Sharma and Sain, 2004) and number of heads per plant (Singh et al., 1986). Epistasis has been reported for many traits in a number of crops: barley (Kularia and Sharma, 2005), maize (Melchinger et al., 1987), sorghum (Finkner et al., 1981), rice (Saleem et al., 2005) and durum wheat (Bnejdi and El Gazzah, 2008).

None of the models explained variation between generation means in the Karim/Ben Bechir cross for grains per spike at Tunis and spikelets per spike at El Kef, suggesting there were higher order interactions or linkage effects. To discover the cause of the model failure further analyses of more generations are necessary.

Trigenic interactions in durum wheat have been reported for grains per spike (Sharma and Sain, 2004). In the present study, mostly the variation in generation means fitted a digenic epistatic model, depending on the cross and site. This indicates that improvements for these traits would be moderately difficult, compared to fitting an additive-dominance model (best from a breeder's point of view); however, this is better than the presence of trigenic interaction.

The majority of traits were largely influenced by dominance × dominance and additive × additive gene effects in both crosses and both sites, with dominance × dominance (l) effects being more pronounced than additive × additive (i) effects (Table 2). This finding is in accordance with those of Dhanda and Sethi (1996), but not Novoselovic et al. (2004) in bread wheat. Mostly the dominance × additive effect did not significantly contribute to the genetic control of these traits.

The non-segregating generations in the present study show that the estimates of genotype by environment interaction components were dependant upon the cross. For Inrat 69/Cocorit 71 the interactive model failed to explain variation of generation mean in tow cases, indicating the presence of more mechanism in the control of this traits. Therefore selection based on the cross Karim/Ben Bechir was better than the cross Inrat 69/Cocorit71.

For Karim/Ben Bechir cross environmental × dominance

(eh) effects are highly significant for spiklets per spike and grain per spike. By contrast, estimates of environment × additive (ed) effects were not significantly different from zero and indicated that for these traits the heterozygote show a greater interaction with the environment than do the homozygote (Table 4). This condition is more favorable than the presence of interaction of homozygote with environment; since homozygous populations are less adaptable than heterozygous populations to varied environmental conditions, as reported by Kaczamarek et al. (2002).

The present study revealed that both additive and non-additive components of genetic variances were involved in governing yield components; with dominance effects and dominance × dominance epistasis more important than additive effects or other epistatic components. However, the heterozygote showed greater interaction with the environment than the homozygote. Maintenance of heterozygosity can give two advantages, the exploitation of epistatic effects and adaptability to varied environmental conditions. Successful methods will be those that can map-up the gene, to form superior gene combinations interacting in a favorable manner and at the same time maintain heterozygosity. This objective can be achieved by restricted recurrent selection (Joshi 1979) and/or di-allele selective mating (Jensen, 1978) methods.

Narrow-sense heritability is important to plant breeders, because effectiveness of selection depends on the additive portion of genetic variation in relation to total variance (Falconer, 1960). In our results, moderate to high values for narrow-sense heritability suggested a considerable participation of genetics in the phenotypic expression of traits and that selection for all traits could be efficient.

ACKNOWLEDGEMENTS

We thank Pr. Monneveux Philippe (CIMMYT Wheat Program, A.P. 6-641, 06600 Mexico D.F., Mexico) for helpful editorial comments and corrections of our manuscript.

REFERENCES

Bnejd F, El Gazzah M (2008). Inheritance of resistance to yellowberry in durum wheat. Euphytica 163: 225-230.
Chalh A, El Gazzah M (2004). Bayesian estimation of dominance merits in noninbred populations by using Gibbs sampling with two reduced sets of mixed model equations. J. Appl. Genet. 43: 471-488.
Dhanda SS, Sethi GS (1996). Genetics and interrelationships of grain yield and its related traits in bread wheat under irrigated and rainfed conditions. Wheat Inf. Serv. 83: 19-27.
Ehdaie B, Waines JG (1989). Genetic variation, heritability and path-analysis in landraces of bread wheat from southwestern Iran. Euphytica 41: 183-190.
Falconer DC (1960). Introduction to quantitative genetics. Ronald Press, New York p. 365.

Finkner RE, MD Finkner, Glaze RM (1981). Genetic control for percentage grain protein and grain yield in grain sorghum. Crop Sci. 21: 139-142.
Grafius JE (1959). Heterosis in barley. Agron. J. 51: 551-554.
Hayman BI (1958). The separation of epistatic from additive and dominance in generation means. Heredity 12: 371-390.
Heidari B, Saeidi G, Sayed-Tabatabaei BE, Suenaga K (2005). The interrelationships of agronomic traits in a doubled haploid population of wheat. Czech J. Genet. Plant Breed 41: 233-237.
Jensen NF (1978). Composite breeding methods and the diallel selective mating system in cereals. Crop Sci. 9: 622-626.
Joshi AB (1979). Breeding methodology for autogamous crops. Indian J. Genet. 39: 567-578.
Kaczamarek Z, Surma M, Adamski T, Jezowski S, Madajewski R, Krystkowiak K, Kuczynska A (2002). Interaction of gene effects with environments for malting quality of barley doubled haploids. Theor. Appl. Genet. 43: 33-42.
Kashif M, Khaliq I (2003). Mechanisms of genetic control of some quantitative traits in bread wheat. Pakistan J. Biol. Sci. 6: 1586-1590.
Kearsey MJ, Pooni HS (1996). The genetical analysis of quantitative traits. 1st edition. Chapman and Hall, London p. 381.
Kularia RK, Sharma AK (2005). Generation mean analysis for yield and its component traits in barley (Hordeum vulgare L.). Indian J. Genet. Pl. Breed. 65: 129-130.
Lewis NL, John D (1999). Epistatic and environmental interactions for quantitative trait loci involved in maize evolution. Res. Camb. 74: 291-302.
Lynch M, Walsh B (1998). Genetics and analysis of quantitative traits. Sinauer Associates Inc., Sunderland, MA p. 980.
Mather K, Jinks JL (1971). Biometrical genetics. 2nd edition. Chapman and Hall, London p. 382.
Mather K, Jinks JL (1982). Biometrical genetics. 3rd edition. Chapman and Hall, London p. 396.
Melchinger AE, Geiger HH, Seitz G (1987). Epistasis in maize (Zea mays L.) III. Comparison of single and three-way crosses for forage traits. Plant Breed 98: 185-193.
Menon U, Sharma SN (1995). Inheritance studies for yield and yield component traits in bread wheat over the environments. Wheat Inf. Serv. 80: 1-5.
Novoselovic D, Baric M, Drenzner G, Lalic A (2004). Quantitative inheritance of some wheat plant traits. Genet. Mol. Biol. 27: 92-98.
Rebetzke GJ, Richards RA, Condon AG, Farquhar GD (2006). Inheritance of carbon isotope discrimination in bread wheat (Triticum aestivum L.). Euphytica 150: 97-106.
Saleem MY, BM Atta, Cheema AA (2005). Detection of epistasis and estimation of additive and dominance components of genetic variation using triple test cross analysis in rice (Oryza sativa L.) Caderno de Pesquisa Sér. Bio., Santa Cruz do Sul 17: 37-50.
SAS (1990). SAS/STAT user's guide. Version 6. 4th ed. SAS institute, Cary, NC. p. 376.
Sharma SN, Sain RS (2004). Genetics of grains per spike in durum wheat under normal and late planting conditions. Euphytica 139: 1-7.
Sharma SN, Sain RS, RK Sharma (2002). Gene system governing grain yield per spike in macaroni wheat. Wheat Inf. Serv. 94: 14-18.
Sharma SN, Sain RS, Sharma RK (2003). Genetics of spike length in durum wheat. Euphytica 130: 155-161.
Singh G, Bhullar GS, Gill KS (1986). Genetic control of grain yield and its related traits in bread wheat. Theor. Appl. Genet. 72: 536-540.
Singh VP, Rana RS, Chaudhary MS, Singh D (1985). Genetics of kernel weight under different environments. Wheat Inf. Serv. 60: 21-25.
Warner JN (1952). A method for estimating heritability. Agron. J. 44: 427-430.
Wright S (1968). Evolution and the genetic of population. Vol 1. Genetic and biometric foundation. University of Chicago Press, Chicago p. 469.

Comparative analysis of genetic and morphologic diversity among quinoa accessions (*Chenopodium quinoa* Willd.) of the South of Chile and highland accessions

Leonardo Anabalón Rodríguez[1] and Max Thomet Isla[2*]

[1]Laboratorio de Mejoramiento Vegetal, Escuela de Agronomía. Universidad Católica de Temuco. Temuco. Chile.
[2]CET Sur. Temuco. Chile.

Quinoa (*Chenopodium quinoa* Willd.) is a widely consumed food crop and a primary protein source for many of the indigenous inhabitants of the Andean region in South America. Identification of quinoa cultivars has been based on phenotypic characters. In the present work, the level of polymorphism and the genetic relationship were studied by means of molecular markers using the amplified fragment length polymorphism (AFLP) technique and twenty morphological characters. Fourteen accessions of quinoa collected in the Araucania and Los Rios Regions, three Andean accessions, and one commercial cultivar were analyzed. Two wild parents were included as outgroup controls. A similarity tree-diagram was made, based on all the AFLP bands generated in the range between 70 and 300 base pairs. With these tools, it was possible to identify molecular differences and similarities that might be associated with important morphological traits such as grain color, panicle color, phenology and geographic distribution.

Key words: *Chenopodium quinoa*, cluster analysis, molecular markers.

INTRODUCTION

Chenopodium quinoa is one of the most important food crops in the Andean highland of South America. We can find it in Chile, Argentina, Ecuador, Perú and Bolivia. It belongs to the Amaranthaceae family (Kadereit et al. 2003), which traditionally includes the economically important species spinach (*Spinacea oleracea* L.) and sugarbeet (*Beta vulgaris* L.). Quinoa is an allotetraploid (2n = 4x = 36), and thus exhibits disomic inheritance for most qualitative traits (Ward 1998; Maughan et al., 2004). The small achene fruits contain an excellent balance of carbohydrates, lipids and protein, making it an excellent food source (Chauhan et al., 1999). This fruit provides an ideal balance of all 20 essential amino acids (Ruas et al., 1999). Quinoa is one of the only few crop plants adapted to the extreme conditions that characterize this region (Prado et al., 2000). There is a great interest for this crop in developing countries since it is considered as one of the most important crops involved in feed condi-

tions improvement of this century. Although increasing quinoa productivity is a primary food-security issue in the Andean Region, limited research on quinoa genetics and plant breeding has been conducted. In general, cultivated quinoa displays a genetic diversity, mainly represented in an ample range of characters like plant coloration, flowers protein content, seeds, saponin content and leaves calcium oxalates content, which allows obtaining a wide range of adaptability to agroecological conditions. Within the diversity centers, the center of Peru (Huan cayo, Ayacucho, Cajamarca), the Ecuadorian Mountain range, the Argentine Northeast, the South of Chile and of Colombia (Grass, Nariño and Cudinamarca)are identified (Jacobsen, 2003). Great ignorance still exists and little investigation has been carried out regarding varieties in the South of Chile, conserved and selected by Mapuche communities and other smallholder farmers. The adaptation capacities of quinoa are huge since we can find varieties developed from sea level up to 4,000 m above, and from 40°S to 2°N of latitude (Jacobsen, 2003). Native Chilean varieties are adapted to other latitudes due to their absence or minor sensitivity to photoperiod during

---

*Corresponding author. E-mail: mthomet@cetsur.org.

the grain filling. In that sense quinoa crops at sea level are less sensitive to damage caused by the conjunction between long high-temperature days, which would explain his extreme adaptability (Bertero, 2001). In spite of the importance that has been attributed to the quinoa crop in the Region of the Andes, only few lines of investigation exist to establish applied genetics and molecular characteristics of this crop. Nevertheless, initiatives to adjust the crops and to open new markets have even made possible to generate some beginning improvement programs outside the original places. These programs aimed to prioritize the increase of grain production, resistance to diseases, tolerance to hydric stress and control of the saponines content (Ochoa et al. 1999). Molecular markers are an effective way to enhance breeding efficiency (Lande, 1991; Patterson et al., 1991; Staub et al., 1996). Up to now, only a few researchers have reported the development and use of molecular markers in quinoa. Wilson (1988a) used data of alozymes in quinoa to confirm the genetic difference between ecotypes of the plateau and valleys. Maughan et al. (2004) made a genetic map. Ruas et al. (1999) used RAPD markers to detect the degree of polymorphism between cultivated and wild species of quinoa. In order to establish genotypic differences between quinoas (Chenopodium quinoa Will.) from the North and the South of Chile, Wilckens et al. (1996) reported to use stored proteins of seed (isoenzimes). Recent studies of characterization of quinoa germplasm have been developed to create a map of microsatellite markers (Mason et al., 2005). In Chile, genetic diversity studies regarding quinoa are only focused on morphologic and biochemical comparison between Andean and local varieties. Nevertheless the molecular markers offer could enable us to obtain a greater and deeper accuracy compared to biochemical markers and therefore, to identify and characterize highly related individuals. Both in Chile and in the rest of the world, molecular markers are considered as excellent and accurate techniques to study genetic diversity and improvement based on DNA analysis and are used for a wide range of crops (Pillay and Myers 1999; Mason et al. 2005; Solano et al. 2007) : Solanum tuberosum (Solano et al., 2007), Lens culinaris (Sharma et al. 1996), Glycine max (Maughan et al., 1996), Lactuca spp. (Hill et al. 1996) and Hordeum vulgare (Becker et al., 1995). Among the molecular markers available with greater ability of accuracy and reproducibility, AFLP (Amplified Fragment Length Polymorphism) analysis represents the most recent technology for taining a great number of molecular markers in prokaryote and eukaryote genomes (Hartings et al., 2008). The AFLP technique (Vos et al., 1995) generally produces between 50 and 100 scorable fragments per polymerase chain reaction (PCR (Maughan et al. 1996). On the other hand, the low cost, easy use, and great quantity of polymorphism of AFLP markers make them very useful for crop analysis investigated in developing countries (Maughan et al. 2004). The use of this method

is also highly relevant since it was used to make the first genetic quinoa map. This technology based on PCR involves three essential steps. First, the digestion of genomic DNA by two restriction enzymes, followed by the ligation of adapter to the extremities of restricted fragments and finally a selective amplification with two conescutive reactions of PCR. The PCR product is denatured and separated in a polyacrylamide gel in which there are usually 60 to 80 bands per DNA sample. The objectives of this study were to (i) analyze the genetic diversity of 14 coastal accessions of (C. quinoa Willd.) using AFLP markers and morphological data, (ii) determine the genetic relationships among southern accessions and (iii) to compare these data with highland accessions and commercial varieties.

MATERIALS AND METHODS

Plant material

The material used (quinoa varieties) came from small-holder farmers of diverse localities in the regions of Araucania and Los Rios. These varieties are ancestral ones, inherited from generation and generation of use and cultivation free from the diverse agricultural modernization programs (Thomet et al., 2003). During this investigation, we evaluated fourteen local varieties (Table 1), three varieties from the Tarapacá Region (Iquique-Universidad Arturo Prat), one enrolled variety (Regalona-Baer) and two outgroup controls (C. album and C. ambrosoide (Figure 1).

Morphological character analysis

Twenty characters (Table 2) of the all varieties and controls, codifying according to description assigned were described and analyzed for quinoa according to IPGRI (1981), then to be exported in a matrix of number-data. The morphologic data were subjected to cluster analysis. A standardization of the data by using Z-scores function was made. Average was applied to the technique of hierarchical conglomerate linkage using a matrix of quadratic similarity applying the Euclidean distances.

DNA isolation

Approximately 100 - 200 mg of material was freeze-dried and ground in liquid nitrogen with a mortar and pestle. Genomic DNA was isolated with Plant DNAzol® following the manufacturer's instructions. RNA was further eliminated by treatment with RNase. The quality and concentration of DNA was evaluated by agarose gel electrophoresis and spectrophotometry. Two independent extractions were performed on each accession.

AFLP analysis

AFLP reactions were carried out using the AFLP Analysis System I kit (Invitrogen Life Technologies) according to the manufacturer's instructions. Each reaction was repeated at least once to verify the AFLP patterns generated.

Approximately 500 ng of genomic DNA was digested for 2 h at 37°C using 2 µl EcoRI/MseI restriction enzyme solution. The AFLP procedure (Vos et al. 1995) was carried out as described by Arens et al. (1998) with slight modifications. Briefly, the entire genomic DNA (400 - 500 ng) was digested with EcoRI and MseI, followed by

Table 1. Material plant included in the analysis.

Sample Nº	ID Accession	Common name	Species	Localities	Altitud masl	Material
1	KM 01	Yellow	*Chenopodium quinoa*	Lautaro	243	Cotiledon
2	KM 02	Red	*Chenopodium quinoa*	Nueva Imperial	51	Cotiledon
3	KM 03	Red	*Chenopodium quinoa*	Ercilla	286	Cotiledon
4	KM 04	Red	*Chenopodium quinoa*	Liquiñe	631	Cotiledon
5	KM 05	Red	*Chenopodium quinoa*	Melipeuco	549	Cotiledon
6	KM 06	Mixture	*Chenopodium quinoa*	Panguipulli	420	Cotiledon
7	KM 07	Red	*Chenopodium quinoa*	Ercilla	290	Cotiledon
8	KM 08	Red	*Chenopodium quinoa*	Vilcún	355	Cotiledon
9	KM 09	Red	*Chenopodium quinoa*	Vilcún	338	Cotiledon
10	KM 10	Yellow	*Chenopodium quinoa*	Vilcún	347	Cotiledon
11	KM 11	Red	*Chenopodium quinoa*	Lautaro	233	Cotiledon
12	KM 12	Yellow	*Chenopodium quinoa*	Temuco	171	Cotiledon
13	KM 13	Red	*Chenopodium quinoa*	Temuco	165	Cotiledon
14	KM 14	Red	*Chenopodium quinoa*	Ercilla	279	Cotiledon
15	R-B	Regalona - Baer	*Chenopodium quinoa*	Temuco	161	Cotiledon
16	R01	Red I	*Chenopodium quinoa*	Iquique- Sector Plomo Loma	3.795	Cotiledon
17	A01	Yellow I	*Chenopodium quinoa*	Iquique- Sector Plomo Loma	3.743	Cotiledon
18	A02	Yellow II	*Chenopodium quinoa*	Iquique- Sector Plomo Loma	3.740	Cotiledon
19	Control I	Quinguilla	*Chenopodium album*	Temuco	161	Cotiledon
20	Control II	Paico	*Chenopodium ambrosoide*	Temuco	161	Cotiledon

ligation of the adapters. Pre-amplification was performed using a single adenine (A) selective nucleotide for each primer. For sele-ctive amplification, an *EcoRI* primer, with three selective nucleo-tides, was used in combination with *MseI* primer with three selective nucleotides. For both pre-amplification and selective amplification, the following amplification profile was used: an initial cycle of 94°C for 30 s, 65°C for 30 s, 72°C for 1 min, followed by 12 touchdown cycles in which the annealing temperature was reduced of 0.7°C per cycle. The annealing temperature was then kept constant at 56°C for the other 23 cycles. Amplification products were separated on a 6% polyacrylamide gel, and made visible with silver staining. Ten primer combinations were tested for their ability to generate reproducible AFLP profiles that could be scored unambiguously. Reproducibility of the

primer combinations was tested by comparing the AFLP profiles of two DNA samples collected from the same individual. Three combinations were specifically chosen for their ability to generate a large number of bands in order to increase accuracy for the identification of possible identical plants.

Data analysis

For each primer combination, the presence or absence of a band in each sample was visually scored. Data were analysis in a binary matrix (Paul et al., 1997; Yee et al., 1999). Genetic similarities were calculated using the Simple Matching coefficient and tree-diagrams obtained by clustering according to the unweighted pair group method with arithmetic average (UPGMA), using the NTSYSpc

2.0.1 program (Applied Biostatistics Inc., NY, USA). The correspondence between the morphological and AFLP similarity coefficient matrices was tested on the basis of correlation analysis for Mantel's test using the MxComp procedure of NTSYS.

RESULTS

Morphological analysis

The grouping analysis of the morphological data made it possible to form three groups (Figure 2). In Group I, two sub-groups were identified. The characteristics which are gathered by the Sub-group (A) accessions are: intense yellow for grain

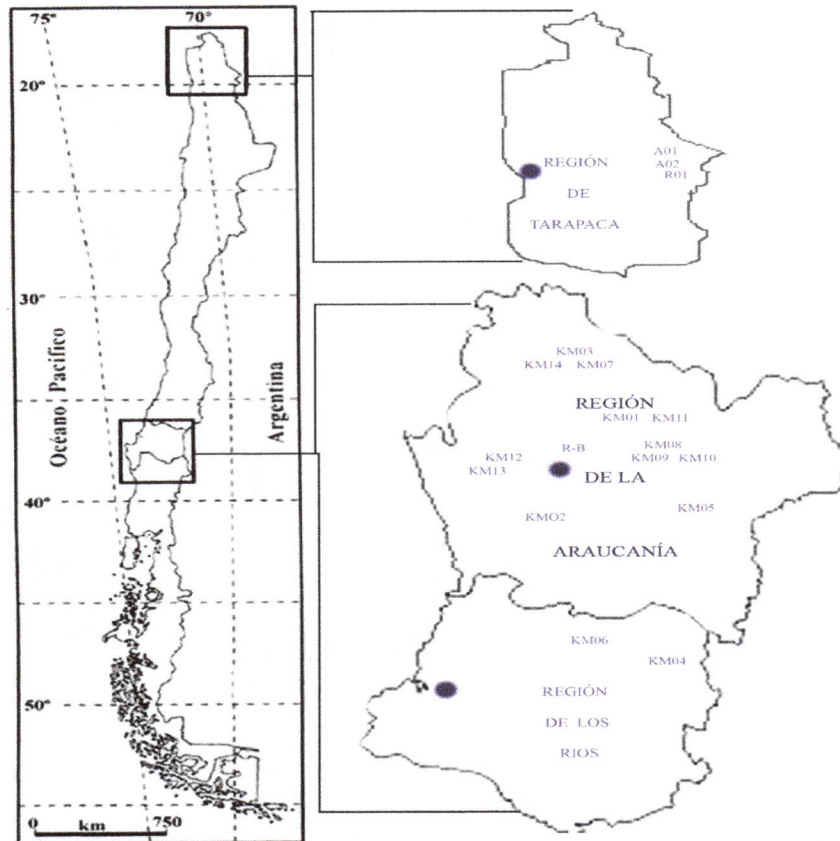

Figure 1. Localities quinoa collections.

Table 2. Descriptive names of the 20 morphological characters examined.

Descriptor list	Morphological characters
Grain	Colour of pericarp
	Shape of fruit edge
Inflorescence	Colour of panicle
	Intensity color of panicle
	Shape of panicle
	Kind of panicle
	Density of panicle
Stem	Colour of stem
	Intensity of colour
	Formation of stem
	Presence of lines
	Colour of lines in stem
Type of growth	Growth habit type
	Number of the primary stem
	Plant height
	Kind of stems
Leaf	Teeth of basal leaf
	Edge of basal leaf
	Colour of basal leaf
Phenological (days)	Emergence Flower buds
	Initiation of flowering
	50% flowering
	Physiological maturity

colour, agglomerated form and yellow colour for the panicle, compact panicle, and precocity. In this group we can find accessions KM01, KM10, KM12 and the registered variety Regalona-Baer. Sub-group (B) included accessions which presented variations in grain colour from intense yellow to brown, red colour, agglomerated form and intermediate density of the panicle. These presented a precocity ranging from early to semi-early. In this group we find accessions KM07 and the accessions which exhibit the same co-efficient of morphological similarity, including accessions KM03, KM04, KM05; KM02, KM08, KM09 and KM11, KM13, KM14 respectively even though they come from different localities. Accession KM06 may be set apart within the sub-group given though it has retained its mixed condition. Group II corresponds to highland accessions of later phenology, yellow to red colour of grain, panicle for are amarantiform and com-pact. The accessions included in this group were A0I, A02 and R01. Group III consisted of two differrent species *C. album* (CH19) and *C. ambro-sioides* (CH20) representing outgroup controls.

AFLP analysis

All of the samples showed a genetic diversity, with a simi-

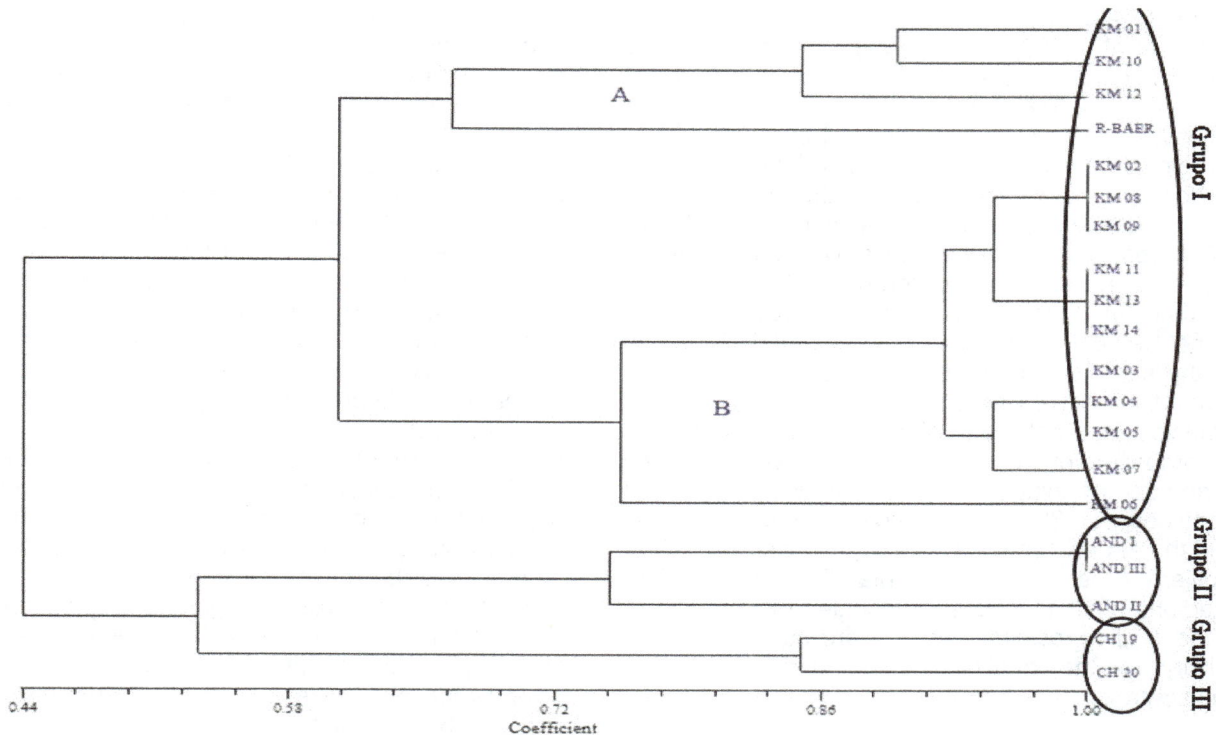

Figure 2. Cluster analysis dendrogram based on morphologic characteristics of the 20 accessions studied

Table 3. Primer combinations used and polymorphic bands generated.

Combination of Primer	Number of Bands	N° of Monomorphic bands	N° of Polimorphic bands	% of polimorphic bands
EcoAAG/MseCAC	64	0	64	100
EcoACA/MseCAA	49	9	40	81.6
EcoACT/MseCAG	37	11	26	70.2
TOTAL	150	20	130	86.6
		13.3 %	86.6%	

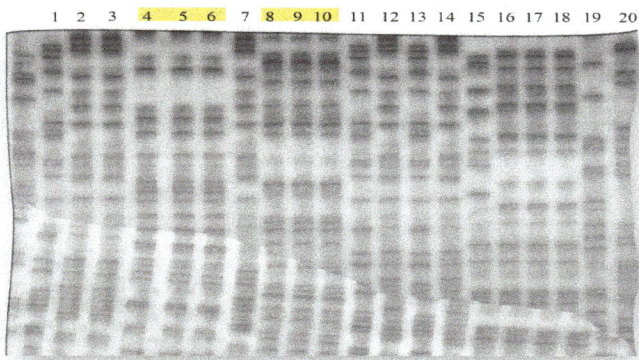

Figure 3. Polymorphisms obtained by AFLP primers EcoAAG/MseCAC.

larity range between 0.54 and 0.97. The tree diagram also throws up a differentiation between the genotypes allocated to Group I, which includes the accessions from the north of Chile, with a coefficient of 0.68; and the Group II accessions, which are samples from the precordillera sector with a coefficient of 0.77. The similarity coefficient between the two groups (I and II) is 0.62. The DNA samples were amplified with10 combinations of primers, using the best three for analysis. Their products generated 150 AFLP bands, of which 130 (86.6%) were polymorphic (Table 3). The bands which presented the same electrophoretic mobility were treated as monomerphic fragments of DNA. The primer pair EcoAAG / MseCAC contributed the highest number of bands, obtaining 100% polymorphic bands that can observe on Figure 3. These results agree with those obtained by Maughan et al. (2004), which indicate that a high level of polymorphism is obtained through analysis using AFLP markers. Moreover, in studies done on other species, AFLP markers are a powerful tool for the analysis of genetic diversity and to ensure a high level of polymer-phism (Kim et al., 1998). Three groups can be identified in the tree diagram (Figure 4). Group I includes eight local accessions, the commercial variety Regalona-characteristics of this group are defined by grain colour, panicle colour, panicle type differentiated and terminal, panicle density and herbaceous growth. Through DNA analysis it could be established that accessions KM13 and KM14 possess the same similarity coefficient (1.0) even though they come from different localities. Group II (Yellow shadow) consists of accessions KM04, KM08, KM09 and KM10 in addition to KM05 and KM6 which present the same similarity coefficient even though they come from different localities.

The principal characteristic of this group is that it is mainly composed of accessions with brown grain, red panicle and geographical location of samples collection at an altitude between 338 m.a.s.l. and 631 m.a.s.l., in the Baer and the Andean accessions A01, A02 and R01. The pre-cordillera zone of the Araucania (IX) and Los Rios (XIV) Regions of Chile. Group III includes the species *C. album* and *C. ambrosioides* which were used to confirm the validity of the analysis, being denominated as outgroup controls.

DISCUSSION

The first objective of this study was to analyze the molecular diversity of 14 coastal accessions of (*C. quinoa* Willd.) using AFLP markers. The characterization of this diversity has improved knowledge relative to the origin of different quinoa accessions conserved by farmers and breeders.The sub group pre-cordillera accessions are differentiated from the rest of the accessions, both from coastal and from the highland accessions, since no intimate genetic relationship exists. The results coincide with previous morphological and iso-enzyme studies, which separate quinoa into two types: a coastal type (Chile) and an Andean plateau type (Wilson 1988b; Risi and Galway 1989). This study represents a complement to the work of Ruas et al. (1999) differentiating the groups of *C. quinoa*, *C. album* and *C. ambrosioides*.

Comparative analysis between AFLP and morphological data

The morphological analysis of the Andean group places it in an independent Group (Group II), however in the AFLP analysis it appears to be integrated into Group I, indicat-ing the existence of a similarity to the genetic material of the accessions collected in the Region. This result is consistent with the existence of common ancestral genes in the crop. The commercial variety Regalona-Baer included in the study is close to Andean eco-types from the molecular point of view, confirming the existence of parental genes originating in Andean material. The Mantel's Test was used to compare the matrices generated from the

AFLP and the morphological data. This test shows a correlation between the morphological and molecular (AFLP) tree diagrams with values of r = 0.09, p = 0.7962, obtained by the MxComp method. Although positive, the concordance value is low, as observed in various other studies regarding varieties of grapes, rice, rye grass and potatoes (Xu et al., 2000; Federici et al., 2001; Roldan-Ruiz et al., 2001; Solano et al., 2007). In addition to this, Spooner et al. (2005), report that DNA digital fingerprint-ing techniques are a better discriminator than morpho-logical data in the analysis of genetic similarity. Our results confirm that DNA analysis is an efficient method for the exploration of genetic diversity in quinoa populations. Although the different genotypes share a common base, the pre-cordillera group differs from the rest of the accessions from the South of Chile and from the varieties coming from the North of the country. Wilson (1988b) hypothesized that ancestral colonization of quinoa in the southern zone of Chile, followed by long periods of gene-tic drift, was the reason for this observed lack of genetic diversity in the Chilean highland populations. Wilson (1988a) also hypothesized that Chilean popula-tions have their origin in the southern Altiplano. This was supported by the data of Christensen et al. (2007), which showed that southern Chilean populations are more similar to Bolivian populations than other quinoas from the Andean highlands. However, the present study indicates that Chilean lowland germplasm is much more genetically diverse than previously believed. Although this observa-tion may potentially shake Wilson's (1988a) Chilean qui-noa origin hypothesis, the most observed diversity at the molecular level found in this study could alternatively be explained by promiscuous outcrossing in the lowland quinoa fields involving abundant weed populations of *C. album* and C. *hircinum*. This natural process combine with the ancestral seed exchange sys-tem, anthropic pressures on selection due to edaphocli-matic and photoperiod factors, would have generated a genetic differentiation in the varieties of the pre-cordillera sector. Molecular analysis using AFLP made it possible to esta-blish the differences and similarities between the mate-rials collected in the South of Chile. So far as we know, this is the first study done in Chile using AFLP markers to analyse genetic diversity in the quinoa germoplasm in the southern zone and in the country.

REFERENCES

Arens P, Coops H, Jansen J, Vosman B (1998). Molecular genetic analysis of black poplar (*Populus nigra* L.) along Dutch rivers. Molecular Ecology. 7(1): 11-18.
Becker J, Vos M, Kuiper M, Salamini F, Heun M (1995). Combined mapping of AFLP and RFLP markers in barley. Molecular Genetic. Genet. 249: 65-73.
Bertero DH (2001). Effects of photoperiod, temperature and radiation on the rate of leaf appearance in quinoa (*Chenopodium quinoa* Willdenow) under field conditions. Annals of Botany 87: 495-502.
Chauhan GS, Eskin Nam, Tkachuk R (1999). Effect of saponin extraction on the nutritional quality of quinoa (*Chenopodium quinoa*

Willd.). J. Food Sci. Technol. 36: 123–126.

Christensen SA, Pratt DB, Pratt C, Stevens MR, Jellen EN, Coleman CE, Fairbanks DJ, Bonifacio A, Maughan PJ (2007) Assessment of genetic diversity in the USDA and CIP-FAO international nurserycollections of quinoa (Chenopodium quinoa Willd.) using microsa-tellite markers. Plant Genet Res 5: 82–95

Federici M, Vaughan D, Tomooka N, Kaga A, Wang Wang X, Doi K, Francis M, Zorrilla G, Saldain N (2001). Analysis of Uruguayan weedy rice genetic diversity using AFLP molecular markers. Electronic J. Biotechnol. 4(3): 130-145.

Hartings H, Berardo N, Mazzinelli GF, Valoti P, Verderio A, Motto M (2008). Assessment of genetic diversity and relationships among maize (Zea mays L.) Italian landraces by morphological traits and AFLP profiling. Theor. Appl. Genet. Jun 27.

Hill M, Witsenboer H, Vos P, Zabeau M, Kesseli R (1996). PCR-fingerprinting using AFLP as a tool for studying genetic relationships in Lactuca spp. Theory. Appl. Genet. 93: 1202-1210.

IPGRI (1981). Descriptores de quínoa. Consejo internacional de recursos fitogenéticos. FAO. Roma. Italia. 27 p.

Jacobsen SE (2003). The worldwide potential for quinoa (Chenopodium quinoa Willd.). Food Rev. Intl. 19 (2): 167-177.

Kadereit G, Borsch T, Welsing K, Freitag H (2003). Phylogeny of amaranthaceae and chenopodiaceae and the evolution of C4 photosynthesis. Int. J. Plant Sci. 164: 959–986.

Kim J, Joung H, Kim Hy, Lim Y (1998). Estimation of genetic variation and relationship in potato (Solanum tuberosum L.) cultivars using AFLP markers. Am. J. Potato Res., 75(2): 107-102.

Lande R (1990). Marker-assisted selection in relation to traditional-methods of plant breeding. In: Stalker HT, Murphy JP (ed) Plant breeding in the 1990s. CAB International, Wallingford, UK. 1991 pp. 437–451.

Mason S, Stevens M, Jellen E, Bonifácio A, Fairbanks D, Coleman C, Mccarty R, Rasmussen A, Maughan P (2005). Development and use of microsatellite markers for germplasm characterization in quinoa (Chenopodium quinoa Willdenow). Crop Science. 45:1618-1630.

Maughan P, Bonifácio A, Jellen E, Stevens M, Coleman C, Ricks M, Mason S, Jarvis D, Gardunia B, Fairbanks D (2004). A genetic linkage map of quinoa (Chenopodium quinoa Willdenow) based on AFLP, RAPD, and SSR markers. Theory Applicated Genetic.109: 1188-1195.Maughan P, Saghai-Maroof M, Buss G, Huestis G (1996). Amplified Fragment Length Polymorphism (AFLP) in soybean: species diversity, inheritance, and near isogenic lines analysis. Theor. Appl. Genet. 93:392-401.

Ochoa J, Frinking H, Jacobs T (1999). Postulation of virulence grups and resistance factors in the quinoa down mildew pathosystem using material from Ecuador. Plant pathology 48: 425-430.

Patterson AH, Tanksley SD, Sorrells M (1991). DNAmarkers in plant improvement. Adv. Agron. 46:39–90.

Paul S, Wachira FN, Powell W, Waugh R (1997). Diversity and genetic differentiation among populations of Indian and Kenyan tea (Camellia sinensis (L.) O. Kuntze) revealed by AFLP markers. Theor. Appl. Gene. 94(2):255- 263.

Pillay M, Myers G (1999). Genetic diversity in cotton assessed by variation in ribosomal RNA genes and AFLP markers. Crop Sci. 39: 1881-1886.

Prado R, Boero C, Gallard M, Gonzalez J (2000). Effect of NaCl on germination, growth, and soluble sugar content in Chenopodium quinoa Willd. seeds. Bot. Bull. Acad. Sci. 41: 27–34.

Risi J, Galwey NW (1989). The pattern of genetic diversity in the Andean grain crop quinoa (Chenopodium quinoa Willd.). I. Associations between characteristics. Euphytica 41: 147–162.

Roldán-Ruiz I, Van Euwijk F, Gilliland T, Dubreuil P, Dillmann C, Lallemand J, De Loose M, Baril C (2001). A comparative study of molecular and morphological methods of describing relationships between perennial ryegrass (Lolium Perenne L.) varieties. Theor. Appl. Genet. 103(8): 1138-1150.

Ruas P, Bonifácio A, Ruas C, Fairbanks D, Andersen W (1999). Genetic relationship among 19 accesions of six species of Chenopodium L., by Randomly Amplified Polymorphic DNA fragments (RAPD). Euphytica 105: 25-32

Thomet M, Sepúlveda J, Palazuelos P (2003). Manejo Agroecológico de la Kinwa. In: La Kinwa Mapuche. Recuperación de un cultivo para la alimentación. Fundación para la Innovación Agraria. Ministerio de Agricultura. Chile. pp. 44-94.

Sharma SK, Knox Mr, Ellis TH (1996). AFLP analysis of the diversity and phylogeny of and its comparision with RAPD analysis. Theor. Appl. Genet. 1996. 93: 751-758

Solano J, Morales D, Anabalón L (2007). Molecular description and similarity relationships among native germplasm potatoes (Solanum tuberosum ssp. tuberosum L.) using morphological data and AFLP markers. Electronic J. Biotechnol. 10(3): 436-443.

Spooner D, Mclean K, Ramsay G, Waugh R, Bryan GA (2005). Single domestication for potato based on multilocus amplified fragment length polymorphism genotyping. Proceedings of the National Academy of Sciences of the United States of America, 2005, 102 (41): 14694-14699.

Staub JE, Serquen FC, Gupta M (1996). Genetic markers, map construction, and their application in plant breeding. Hort. Sci. 31: 729–741.

Vos P, Hogers R, Bleeker M, Reijans M, Van De Lee T, Hornes M, Friters A, Pot J, Paleman J, Kuiper M, Zabeau M (1995). AFLP: a new technique for DNA fingerprinting. Nucleic Acids Res., 23 (21): 4407- 4414.

Ward S (1998) A new source of restorable cytoplasmic male sterility in quinoa. Euphytica 101(2):157-163.

Wilckens R, Hevia F, Tapia M, Albarrán R (1996). Caracterización de dos genotipos de Quinoa (Chenopodium quinoa Willd.) chilena. I. Electroforesis de proteínas de la semilla. Agro-Ciencia. 12(1): 51-56.

Wilson HD (1988a). Quinoa biosystematics I: domesticated populations. Econ. Bot. 42: 461–477.

Wilson HD (1988b). Quinoa biosystematics II: free living populations. Econ. Bot. 42: 478–494.

Xu R, Tomooka N, Vaughan DA (2000). AFLP markers for characterizing the azuki bean complex. Crop Sci., 40 (3): 808-815.

Yee E, Kidwell KK, Sills GR, Lumpkin TA (1999). Diversity among selected Vigna angularis (Azuki) accessions on the basis of RAPD and AFLP markers. Crop Sci., 39(1) :268-275.

Review article: Quality protein maize (QPM): Genetic manipulation for the nutritional fortification of maize

P. A. Sofi[1*], Shafiq A. Wani[1], A. G. Rather[2] and Shabir H. Wani[3]

[1]Directorate of Research, SKUAST-K, Shalimar, 191121, J and K, India.
[2]Rice Research Station, SKUAST-K, Khudwani, India.
[3]Department of Plant Breeding, Genetics and Biotechnology, PAU, Ludhiana, India.

Cereals are the only source of nutrition for one-third of the world's population especially in developing and underdeveloped nations of Sub-Saharan Africa and South-east Asia. The three major cereals, rice, wheat and maize constitute about 85% of total global cereals production amounting to about 200 million tonnes of protein harvest annually at an average of 10% protein content, out of which a sizeable proportion goes into human consumption (Shewry, 2007). A major concern in case of developing nations is that in most cases, a single cereals crop is the major food staple and as such the nutritional profile of cereal crops assumes great significance. Grain protein content of cereals has a very narrow range with rice (5.8-7.7%), maize (9-11%), barley (8-15%) and wheat (7-22%) as reported by various workers. In many developing countries of Latin America, Africa and Asia, maize is the major staple food and often the only source of protein. At global level, maize accounts for 15% of proteins and 20% of calories in world food diet. But unfortunately, the nutritional profile of maize is poor as it is deficient in essential amino acids such as lysine, tryptophan and methionine due to a relatively higher proportion of prolamines in maize storage proteins which are essentially devoid of lysine and tryptophan. The reason concerning this is that lysine, tryptophan and threonine are the limiting amino acids in human beings and non-ruminants. Maize is also an important component of livestock feed especially in developed nations where 78% of total maize production goes into livestock feed. Therefore, breeding strategies aimed at improving the protein profile of maize will go a long way in reducing prevalence and persistence of malnutrition in developing world.

Key words: *Mucronate*, protein, maize, zein.

INTRODUCTION

Storage proteins in maize

The maize grain largely consists of endosperm that is rich in starch (71%). Both the embryo and endosperm contain proteins but the germ proteins are superior in quality as well as quantity. Zeins are a class of alcohol soluble proteins that are specific to endosperm of maize (Prassana et al., 2001) and are not detected in any other plant part. The maize endosperm consists of two district regions having different physical properties. The aleurone layer is the outer most layer rich in hydrolytic enzymes secreted by specialized cells. Within the aleurone layer is the starch rich endosperm having vitreous and starchy regions. The zein proteins found in vitreous region form insoluble accretions called protein bodies in the lumen of rough endoplasmic reti-culum and towards maturation are densely packed between starch grains (Gibbon and Larkins, 2005). These zeins consist of albumins, globulins, glutelins and prolamins and constitute about 50-60% of maize proteins. The prolamins are rich in proline and amide nitrogen derived from glutamine. All prolamins are alcohol soluble (Shewry and Halford, 2002). The prolamins of maize grain are called zeins and consist of one major class (α-zeins) and three minor classes (β, γ and δ). The zein fraction α is rich in cystein while β- and γ-fractions are rich in methionine. These four types α, β γ and δ constitute about 50-70% of maize endosperm and are essentially rich in glutamine, leucine and proline and poor in lysine and tryptophan. Other proteins such as globulins (3%), glutelins (34%) and albumins (3%) are collectively called non-zeins (Figure 1). The zein fraction in normal maize normally contains higher proportion of leucine (18.7%), phenyla-lanine (5.2%) isoleucine (3.8%), valine (3.6%) and tyrosine (3.5%), but smaller amounts of other essential

*Corresponding author. E-mail: phdpbg@yahoo.com.

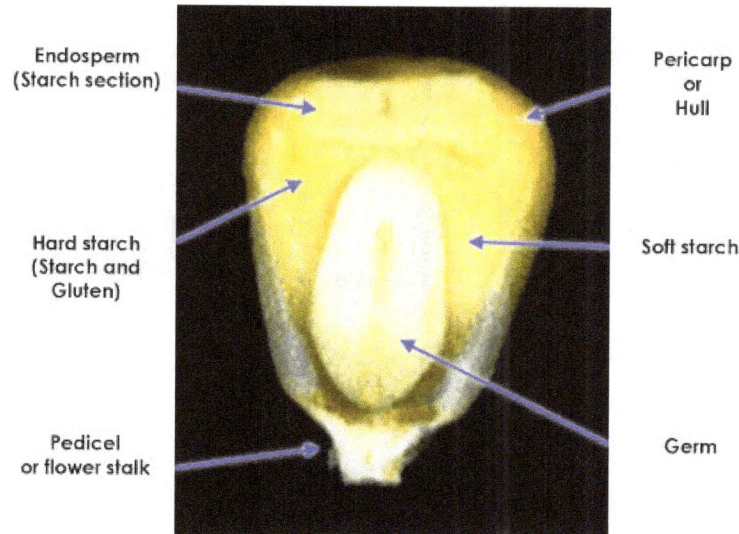

Figure 1. Structure of maize kernel (Source: www.fao.org).

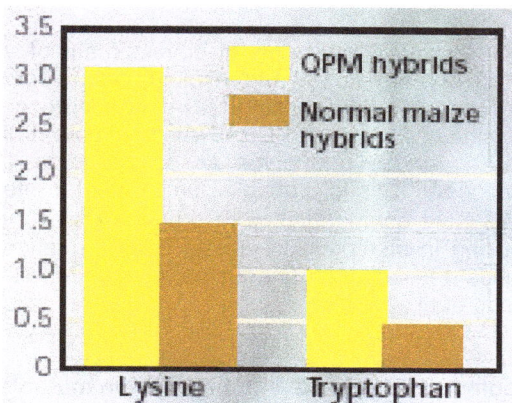

Figure 2. Comparative composition of normal and *o2* maize for lysine and tryptophan Genetic manipulation of protein quality.

amino acids such as threonine (3%), histidine and cysteine (1%), methionine (0.9%), lysine (0.1%) and is essentially devoid of tryptophan as it is absent from the major prolamin fraction (α-zeins) of maize kernel. The non-zein protein f raction is balanced and rich in lysine and tryptophan (Vasal, 2000). The zeins are synthesised on rough endoplasmic reticulum membrane and accumulate as protein bodies in its lumen (Figure 2).

With an objective to screen out maize lines containing improved amino acid balance in endosperm to enhance its biological value, efforts were directed in early 1900 to identify such genotypes. Two major problems were faced: one that no specific genes were identified that governed the amino acid profile of maize proteins, and could be used in breeding programmes. Secondly the lack of a simple genetic system precluded the use of backcross programmes to improve upon protein quality in maize. However, in 1920, a naturally

occurring maize mutant was identified in Connecticut maize fields in USA that had soft and opaque grains and was named as opaque 2 (*o-2*) (Singleton, 1939). In 1960, Nelson and Mertz worked with the mutant lines at Connecticut Experiment station to identify maize lines with improved protein profile (Krivanek et al., 2007). In 1961, Researchers (researchers) at Purdue University observed that mutant lines that were homozygous for *o2* allele had significantly higher lysine (almost double) in endosperm compared to normal maize. These discoveries aroused great enthusiasm and hope among researchers towards genetic manipulation of protein quality in maize and resulted in discovery of various other mutant types that had altered amino acid com-position. These include the *floury-2 (fl-2)*, *Mucronate (Mc)* and *Defective endosperm B30 (DE B30)*. The opaque mutants are recessive *(o1, o2, o5, o9-11, o13, o17)*, the floury mutation is semidormant *(fl-1, fl-2 and fl-3)* where as *Mucronate* and defective endosperm are dominant mutations. Figure 3 shows the position of zein genes and the mutations on maize gene map. The opaque mutations affect the regulatory network whereas floury; *Mucronate* and defective endosperm affects the storage proteins (Gibbon and Larkin, 2005). The improved protein quality of such mutants was apparently due to increase in proportion on non-zein fraction that is rich in lysine and tryptophan and repression of zein synthesis. Each of the zein polypeptide is a product of differential structural gene (Zp). These zp genes are simply inherited and are members of a large group of genes (upto 150). In terms of zein repression o7 >o2 > fl-27> De B30. Epistatic interactions have also been reported among various regulatory mutants (Prassana and Sarkar, 1991). Thus o2 and o7 are epistatic over fl-2, whereas o2 and Mc have synergistic effect. The reduction in levels of zeins by various mutant allele is accomplished either by reduction in levels of various zein sub-units, rate of

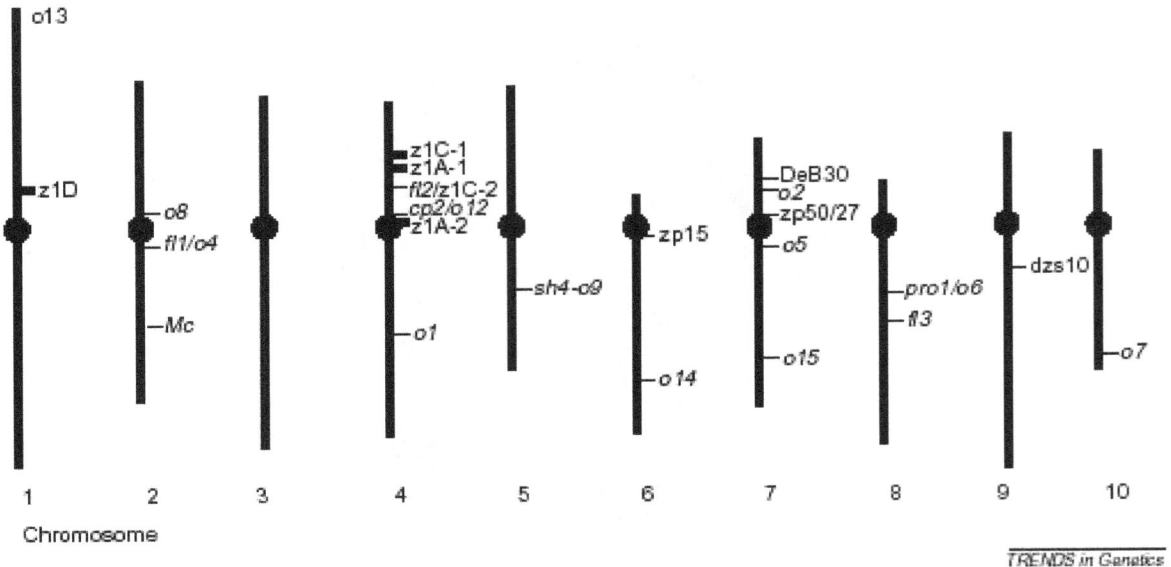

Figure 3. Chromosomal locations of various zein genes (source: Gibbon and Larkin, 2005).

accumulation of zeins, increase in methionine content and effect on timing and pattern of storage protein accumulation. The only mutant gene conditioning protein quality in maize is O2 which encodes a defective basic-domain-leucine-zipper transcription fact or and has been mapped to short arm of chromosome 7. In its dominant form o2 regulates the expression of 22 KDa α-zeins (Damerval and Eevienne, 1993) and other genes including lysine ketoglutarate reductase the mutant allele 02 typically causes a significant increase in non-zein fraction associated with decrease in lysine-poor α-zein proteins. Moreover, the defective LKR enzyme causes increased levels of free lysine. In normal maize, the free amino acids account for a low proportion of total amino acids in normal grains. The amino acids lysine threonine and metheonine are synthesized from asparatic acid (Shewry, 2007). There are complex regulatory networks that maintain low levels of free amino acids in grains by feedback inhibition. The increased levels (level) of lysine in o2 mutants is due to higher levels of an elongation factor of protein sysnthesis (eEF1A). This factor though itself being rich in lysine (10%) but accounts for only 2% of lysine in the endosperm. Therefore, it is evident that the higher expression of eEF1A is accompanied by a number of transcription factors. QTL mapping studies have revealed linkage between eEF1A and genes encoding zein storage proteins (Valenzualla et al., 2004). The floury-2 mutants (fl-2) encode a 22KD α-zein gene with a defective signal peptide that causes mutant polypeptide to accumulate in the membrane of endoplasmic reticulum (Copleman et al., 1995). Similarly, in case of defective endosperm mutants (De B30), a 19 KDa α-zein is produced that also has of a defective signal peptide (Gilkin et al., 1997). However, in case of Mucronate (Mc) mutants, there is a frame shift mutation in 16 KDa γ-zein. All these mutants

together cause disruption in development of vitreous texture of maize kernel and result in opaque phenotype. The recent studies using RNA interference based silencing of 22 and 19 KDa RNAi lines more profoundly caused opaque phenotype as compared to 19 KDa component. This is probably due to greater interaction of 22 KDa components with β and γ-zeins (γ-zeins) resulting in disruption in protein body formation which causes the opaque phenotype (Segal et al., 2003; Huang et al., 2004).

Problems associated with high lysine mutants

Even though high lysine mutants aroused tremendous interest and enthusiasm for their possible use in developing maize with superior protein quality, but rapidly the negative plieotropic (pleiotropic) effects of such mutations began to be recognised. These undesirable features were limiting factors to its widespread use and adoption (Lauderdale, 2002). Even though the endosperm protein mutants such as o2 and fl-2 favourably change the amino acid profile of maize grain, they also cause certain undesirable consequences, as is expected of most mutants. The undesirable characteristics include reduced yield than normal maize, low grain consistence and a farinaceous endosperm that retains water (Toro et al., 2003). These features result in a soft, chalky endosperm that dried slowly making it prone to damage, a thick pericarp, more susceptibility to diseases and pests, higher storage losses and also affects harvest ability. Since the kernel weight is reduced due to less density per unit volume as starch is loosely packed with lot of air spaces, there is corresponding decline in the yield (Singh and Venkatesh, 2006) which can be almost to the tune of 10 percent or above. Especially in developing countries, where farmers are accustomed to hard flint and dent grains, the kernel appearance of such mutants made it less ideal for large scale use and adoption in target areas. The mutations that alter grain protein synthesis cause changes in texture of grains. The early opaque-2 (o2) mutants had reduced levels of α-zeins resulting in small unexpanded protein bodies (Geetha et al., 1991), whereas, o15 that reduces γ -zeins leads to smaller number of protein bodies. Other mutations such as floury-2 (fl-2), Mucronate (Mc) and defective endosperm (De B30) result in irregularly shaped protein bodies.

In order to overcome these apparent limitations for large scale use of such mutants, efforts were directed towards identification of alternative mutants that did not carry such disadvantages. This resulted in identification of additional mutants of opaque and floury series, even though none of them eventually could get to farmers fields. This dampened the spirit with which high lysine maize research was pursued as the complexity of the coordinate gene action governing endosperm protein profile became more and more evident. Another major set back to opaque-2 mutant research came in 1973 when WHO and UN revised the energy and protein recommendations. The energy was given more priority over protein for defining them as major limiting factors in malnutrition (Lauderdale, 2002). As a result, the interest of researchers got distinctly polarised towards high energy rather than high protein, therefore, a renewed focus towards higher yields to meet energy demands.

From opaque-2 to QPM: Endosperm modifiers

The undesirable traits associated with opaque-2 mutation coupled with the de-emphasis of protein requirements, the interest in o-2 was virtually off the ground but researchers at CIMMYT and elsewhere continued their efforts to develop maize that had high lysine and tryptophan and could favourably compete with normal maize for yield and grain characteristics. Under the dynamic leadership of Dr. S.K. Vasal, Researchers at CIMMYT, and University of Natal, RSA, various endosperm modifier genes were identified that could favourably alter the grain characteristics, thereby overcoming an important obstacle in popularisation of high lysine o-2 maize. These modifier genes do not have any effect of their own as such but interact to improve the kernel hardiness and appearance and increase kernel weight and density. A large number of reports followed wherein varying degrees of endosperm modifications were observed (Paez et al., 1969; Anapurna et al., 1971). This ends up with the idea that such endosperm modifier genes could be used along with o-2 gene either singly or in combination with other mutants such as sugary-2 (Su-2) in order to have acceptable characteristics in the final product. Such combinations resulted in maize lines that possessed high lysine, vitreous grains and better protein digestibility even though yield was affected.

The mechanism by which the endosperm modifiers change the grain structure from chalky to vitreous in modified opaque-2 (mo2) is not clearly understood,

neither have the genes been characterised in terms of their number, chromosomal locations, nor expression levels in different genetic backgrounds, even though Lopes et al. (1995) linked two chromosomal regions in chromosome 7 with endosperm modification. The o2 mutants that have reduced levels of 22 KDa α -zeins, mo-2 mutants have 2-3 times higher levels of 27 KDa γ -zeins (Geetha et al., 1991). The higher level of γ -zeins is thought to initiate the formation of protein bodies, due to disulphide bond mediated cross linking of 27 KDa γ -zeins with other cysteine-rich proteins. The observation was that Moro et al. (1995) the modifier genes increased the levels of 27 KDa γ -zeins in opaque-2 backgrounds only, indicating thereby that such genes do not, per se, improve the protein quality of maize (Moro et al. 1995). Moreover, they found that higher level of γ -zeins increased the hardness to a particular level only. A scale from 1-5 has been revised for endosperm modification in maize. Score 1 is fully modified that my or may not be QPM, while as score of 5 means chalky, opaque and soft kernels. Breeders use a scale of 2-3 in early breeding generations but full modification is sought in later cycles in QPM converted back ground. The aim is to achieve complete grain modification without losing the o2 phenotype.

At CIMMYT, a conservative approach was adopted in developing mo2 genotypes to strike a balance between proteins levels and grain quality and competetive yield levels. Therefore, emphasis was laid on maintaining its lysine content while making it competitive with conventional maize varieties. This improved o2 maize was renamed as quality protein maize (QPM). QPM essentially has about twice the levels of lysine and tryptophan than normal maize and also increased levels of histidine, arginine, aspartic acid and glycine. It also has reduced levels of glutamic acid, alanine, leucine, especially, the lower levels of leucine is an added advantage as it results in a more balance leucinesoleucine ratio that helps to liberate more tryptophan.

In addition to these two genetic systems that is, o-2 and endosperm modifiers, a third genetic system consist of amino acid modifier genes affecting relative levels of lysine and tryptophan (Krivanek et al., 2007). As against the lysine and tryptophan levels of 2 and 0.4% in normal maize, the corresponding values in QPM are 4 and 0.8% respectively. This is largely due to a decrease in the zein fraction from 47.2% in normal maize to 22.8% in o2 mutants. Genes governing the levels of lysine have been mapped to the chromosome2, 4 and 7 (Wang et al., 2001; Wu et al., 2002). Such genes have been identified and include eEF1A (7L), eEF1A (4S), eEF1A (2S), FAA (1L), FAA (2S), FAA (2L), FAA (3S), FAA (4L), FAA (5L), FAA (7L), FAA (8S) and FAA (9S). It is now quite evident that from a simple genetic system in o-2, QPM has evolved by combination of various genetic systems, behaving to be handled like classical quantitative trait.

Breeding efforts and achievements: CIMMYTs role

Global research on QPM had faded elsewhere, however,

Table 1. CIMMYT gene pools/populations of QPM.

Pool/population	Adaptation	Maturity	Seed colour	Seed texture
Pool 15 QPM	Tropical	Early	White	Flint-dent
Pool 17 QPM	-do-	-do-	Yellow	Flint
Pool 18 QPM	-do-	-do-	-do-	Dent
Pool 23 QPM	-do-	Late	White	Flint
Pool 24 QPM	-do-	-do-	-do-	Dent
Pool 25 QPM	-do-	-do-	Yellow	Flint
Pool 26 QPM	-do-	-do-	-do-	Dent
Pool 27 QPM	Sub-tropical	Early	White	Flint-dent
Pool 29 QPM	-do-	-do-	Yellow	-do-
Pool 31 QPM	-do-	Medium	White	Flint
Pool 32 QPM	-do-	-do-	-do-	Dent
Pool 33 QPM	-do-	-do-	Yellow	Flint
Pool 34 QPM	Tropical	-do-	-do-	Dent
Population 61	Tropical	Early	Yellow	Flint
Population 62	-do-	-do-	White	Semi-flint
Population 63	-do-	-do-	White	Dent
Population 64	-do-	-do-	White	-do-
Population 65	-do-	-do-	Yellow	Flint
Population 66	-do-	-do-	-do-	Dent
Population 67	Sub-tropical	-do-	White	Flint/semi-flint
Population 68	-do-	-do-	-do-	Dent/semi dent
Population 69	-do-	Medium/late	Yellow	Flint
Population 70	-do-	-do-	-do-	Dent

Source : Vasal et al. 1993 a, b.

at a wide array of tropical and sub-tropical QPM gene pools and populations with hard endosperm have been developed (Bjarnson and Vasal, 1992). Initially the emphasis was on development of donor stock by selection for modified grain texture in QPM backgrounds using various selection schemes. The donor stocks were isolated from soft *o2* materials that showed varying degrees of kernel modification due to differential accumulation of modifier genes. The development was accomplished by intra-population selection of genetic modifiers in *o2* background followed by grouping of modified *o2* sources into pools which were recombined. The materials from these two approaches were used as donor for conversion programmes. These efforts even though painstaking led to identification of four tropical hard endosperm populations (composite K; Ver 181-Ant gp venezula-1; Thai composite; PD 9MS6) and one highland (Composite-1). Later on, these donor stocks were used for large scale conversion of non-QPM materials into QPM into a wide array of genetic background from different agro-climatic zones. Table 1 presents various germplasm pools and populations developed by CIMMYT for use in various national maize programmes. This resulted in release of a larger number of QPM varieties by conversion of elite tropical, subtropical and highland lines/populations into QPM versions. The yield gaps were progressively narrowed down to increase the acceptance environments (Table 2).

In 1985, when CIMMYT began its hybrid maize programme, there was corresponding shift of focus in case of QPM germplasm development from OPV's to hybrids for different target environments. Consequently, population improvement was replaced by hybrid breeding. Efforts were directed towards characterization of QPM germplasm in terms of combining ability and heterotic patterns through a serves of dialled studies world over. The result was release of a large number of QPM hybrids that out yielded the local non-QPM check in regional trails (Table 3).

Adoption and impact of QPM

The target countries for large scale cultivation of QPM have been those where maize finds substantial use for human consumption and animal feed. These countries have different levels of development ranging from developed nations like Mexico and Brazil to developing /underdeveloped nations of Africa and Asia. Among humans, women and children have been major targets while as in case of animals, pig and poultry are major targets. In 1977, only four countries grew QPM but in 2003, more than 23 countries have released QPM varieties for large scale cultivation on area over 3.5 million hectares with Mexico alone accounting for about 2.5 million hectares. Presently, the area under QPM is about 2.5 million hectares. This has been largely possible due to the finding of QPM research at CIMMYT by NIPPON foundation Japan and Canadian International Development Agency (GDA). The Nippon funded project "The improvement of quality protein

Table 2. QPM varieties released since 1996.

Variety	Pedigree	Country
NB-nutrina	Poza Rica 8763	Nicaragua
Susma	Across 8363SR	Mozambique
Obtamapa	-do-	Mali
Nalongo	-do-	Uganda
Obatampa	-do-	Benin
BR-473	-do-	Brazil
BR-451	-do-	Brazil
Assume preto	-do-	Brazil
Obatampa	Across 8363SR	Burkina Faso
Obatampa	-do-	Guinea
VS-537C	Poza Rica 8763	Mexico
VS-538 C	Across 8762	Mexico
Espoir	-do-	Burkina faso
Obatampa	Across 8363SR	Cameroon
Mamaba	-do-	Guinea
CMS 475	-do-	Guinea
K 9101	-do-	Guinea
CMS 473	-do-	Guinea
WSQ104	Pool 15	Kenya
EV 99 QPM	-do-	Nigeria
EV 99 QPM	-do-	Senegal
DMRESR WQPM	-do-	Senegal
Susma	Across 8363SR	Senegal
Lishe-K1	Across 8363SR	Tanzania
Ev 99 QPM	-do-	Togo

Source : Cordova, 2000; Krivanek et al., 2007.

maize in selected developing countries", focussed on promotion of QPM in countries, where maize is a staple and where the probability of adoption and impact is high. In sub-saharan Africa, 17 countries are growing QPM on around 200000 hectares with Ghana alone accounting for about 70000 hectares, Obatampa being the major cultivar. NIPPON foundation, CIDA and Rockefeller foundation have been instrumental in promoting the development and dissemination of QPM in Africa (CIMMYT, 2005). The emphasis has been on conversion of elite maize OPV's and hybrids into QPM's. In India, QPM research was initiated under AICMIP in 1966 and resulted in three o2 composites namely Shakti, Rattan and Protina. Later on using endosperm modification system, mo-2 composite "Shakti-1" was released in 1998. Later on under NATP, two hybrids "Shaktiman-1" and "Shaktiman-2" were also released using CIMMYT inbreds as parental lines (Prasanna, et al., 2001). In China, a number of high yielding QPM hybrids are under cultivation covering an area of about 1000 hectares. It is expected that by 2020, about 30% of maize area in China will be under QPM cultivars (Gill, 2008).

Impact dietary assessment studies

Most important target group for QPM consumption are children. Out of the 189 nations listed by UNICEF, Malawa, Mozambique, India and Bolivia have highest under five mortalities. The impact of QPM can be highly significant in such target group in these vulnerable nations. An early study on Indian children fed with food supplemented with normal maize, QPM and milk (Singh, et al., 1980). The results from children taking QPM were encouraging as the growth parameters recorded were comparable to those of milk. A more comprehensive study was carried out in Ghanaian children (0-15 months) given food supplemented with QPM and normal maize (Afriyiet et al., 1998). The QPM fed children were healthier, suffered fewer fatalities and had better growth rates. The impact of QPM in human nutrition will however, depend on a number of factors. It needs to be established whether the intake of QPM essentially results in enhanced protein utilization in children and adults. Moreover, the results have to be validated especially in the areas where malnutrition is prevalent and maize is an important component of diet.

Table 3. QPM hybrids released since 1996.

Hybrid	Pedigree	Country
HQINTA-993	(CMS-144 x CML-159) CML-176	Nicoragua
HB-PROTICTA	-do-	Guatemala
HQ-61	-do-	El-Savador
HQ-31	-do-	Honduras
Zhongdan 9409	Pool 33 x Temp QPM	China
Zhongdon 3850	-do-	China
QUIAN 2609	(Tai 19 x Tai 02) CML-171	China
Shaktiman-1	(CML-142 x CML-150) CMS -176	India
Shaktiman-2	CML-176 x CML-186	India
QS7705	-do-	South Africa
GH-132-28	P62 x P63	Ghana
BHQP-542	(CML-144 x CML-159) CML-176	Ethiopia
INIA	CML-161 x CML-165	Peru
FONAIAP	(CML-144 x CML-159) CML-176	Venezuala
HQ-2000	CML-161 x CML-165	Vietnam
H-441C	CML-186 x CML-142	Mexico
H-367C	CML-142 x CML-150	Mexico
H-553C	(CML + 42 x CML-150) CML-176	Mexico
H-519C	(CML-144 x CML-159) CML-176	Mexico
H-368C	CML-186 x CML-149	Mexico
H-469C	CML-176 x CML-186	Mexico
KH500Q	(CML-144 x CML-159) CML 1-81	Kenya
KH631Q	(CML + 44 x CML-159) CML-182	Kenya
Lishe-H1	(CML-4 x CML-159) CML-176	Tanzainia
Lishe-H2	Obatampa (CML-144 x CML-176	Tanziania
ZS261Q	(CZL01006 x CML-176 (CZLO1005 x CML-181)	Zimbabwe

Source : Cordova, 2000; Krivanek et al. 2007.

Another application of QPM is as animal feed, especially for monogastric animals such as pigs and poultry, which require a more complete protein than cereals alone can provide, as is case with normal maize that is deficient in lysine and tryptophan. A number of studies have proved that the more potential impact of QPM can be its use in commercial feeds for pigs and poultry as it results in improved growth. In Brazil and El-salvador, use of QPM in animal feed reduced use of soybean meal by about 50% besides reducing use of synthetic lysine substantially (Pereira, 1992). It also resulted in saving of 3-5% in development of feed for pig and poultry. In China, the QPM variety Zhong Dan-9409 has been used in animal feed. It has 8-15 per cent yield advantage and about 80 per cent more lysine and tryptophan. Zhai (2002) analysed the effect of replacing normal maize with QPM and found that in case of poultry, there was no apparent increase in amino acid digestibility but in case of pigs, there was a significantly higher amino acids and weight gain in case of pigs at various growth stages. Similarly, in Ghana, the QPM variety "Obatampa" has been used in both human nutrition as well as `animal feed. The potential impact of QPM as animal feed is restricted to Mexico, Brazil, China, Bolivia, which have high per capita meat supply. Contrarily, in African countries, due to the low meat supply market it a less potential marked for QPM as animal feed component

Genetic engineering for improving protein quality

Genetic engineering has targeted all the traits amenable to manipulation and quality is not exception to it. The use of genetic engineering in improving protein quality of maize is a very potential area of application of biotechnology since conventional breeding suffers from various draw backs. In fact the earlier breeding efforts have been some times frustrating with either no results or limited success. Grain protein quality is governed by various biochemical pathways and once those pathways are elucidated, it is easier to manipulate them for bringing about desired changes in grain protein profile. Various biochemical pathways are potential targets for manipulation using genetic engineering (Shewry, 2007).

Increasing lysine rich proteins

Introduction of proteins that carry higher proportion of lysine residues is a unique approach to improve grain protein. This approach offers advantages as it is not associated with any adverse effect on grain texture as in case of *o2* maturation. Furthermore, since a number of such proteins have been identified among cereals it will not attract too many regulatory controversies. A number of proteins have been found to have higher lysine content including β-amylase (5%), protein Z (7.1%), chrymotrypsin inhibitors CI-I (9.5%) and CI-2 (11.5%) and hordeothionin of barley. Yu et al. (2004) transformed maize with a lysine rich protein from potato pollen resulting in 50% increase in grain protein and lysine. But lack of information about its biological functions is a major concern for using it on large scale.

Reducing zein synthesis

In maize, RNA interference induced downregulation of 22 KDa α-zeins (Segal et al., 2003) and 19 KDa α-zeins (Huang et al., 2005) resulted in higher lysine content (upto 16-20% more lysine) much below the 02 (02) mutants. Double stranded RNA (ds RNA) had been used as a refined approach to simultaneously down regulate both 22 KDa and 19 KDa α -xeins resulting in increase in lysine from 2.83 to 5.62% and tryptophan from 0.69 to 1.22%. The advantage of this approach is based on the fact that the dominant nature of transgene ensures maintenance of quality in farmers fields under varying degrees of contamination fro pollen of normal maize, which is a major problem in case of recessive O2 mutants.

Increasing the level of free amino acids

Even though free acids account for small proportion of grain proteins, nevertheless they are also a potential target for genetic manipulation. The essential amino acids such as lysine, threonine and metheoine, in all higher plants, are synthesised from aspartic acid via a pathway that is highly branched and under complex feed back inhibition (Zhu et al., 2007). The key enzymes of pathway are aspartate kinase (AK) and dihydropicolinate synthase (DHPS).The former (AK) is important at early steps of pathway and is inhibated by both lysine and threonine while as latter is inhibited by lysine only. Mazur et al. (1999) expressed a *Corynebacterium* DHPS gene driven by globulin-1 promoter in aleurone and embryo of maize and observed 50-100% increase in free lysine. However, no increase in free lysine was observed when the same gene was expressed in endosperm (driven by glutelin-2 (γ -zein) promoter). Zhu and Galili (2003) expressed a becterial (*E. coli* or *Coryneoacterium*) feed back insensitive, DHPS in *Arabidopsis* and achieved 12 fold increase in grain lysine content. The expression of DHPS in maize increase the free lysine content in grain from 2 to 30% of total amino acid pool (Zhu et al., 2007). Monsanto in 2006 has also released transgenic maize with high lysive (lysine) by expressing feed back insensitive DHPS gene from *Corynebacterium* drive by globulin-1 gene promoter. The free lysive in grain (dry weight

basis) increased from 2500-2800 ppm to 3500-5300 ppm.

Marker-assisted selection for QPM breeding

The development of QPM requires manipulation of various genetic systems such as O2, endosperm modifiers and amnic acid modifiers and as such conventional breeding procedures are quite laborious and the results some times frustrating. It is very tedious to continuously select for optimum level of one trait while maintaining desired level of other. The conventional backcross strategy for conversion of normal maize to QPM suffers from two major problems. One that *o2* being a recessive trait, selection has to be carried out at each backcross in order to fix the recessive *o2* allele, prior to selection for endosperm modification, thereby extending the time period for line conversion. While as, latter is inhibited by lysine only. Moreover, the quality traits such as grain protein content cannot be selected prior to seed formation making screening very difficult. Besides, low cost and reliable methods of screening are not available.

Marker assisted selection is an appropriate technology for traits such as high lysine in maize and can be a cost effective procedure for selecting *o2* locus in breeding populations (Dreher et al., 2003). With sequencing of maize genome being finished, a large number of market system are now available that are associated with *o2* and endosperm modification phenoltype (Lopez et al., 2004; Bantle and Prasanna, 2003). An appropriate application of such markers will greatly enhance the efficiency of selection for improvement of grain protein in maize besides cutting down at cost and time. Both foreground MAS and background MAS can be effectively employed for selecting *o2* phenotype besides ensuring maximum recovery of recurrent parent. Babu et al. (2005) used MAS for development of QPM parental lines of Vivek-9 hybrid and could developed QPM hybrid in less than half the time required through conventional breeding. Danson et al. (2006) used various markers to introgress *o2* gene into herbicide tolerant elite maize inbred lines. They found that using marker for QPM and endosperm modification in tonden can greatly enhance the selection efficiency for isolating fully modified kernels in QPM background. reviewed enthusiasm in our endeavour to make QPM of real potential use."

Conclusion

QPM research has witnessed early enthusiasm, a phase of near abandonment and a phase of renewed interest. The key feature of all these phases has been the commitment of QPM breeders at CIMMYT and elsewhere and the results are quite obvious. Today improve (improved) QPM varieties and hybrids have been released in about 40 countries with African countries being the major target. CIMMYT has done all the good work to make QPM research a worthwhile venture. It is in recognition of this commitment that S.K. Vasal (Maize breeder) and E. Villegas (Cereal chemist) were confessed upon the prestigious world food prize in 2000 for their efforts at CIMMYT to develop QPM. But the overall success of QPM has been due to painstaking efforts of a number of scientists at CIMMYT and elsewhere notably Magni Bjarnson, Kevin Pixley, Brain Larkins, Norman Borlaug, Hugo Cordovo, Timothy Reeves and S. Pandey, various agencies such as Nippon foundation, CIDA, Rockefeller foundation and Sasakann Global 2000 and various partner national maize research systems.

The most important goal of QPM research has been to reduce malnutrition in target countries through direct human consumption, even though, the impact, as of now, has now been as great perceived, it is expected that greater impact will accrue out of development and dissemination of improved hybrids, OPV's and Synthetics in Africa and Mesoamerica. The new QPM synthesis have desirable features such as low uniform ear placement, resistance to ear rot and root lodging and better grain protein profile (Cordovo, 2001). Efforts are on at CIMMYT and elsewhere to transfer high lysine trait in elite stress tolerant genotypes through marker assisted selection. The sequencing of maize genome and identification of markers associated with protein and grain modification will help rapid identification of genes responsible for such traits and thereby accelerate the development of QPM varieties for target countries where probability of impact in high (Gibbon and Larkins, 2005). In terms of feed component, China, Brazil, Mexico and Vietnam are potential targets as maize is extensively used as component of animal feed.

For better dissemination and adoption of QPM in target countries we need to understand factors affecting is adoption/disadoption, agronomic characteristics, awareness about nutritional quality, availability of seed and ease of recycling (Kravinek et al., 2007). The words of S.K. Vasal are worthwhile to mention, "Good science is not free from difficulties, frustration and criticism. This should be viewed to generate creativity, revisiting different approaches and activities, and making constant adjustments for efficient use of resources at all levels. Periodic reviews help to de-emphasize some aspects while expanding others, if necessary and introducing new initiatives, while were not in place already, therefore, it is very essential to make new initiatives, try novel ideas and pump reviewed enthusiasm in our endeavour to make QPM of real potential use."

REFERENCES

Prasanna B, Sarkar K (1991). Coordinate genetic regulation of maize endosperm. Maize genetics Perspectives ICAR pp. 74-86.

Bantle K, Prasanna B (2003). Simple sequence repeat polymorphium in QPM lines. Cuphytica 129: 337-344.

Babu R, Nair S, Kumar A, Venkatesh S, Shekhar J, Singh NN, Gupta H (2005). Two generation marker aided backcrossing for rapid conversion of normal maize lines to quality protein maize. Thero. Appl. Genet. 111: 888-897.

Bjarnson M, Vasal SK (1992). Breeding for quality protein maize.

Plant Breed. Rev. 9: 181-216.

Cordova H (2001). Quality protein maize: improved nutrition and livelihoods for the poor. Maize Research Highlights 1999-2000, CIMMYT pp. 27-31.

CIMMYT (2005). The development and promotion of QPM in sub-saharan Africa, Progress Report Submitted to Neppon Foundation. 6: 312-324.

Damerval C, Devienee D (1993). Qualification of dominance for proteins plieotropically affected by opaque-2 in maize. Heredity 70: 38-51.

Danson J, Mbogori M, Kimani M, Lagat M, Kuria A, Diallo A (2006). Marker-assisted introgression of opaque 2 gene into herbicide tolerant elite maize inbred lines. African J. Biotech. 5: 2417-2422.

Gibbon B, Larkin B (2005). Molecular genetic approaches to developing quality protein maize. Trends Genet. 21: 227-233.

Geetha K, Lending C, Lopes M, Wallace J, Larkins B (1991). Opaque-2 modifiers increase α-zein synthesis and alter its spatial distribution in maize endosperm. Plant Cell 3: 1207-1219.

Gill G (2008). Quality protein maize and special purpose maize improvement. In "Recent Advances in crop improvement" CAS training at PAU from 05-25 Feb, 2008. pp. 377-385.

Huang S, Frizzi A, Florida C, Kruger D (2006). High lysine and high tryptophan transgenic maize resulting from reduction of both 19- and 22 KDa α-zeins. Plant Mol. Biol. 6: 525-535.

Huang S, Kruger D, Grizz A, Ordene R, Florida C, Adams W, Brown W, Luethy M (2005). High lysine corn produced by combination of enhanced lysine biosysnthesis and reduced zein accumulation. Plant Biotech. J. 3: 555-569.

Huang S, Adams W, Zhou Q, Malloy K, Voyles D, Anthony J, Kriz A, Luethy M (2004). Improving nutritional quality of maize proteins by expressing sense and antisense gents. J. Agric. Food Chem. 52: 1958-1964.

Krivanek A, Groote H, Gunaratna N, Diallo A, Freisen D (2007). Breeding and disseminating quality protein maize for Africa. Afr. J. Biotech. 6: 312-324.

Lauderdale J. (2002). Issues regarding targeting and adoption of quality protein maize. CIMMYT working paper p. 31.

Lopez M, Gloverson L, Larkins B (2004). Genetic mapping of opaque-2 modifier genes. Maize Genet. Newsletter 69: 165.

Mazur B, Krebbers E, Tingey S. (1999). Gene discovery and product development for grain quality traits. Sci. 285-372.

Prasanna B, Vasal S, Kasahun B, Singh N.N. (2001). Quality protein maize. Curr. Sci. 81: 1308-1319.

Paez A, Helm J, Zuber S (1969). Lysine content of opaque-2 kernels having different phenotypes. Crop Sci. 9: 251-252.

Segal G, Song R, Messing J (2003). A new opaque variant of maize by a single dominant RNA-interference – inducing transgene. Genet. 165: 387-397.

Segal G, Songh R, Messing J. (2003). A new opaque variant of maize by single dominant RNA-I inducing transgene. Genet. 165: 387-397.

Singh J, Koshy S, Agarwal K, Singh NN, Lodha M, Sethi A (1980). Relative efficiency of opaque-2 maize in growth of preschool children. Ind. J. Nutr. Dietet. 17: 326-334.

Singh NN, Venkatesh S (2006). Development of quality protein maize inbred lines. In: Heterosis in Crop Plants. Ed. Kaloo G, Rai M, Singh M, Kumar S. Res. Book Center, New Delhi, pp. 102-113.

Shewry P (2007). Improving protein content and composition of cereal grain. J. Cereal Sci. 46: 239-250.

Shewry P, Halford N (2002). Cereal seed storage proteins : structure, properties and role in grain utilisation, J. Expt. Bot. 53: 947-958.

Toro A, Medici L, Sodek L, Lea P, Azevedo R (2003). Distribution of soluble amino-acids in maize endosperm mutants. Scientia Agricola 60: 91-96.

Vasal SK (2001). Quality protein maize development : An exciting experience. Seventh Eastern and South Africa Regional Maize Conference pp. 3-6.

Vasal SK, Srinivasan G, Pandey S, Ganzalez F, Crossa J, Beck DF (1993a). Heterosis and combinign ability of CIMMYT's Quality protein maize germplasm. I. lowland tropical. Crop Sci. 33: 46-51.

Vasal SK, Srinwasan G, Gonzalez C, Beck D, Crossa J (1993b). Heterosis and combining ability of CIMMYT's quality protein maize germplasm. II. Sub-tropical. Crop Sci. 33: 51-57.

Vasal SK (2000). Quality protein maize story. Proceedings of workshop on Improving Human nutrition through agriculture. The Role of International Agricultural Research, IRRI. pp. 1-16.

Wang X, Woo Y, Kim C, Larkins B (2001). Quantitative trait locus mapping of loci influencing elongation factor 1 alpha content in maize endosperm. Plant Physiol. 125: 1271-1282.

Wu R, Lou X, Ma C, Wang X, Larkins B, Casella G (2002). An improved genetic model generates high resolution mapping of QTL for protein quality in maize. PNAS 99: 11281-11286.

Yu J, Peng P, Zhang X, Zhao Q, Zhy D, Su X, Liu J, Ao G (2004). Seed specific expression of lysine rich protein Sb401 gene significantly increases both lysine and total protein in maize seeds. Mol. Breed. 14: 1-7.

Zhai S (2002). Nutritional evaluation and utilization of QPM Zhang Dan 9404 in laying ren feeds. M.Sc. Thesis. North Western Agri. & Forestry Univ. of Sci. & Tech, China, pp. 13-21.

Zhu X, Galali G (2003). Increased lysine synthesis coupled with a knock out of its catabolism synergistically boosts lysine content an also trans regulators metabolism for other amino acids. Plant Cell 15: 845-853.

Zhu C, Naqvi S, Sonia G, Pelacho A, Capell T, Christou P (2007). Transgenic strategies for nutritional enhancement of plant. Trends Plant Sci. 12: 548-555.

4

Genetic variation of 1RS arm between sibling wheat lines containing 1BL.1RS translocation

Tan FeiQuan[1], Fu ShuLan[1], Tang ZongXiang[1], Ren ZhengLong[1,2*] and Zhang HuaiQiong[1]

[1]State Key Laboratory of Plant Breeding and Genetics, Sichuan Agriculture University, Ya'an, Sichuan 625014, China.
[2]School of Life Science and Technology, University of Electronic Science and Technology of China, Chengdu 610054, China.

3 sets of sibling 1BL.1RS translocation wheat lines (each set containing 2 sibling lines) derived from 3 different F_4 single plant. Between 2 sibling wheat lines, different resistance to powdery mildew was observed. In addition, different bands in α-gliadin mobility zone between 2 sibling wheat lines were also significantly observed by acid polyacrylamide gel electrophoresis (A-PAGE). However, the nucleotide sequence of α/β-gliadin precursor gene cloned from 2 sibling wheat lines were 100% similarity. Polymerase chain reaction (PCR) analysis using 180 wheat microsatellite markers dispersed on 7 homeologous groups of wheat indicated that little genetic variation occurred in wheat A, B and D genomes of these sibling wheat lines. Structural variation of 1BL.1RS chromosome arm was detected by PCR analysis. Different rye-specific repetitive DNA exhibited different variation. The different structure of 1RS chromosome arms may be related to the genetic variation between the 2 sibling wheat lines. The results lead us to think that the variation of 1RS arm within a species should be utilized in wheat breeding program. To detect variation of repetitive DNA can provide better understanding of the genetic variation of 1BL.1RS translocation and will be useful for utilization of 1RS arm in wheat breeding program.

Key words: 1BL.1RS translocation, repetitive DNA, sibling wheat lines, gentic variation.

INTRODUCTION

The introduction of alien chromosomes from related species into wheat (Triticum aestivum L.) has proven to be useful for increasing the genetic diversity available to wheat breeders. Rye (Secale cereale L.) has been used extensively as a source of elite genes introduced into wheat genome. Translocation 1BL.1RS involving the short arm of rye chromosome 1R and the long arm of wheat chromosome 1B is widely used in wheat breeding program. The 1BL.1RS translocation has been introduced into hundreds of wheat cultivars worldwide through the Russian wheat cultivars 'Kavkaz', 'Aurora' and their deri-

vaties (Raiaram et al., 1990; Rabinovich, 1998). Because the 1RS arm carries genes for resistance to diseases and positively affects agronomic traits including yield performance, yield stability and wide adaptation (Schlegel and Meinel, 1994; Moreno-Sevilla et al., 1995; McKendry et al., 1996; Kim et al., 2004), wheat-rye 1BL.1RS translocation lines will be still useful in wheat breeding program. It has already reported that the effect of source of rye chromatin is greater than its position effect in wheat genome and selection of 1RS source is important in producing constantly higher grain yield in 1RS translocation lines (Kim et al., 2004). Approaches have been made to the production of new 1BL.1RS translocation line from different origins of cultivated rye, with aim to improve the genetic polymorphism of the 1RS chromosomes for better diseases resistance (Lukaszewski et al., 2001; Ko et al., 2002a). With the diversified 1BL.1RS translocation lines developed, the cytogenetic and biochemical techno-

*Corresponding author. E-mail: auh5@sicau.edu.cn.

Abbreviations: A-PAGE, acid polyacrylamide gel electrophoresis; PCR, polymerase chain reaction.

'MY11'×rye
↓
F1 (treated with colchicines) ×'MY11'
↓
'MY11'×BC1F1
↓
BC2F1
↓
BC2F4 (single plant)
↓
BC2F5 population
↓
wheat lines

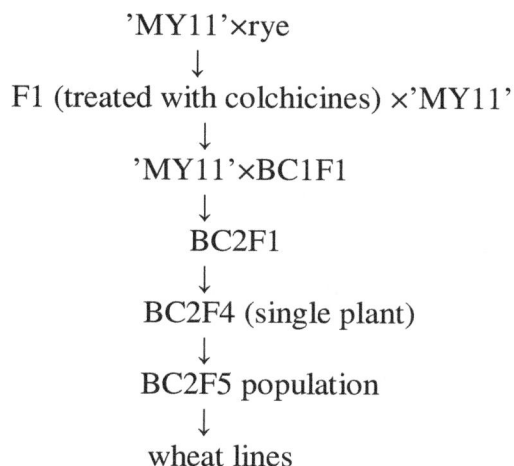

Figure 1. Steps for obtaining wheat lines.

logies were exploited to characterize the genetic variation of 1BL.1RS translocation in wheat background (Li et al., 2002; Yan et al., 2005). To better understand the genetic variation of 1BL.1RS translocation will be useful for utilization of 1RS arm in wheat breeding program.

However, besides cytogenetic and biochemical technologies, DNA sequences can also be used to investigate the genetic diversity of 1BL.1RS translocation. It has reported that the major determinants of genome system architecture are the repetitive elements in the genome, such as tandem repeats and dispersed repeats (Shapiro, 2005). Species in the tribe *Triticeae* hold large genomes and the majority of their genomes consist of repetitive DNA sequences (Flavell and Smith, 1976; Flavell, 1986). Taken these reports together, it is possible and reasonable to use repetitive DNA sequences to detect genetic diversity induced by 1BL.1RS translocation. Furthermore, some sibling wheat lines/cultivars which were developed in wheat breeding program exhibit diversity in some traits. These sibling wheat lines are good materials for studying the mechanism by which repetitive DNA sequences induce genetic variation.

In this study, we selected 3 sets of sibling 1BL.1RS translocation lines from 76 wheat lines. Each set of sibling 1BL. 1RS translocation line was derived from a F_4 single plant. However, some different traits were observed between 2 sibling translocation lines. We mainly investigated the structure of 1BL.1RS translocation chromosome using repetitive DNA to discover the probable molecular mechanisms which in-duce the genetic variation in sibling 1BL.1RS translocation lines.

MATERIALS AND METHODS

Plant materials and pedigree

Common wheat (*T. aestivum* L.) 'Mianyang11' ('MY11') was used as

recipient and rye (*Secale cereale* L.) inbred lines R3, R12 and Baili were used as donors. 'MY11' was released in 1981 and was a widespread high-yielding wheat cultivar in Southwest China. It has already been susceptible to several diseases such as stripe rust, powdery mildew and *Fusarium* head blight for many years. Wheat lines 96-132-1 and 96-132-2 were obtained from the cross between 'MY11'×R3. Wheat lines 96-137-2 and 96-137-3 were derived from 'MY11'×R12. Wheat lines 96-212-1 and 96-212-2 were derived from 'MY11'×Baili. These wheat lines were obtained by following steps (Figure 1).

The same single plant selected from each wheat line was used for each analysis in this study, such as disease resistance observation, cytological analysis, seed storage protein electrophoresis, southern blot analysis and PCR analysis. Wheat cultivars Chinese spring ('CS') and rye inbred line L155 were also used in this study.

Disease resistance observation

The 3 sets of sibling wheat lines were evaluated for resistance to powdery mildew. Seedlings were inoculated by the epidemic predominant race of powdery mildew in southwest China. 6 infection types were recorded on a scale of 0, 0;,1, 2, 3 and 4. 0 is no visible symptoms, 0; is necrotic flecks, 1 is highly resistant, 2 is moderately resistant, 3 is moderately susceptible and 4 is highly susceptible.

Cytological analysis and seed storage protein electrophoresis

A-PAGE and sodium dodecyl sulphate-polyacrylamide gel electrophoresis (SDS-PAGE) were used to separate endosperm gliadin proteins and glutenin subunits, respectively. The procedures were described by Yang et al. (2001). For the identification of the chromosomes, the C-banding techniques described by Ren and Zhang (1995) were applied. Meiotic behavior of the 6 wheat lines was investigated according to Li et al. (2002).

Selection and design of primer

3 sets of primer pairs were designed according to original pSc119.1 sequence (McIntyre et al., 1990), pSc20H sequence (GenBank accession No.AF305943) and alpha-/beta-gliadin storage protein precursor gene (GeneBank Accession NO. DQ166376). This gene was selected randomly. The primer pairs of sequences pSc119.1, pSc20H and the gene are named as Pr119.1 (5'TTGGC CCTCA TGCCT TTAGT CCTTG C3'; 5'CTTGG CCCTC TCCGC TTGAC CGTTG CTC3'), Pr20H (5' GTTGG AAGGG AGCTC GAGCT G 3'; 5'GTTGG GCAGA AAGGT CGACA TC3') and Prαβ-gli (5'CATCC TTGCC CTCCTT GCTA3'; 5'TGGTA CCGAA GATGC CAAAT3'), respectively. Sequences pSc119.1 and pSc20H are all rye-specific interspersed repetitive DNA (McIntyre et al., 1990; Ko et al., 2002b). Additionally, one microsatellite marker *SCM 9* mapped on chromosomal arm 1RS (Saal and Wricke, 1999) and 180 microsatellite markers dispersed on 7 homeologous groups of wheat (Röder et al., 1998) were also used in this study.

PCR analysis, cloning and sequencing

Genomic DNAs were extracted from each single plant according to the method described by Zhang et al. (1995). Each PCR reaction (25 µL) using simple sequence repeat (SSR) primers, Pr119.1 and Pr20H contained 50 mM KCL and 10 mM Tris-HCL (pH 8.8), 1.5 mM $MgCL_2$, 200 µM of dNTP, 200 µM of each primer, 1.0 unit of *Taq* polymerase (Promega), and 40-60 ng of genomic DNA. The annealing temperature of SSR markers was according to Röder et al. (1998) and Saal and Wricke (1999). The annealing temperature

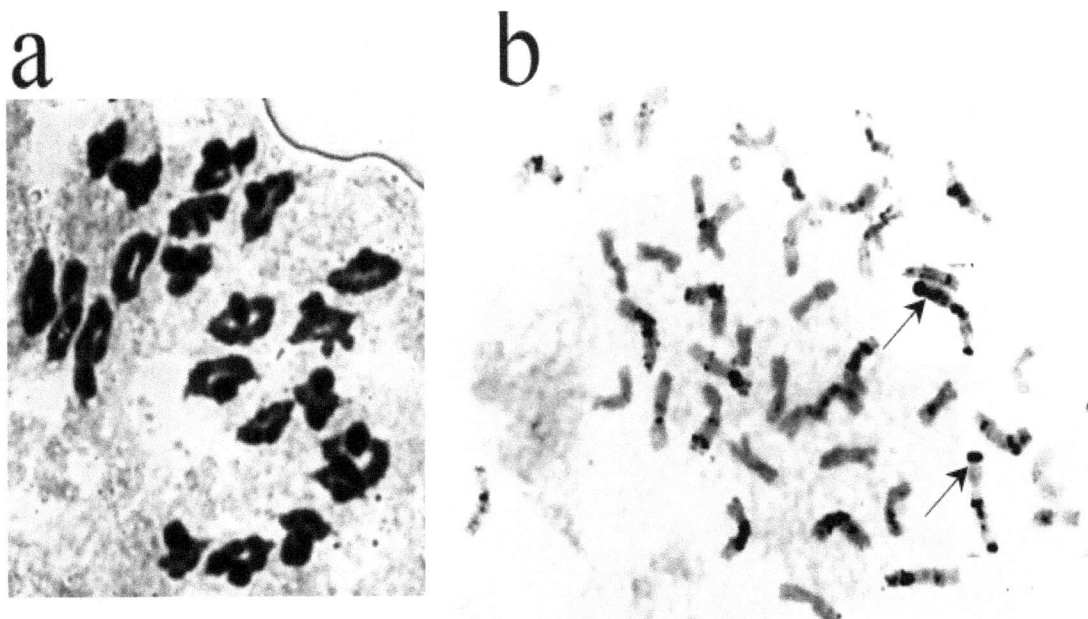

Figure 2. Cytological analysis of 6 wheat lines. (a) 21 cycle bivalents of 6 wheat lines at meiotic metaphase I (Only 96-132-1 is shown); (b) C-banding of 6 wheat lines (Only 96-132-1 is shown). Arrows show the 1BL.1RS chromosomes.

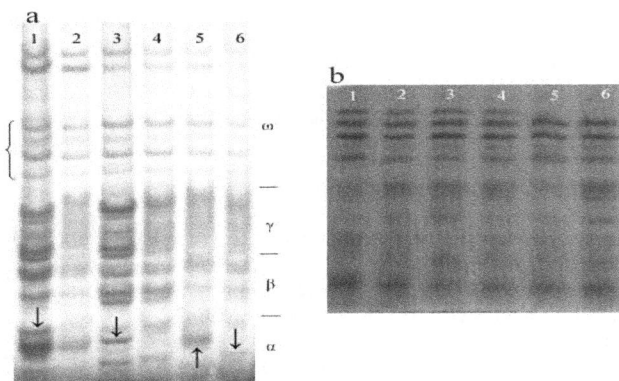

Figure 3. Storage protein electrophoresis of 6 wheat lines. (a) Gliadin storage protein separated by A-PAGE. (b) Glutenin subunits separated by SDS-PAGE. Lane 1, 96-132-1; Lane 2, 96-132-2; Lane 3, 96-137-2; Lane 4, 96-137-3; Lane 5, 96-212-1; Lane 6, 96-212-2. Bracket indicates bands of rye secalins. Arrows indicate different bands in α-gliadin mobility zone between two sibling wheat lines.

of Pr119.1 and Pr20H was 60°C. Each PCR reaction (25 μL) using Praβ-gli contained 80 ng of template, 2.5 μL 10×PCR buffer, 1.5 mM MgCL$_2$, 200 μM of dNTP, 200 μM of each primer, 1.25 unit of Ex *Taq* polymerase (TaKara, Japan), the annealing temperature was 60°C. PCR amplifications were carried out on an MJ research PTC-200 thermocycler, using a program that consisted of initial denaturation for 3 min at 94°C, followed by 35 cycles of 1 min at 94°C, 30 s at annealing temperature, 2 min at 72°C and final extension for 10 min at 72°C. The amplified products were analyzed on on 2% agarose gels (FMC brand, Spain). Using 0.5 Tris-borate-

EDTA buffer with constant power 120 v (4 v/cm). The PCR fragments were visualized using ethidium bromide staining methods.

Target PCR product of Praβ-gli was recovered using gel extraction kit (Omega E. Z. N.A., USA) and products were cloned into pMD18-T simple vector (Takara, Japan). Clones were sequenced by the commercial company Invirtrogene biotechnology (Shanghai) Co., Ltd. Sequence analysis was performed with the software DNAMAN Version 4.0.

Southern blot analysis

The procedure of southern blot was described as Tang et al. (2006). Genomic DNAs of wheat lines were digested by *Bam*HI. The plasmid pSc119.1 (gifted by Dr. Gustafson, University of Missouri Columbia, USA) was used as probe. The probe was labeled with [^{32}P] dCTP using the random primed DNA labeling kit (TaKaRa Biotechnology Co., Ltd, Dalian).

RESULTS

Characteristics of the 3 sets of sibling wheat lines

The meiotic behavior of the 6 wheat lines was investigated at metaphase I and 21 cycle bivalents were observed (Figure 2a). The regular meiotic behavior reveals their cytological stability. C-banding analysis indicates that all the 6 wheat lines are 1BL.1RS translocations (Figure 2b). The resistance to powdery mildew is different between 2 sibling wheat lines of each set. Wheat lines 96-132-1, 96-137-2 and 96-212-1 are highly susceptible to powdery mildew, however, wheat lines 96-132-2, 96-137-3 and 96-212-2 showed highly resistant to powdery mildew. The seed gliadin A-PAGE patterns of the 6 wheat lines are shown in Figure 3a, where one can clearly observe that bands in α-gliadin mobility zone between 2 sibling wheat lines are significantly different. However, the glutenin composition between 2 sibling wheat lines is same (Figure 3b). Furthermore, the nucleotide sequence of α/β-gliadin

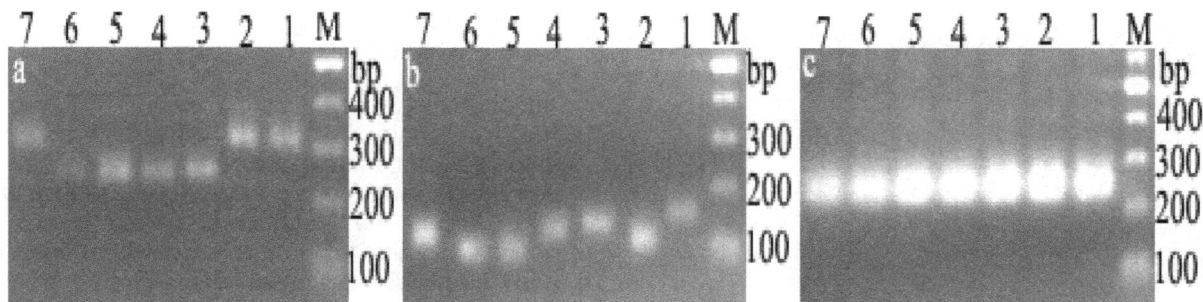

Figure 4. PCR amplification with wheat SSR markers. (a) Amplification products of X*gwm*140; (b) Amplification products of X*gwm*359; (c) Amplification products of X*gwm*4. Lane 1, 96-132-1; Lane 2, 96-132-2; Lane 3, 96-137-2; Lane 4, 96-137-3; Lane 5, 96-212-1; Lane 6, 96-212-2. Lane 7, 'MY11'; M, DNA size marker.

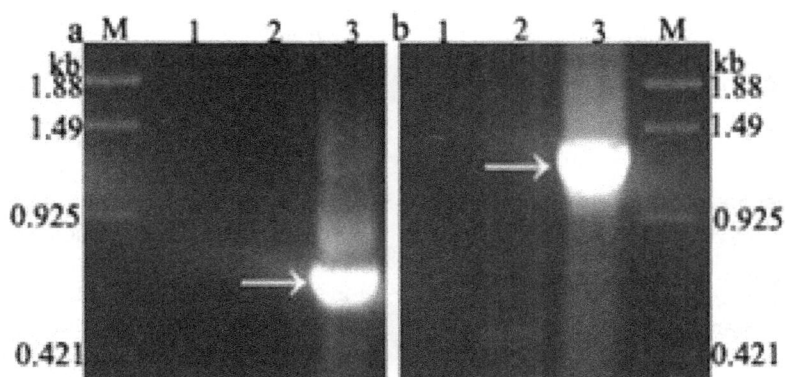

Figure 5. Specialty testing of primer pairs. (a) Amplification product of primer Pr119.1. (b) Amplification product of primer Pr20H. Arrows indicate the target fragment of each primer pair. Lane 1, 'CS'; Lane 2, 'MY11'; Lane 3, rye L155; M, DNA size marker.

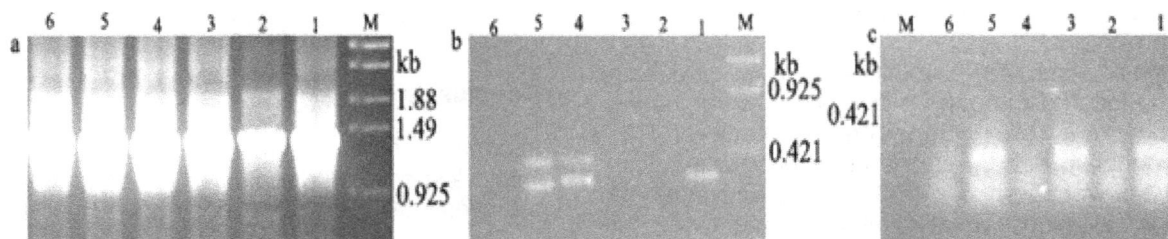

Figure 6. PCR amplification of 6 wheat lines using primers Pr20H (a), Pr119.1 (b) and *SCM*9 (c). Lane 1, 96-132-1; Lane 2, 96-132-2; Lane 3, 96-137-2; Lane 4, 96-137-3; Lane 5, 96-212-1; Lane 6, 96-212-2. M, DNA size marker;

precursor gene cloned from 2 sibling wheat lines from each set showed 100% similarity among them. The gene sequences cloned from the three sets of sibling wheat lines were submitted to the public database of NCBI GenBank. Sequence with Genbank Accession NO. EF165554 was cloned from 96-132-1 and 96-132-2, sequence with Genbank Accession NO. EF165555 was cloned from 96-137-2 and 96-137-3 and sequence with Genbank Accession NO. EF165556 was cloned from 96-212-1 and 96-212-2. Out of the 180 wheat SSR markers, only 5 markers displayed polymorphism among 'MY11' and the 6 wheat lines (Figures 4a, 4b). These SSR markers are X*gwm*140, X*gwm*408, X*gwm*469, X*gwm*497 and X*gwm*539. The other wheat SSR markers amplified the same band pattern in 'MY11' and the analyzed 6 wheat lines (Figure 4c).

Rye-specificity of primers Pr119.1 and Pr20H

The genomic DNA extracted from rye L155, 'MY11' and 'CS' were amplified with the two sets of primer pairs Pr119.1 and Pr20H. All the primer pairs amplified the target products only from the genomic DNA of L155, however, no products were amplified from the genomic DNA of 'MY11' and 'CS' (Figure 5), thus showing that the 2 primer pairs are rye-specific.

Structural variation of 1BL.1RS translocation chromosomes

Primer pairs Pr20H amplified the target fragments from the genomic DNA of all the 6 wheat lines (Figure 6a). However, primer pairs

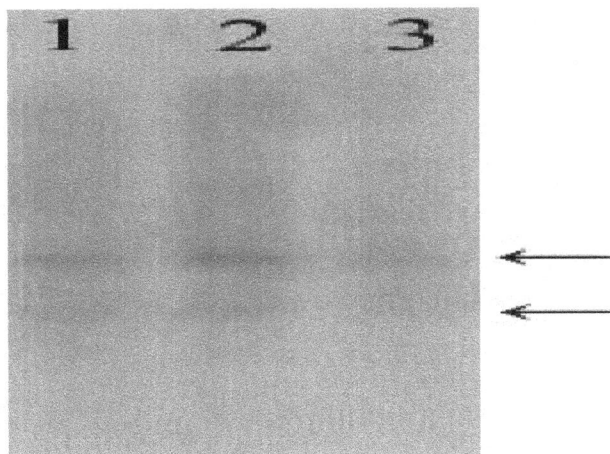

Figure 7. Southern blot of the genomic DNA extracted from 96-132-2, 96-137-2 and 96-212-2 using pSc119.1 as probes. Lane 1, 96-132-2; Lane2, 96-137-2; Lane 3, 96-212-2. Arrows indicate the hybridization signals.

Pr119.1 amplified only the target fragments from the genomic DNA of wheat lines 96-132-1, 96-137-3 and 96-212-1, but not from the wheat lines 96-132-2, 96-137-2 and 96-212-2 (Figure 6b). Furthermore, a non-target fragment was also amplified from the genomic DNA of wheat lines 96-137-3 and 96-212-1 using primer pair Pr119.1 (Figure 6b). The PCR-amplified bands using rye microsatellite markers SCM9 were different between 2 sibling wheat lines (Figure 6c). Southern-blot analysis using pSc119.1 as probe indicated that the sequence pSc119.1 are not eliminated from wheat lines 96-132-2, 96-137-2 and 96-212-2 (Figure 7).

DISCUSSION

Structure variation of 1BL.1RS chromosomes

Rye-specific dispersed repetitive DNA sequence pSc119.1 is located throughout the 14 rye chromosome arms (McIntyre et al., 1990). The difference of PCR amplification using Pr119.1 between 2 sibling wheat lines discovers the variation of pSc119.1 and indicates the structural variation of 1RS arm at the same time. Furthermore, the different amplification products of SCM9 between 2 sibling wheat lines detects the variation of flanking regions of microsatellite and also suggests the structural variation of 1RS arm. It has already been reported that variation of repetitive elements will occur immediately following wide hybridization (Ma et al., 2004; Salina et al., 2004; Han et al., 2005; Ma and Gustafson, 2006). In this study, the 6 wheat lines were also derived from wheat-rye wide hybridization. The variation of repetitive DNA sequences has still been observed in these wheat lines although they are in F_5 generation, indicating that variation of repetitive element may be continual in progeny derived from wide hybridization. PCR analysis did not discover the variation of repetitive sequence pSc20H, indicating that different interspersed repetitive elements play different roles in genomic variation.

Mechanism inducing genetic divergence between 2 sibling wheat lines

Although the 3 sets of sibling wheat lines were all derived from a single F_4 plant, some differences of traits such as resistance to powdery mildew and seed gliadin composition was observed between 2 sibling wheat lines. Whether did the mutation of genes induce the divergences of traits? The cases that wheat lines are sibling and α/β-gliadin precursor gene cloned from 2 sibling wheat lines were 100% similarity lead us to think divergences of traits probably did not result from mutation of genes. Shapiro (2005) indicated that it is possible for 2 genomes in different species to have identical coding sequences but distinct signals and genome system architectures and the different patterns of coding sequence expression will lead to phenotypic and ecological diversity. Madlung et al. (2002) have described the mechanisms of epigenetic gene regulation resulting in phenoltype instability among sibling allotetraploids of Arabidopsis. Furthermore, the PCR analysis using 180 wheat SSR markers does indicate that the wheat A, B and D genomes are high genetic similarity between 2 sibling wheat lines. The case that the glutenin composition between 2 sibling wheat lines is same also indicates that 2 sibling wheat lines are genetic identity. Taken together, it is reasonable to presume that the divergences of traits between 2 sibling wheat lines probably were induced by the variation of 1RS arm.

As well known, in wheat breeding process, heterozygosity still exists in wheat lines even though they are in high generation. The results in this study lead us to think that using repetitive DNA sequences may elucidate the reasons why the heterozygosity exists in wheat lines when they are in high generation.

Conclusions

In conclusion, the genetic variation of 1RS could occurred within a species and repetitive DNA sequences are useful for investigating the genetic variation

ACKNOWLEDGEMENTS

This work was supported by national natural science foundation of China (Grant No.30730065) and youth foundation of Sichuan Agriculture University (Grant No. 00131300). We also want to thank Hanmei Liu for her technical assistance.

REFERENCES

Flavell RB (1986). Repetitive DNA and chromosome evolution in plants. Philos Trans R Soc Lond B 312: 227–242.
Flavell RB, DB Smith (1976). Nucleotide sequence organisation in the wheat genome. Heredity 37: 231-252.
Han FP, Fedak G, Guo W, Liu B (2005). Rapid and repeatable elimination of a parental genome-specific DNA repeat (pGc1R-1a) in

newly synthesized wheat allopolyploids. Genet.170: 1230-1245.

Kim W, Johnson JW, Baenziger PS, Lukaszewski AJ, Gaines CS (2004). Agronomic effect of wheat-rye translocation carrying rye chromatin (1R) from different sources. Crop Sci. 44:1254-1258.

Ko JM, Seo BB, Suh DY, Do GS, Park DS (2002a). Production of new wheat line prossessing the 1BL.1RS wheat-rye translocation derived from Korean rye cultivar Paldanghomil. Theor Appl Genet. 104:171-176.

Ko JM, Do GS, Suh DY, Seo BB, Shin DC, Moon HP (2002b). Identification and chromosomal organization of two rye genome-specific RAPD products useful as introgression markers in wheat. Genome 45:157-164.

Li YW(not cited. Provide or delete), ZS Li, Jia X (2002). Meiotic Behavior of 1BL/ 1RS Translocation Chromosome and Alien Chromosome in Two Tri-genera Hybrids. Acta. Bot. Sin. 44:821-826.

Lukaszewski AJ, Porter DR, Baker CA, Rybka K, Lapinski B (2001). Attempts to transfer Russian wheat aphid resistance from a rye chromosome in Russian triticales to wheat. Crop Sci. 41:1743-1749.

Madlung A, Masuelli RW, Watson B, Reynolds SH, Davison J, Comai L (2002). Remodeling of DNA methylation and phenotypic and transcriptional changes in synthetic Arabidopsis allotetraploids. Plant Physiol. 129: 733-746.

Ma XF, Fang P, Gustafson JP (2004). Polyploidization-induced genome variation in triticale. Genome 47: 839-848.

Ma XF, Gustafson JP (2006). Timing and rate of genome variation in triticale following allopolyploidization. Genome 49: 950-958.

McIntyre CL, Pereira S, Moran LB, Appels R (1990). New Secale (rye) DNA derivatives for the detection of rye chromosome segments in wheat. Genome 33: 635-640.

McKendry AL, Tague DN, KE Miskin (1996). Effect of 1BL.1RS on agronomic performance of soft red winter wheat. Crop Sci. 36: 844-847.

Moreno-Sevilla B, Baenziger PS, Shelton DR, Graybosch RA, Peterson CJ(1995). Agronomic performance and end-use quality of 1B vs. 1BL/1RS genotypes derived from winter wheat 'Rawhide'. Crop Sci. 35:1607-1612.

Rabinovich SV (1998). Importance of wheat-rye translocation for breeding modern cultivars of Triticum aestivum L. Euphytica 100: 323-340.

Ren ZL, Zhang HQ (1995). An improved C-banding technique for plant chromosomes. J. Sichuan Agri. Univ. 13:1-5.

Röder MS, Korzun V, Wendehake K, Plaschke J, Tixier MH, Leroy P, Ganal MW (1998). A Microsatellite map of wheat. Gene 149: 2007-2023.

Saal B, Wricke G (1999). Development of simple sequence repeat markers in rye (Secale cereale L.).Genome 42: 964-972.

Salina EA, Numerova OM, Ozkan H, Feldman M (2004). Alterations in subtelomeric tandem repeats during early stages of allopolyploidy in wheat. Genome 47: 860-867.

Schlegel R, Meinel A (1994). A quantitative trait locus (QTL) on chromosome arm 1RS of rye and its effect on yield performance of hexaploid wheats. Cereal Res. Commun. 22: 7-13.

Shapiro JA (2005). A 21st century view of evolution: genome system architecture, repetitive DNA, and natural genetic engineering. Gene 345: 91-100.

Tang ZX, Ren ZL, Wu F, Fu SL, Wang XX, Zhang HQ (2006). The selection of transgenic recipients from new elite wheat cultivars and study on its plant regeneration system. Agri. Sci. in China 5: 417-424.

Yan BJ, Zhang HQ, Ren ZL (2005). Molecular cytogenetic identification of a new 1RS/1BL translocation line with secalin absence. Hereditas(Beijing), 27(4): 513-517.

Yang ZJ, Li GR, Jiang HR, Ren ZL (2001). Expression of nucleolus, endosperm storage proteins and disease resistance in an amphiploid between Aegilops tauschii and Secale silvestre. Euphytica 119: 317-321.

Zhang HB , ZhaoXP , Ding X , Paterson AH , Wing RA (1995). Preparation of megabase-sized DNA from plant nuclei. Plant J. 7: 175-184.

5

Influence of temperature and genotype on *Pythium* damping-off in safflower

M. H. Pahlavani[1]*, S. E. Razavi[2], F. Kavusi[3] and M. Hasanpoor[3]

[1]Department of Plant Breeding and Biotechnology, Gorgan University of Agricultural Sciences and Natural Resources, P.O. Box 386, Gorgan, Iran.
[2]Department of Plant Protection, Gorgan University of Agricultural Sciences and Natural Resources, P.O. Box 386, Gorgan, Iran.
[3]BS students of Plant Breeding; Gorgan University of Agricultural Sciences and Natural Resources, P.O. Box 386, Gorgan, Iran.

Improvement of genetic potential in safflower (*Carthamus tinctorius*) against Pythium species would be an efficient means of control of this major seed and seedling fungal pathogen. The type and content of reaction for plant to pathogen could be severely affected by environmental conditions such as temperature. In this study seed rot and seedling damping-off of fourteen safflower genotypes that came from different origins, were evaluated using *Pythium ultimum* infected and sterile paper towels at temperatures 10, 15, 20, 25 and 30°C. Both factors including the temperatures and the genotypes and their interaction affected seed germination of safflower. The results showed that temperature had a significant effect on number of normal and diseased seedlings in Pythium-infected media. Among the five different levels of treated temperatures, the lowest number of normal seedlings occurred at 25 and 30°C, and the lowest number of diseased seedlings were also observed at 10 and 15°C. There was a considerable difference among the fourteen studied genotypes for number of normal seedlings and number of diseased seedlings in infected media under laboratory conditions. The effect of genotype × temperature interaction on both number of normal seedlings and number of diseased seedlings was no significant. Cultivar CW-74 had the lowest, and cultivars LRV-51-51 and LRV-55-259 had the highest number of normal seedlings under *Pythium*-infected conditions. And also, Line 34072 had the lowest, and cultivar CW-74 had the highest number of diseased seedlings in Pythium-infected media. In fields infesting with *P. ultimum*, sowing safflower seed when temperature is more than 15°C is likely to have poor stand establishment due to seed rot and seedling damping-off. Therefore it is advisable to plant safflower early when soil temperature is cool.

Key words: *Pythium ultimum*, zoospore, seed, seedling, rot.

INTRODUCTION

Safflower, *Carthamus tinctorius* L., is an annual, broad leaf crop which belongs to the family of Compositeae. Safflower is cultivated worldwide as an oilseed or ornamental crop. In Iran, this crop is grown for its seeds to extract oil or feed home birds, and also for its flowers to use in medicine or ornamental purposes, and is being cultivated on approximately 1000 ha annually (FAO, 2008). Safflower suffers severely from soil pathogens, which may attack seed, germinating seed, and young seedlings or at time of seed formation, causing directly or indirectly yield and quality losses. Seed and seedling rots by *Pythium* as well as *Phytophthora* rots are among the more devastating soil borne diseases of safflower (Heritage et al., 1984; Huang et al., 1992). Studies showed that *Pythium ultimum* Trow. is the causal agent of seed rot and seedling damping-off of safflower in Iran and other countries (Ahmadi et al., 2008; Ahmadinejad and Okhovat, 1976; Huang et al., 1992, Mundel et al., 1995). It is not only made some limitations for safflower production in Iran, but also for the other producing areas in the world. The pathogen parasitizes seeds and invades the hypocotyl or first internode tissues of safflower seedlings

*Corresponding author. E-mail: hpahlavani@yahoo.com.

and causes rotting and collapse of infected tissues and finally decays seeds and seedlings (Kolte, 1985; Thomas, 1970). Although different chemical fungicides are used to control damping-off, but similar to other fungi diseases the best way for decrease the losses is planting the resistant cultivars.

The condition of infection, seed decay and seedling death caused by Pythium has been studied in safflower and other crop plants (Ben-Yephet and Nelson, 1999; Fortnum et al., 2000; Mundel et al., 1995; Thomas, 1970). Like other soil borne pathogenic fungi, severity of infection by Pythium, incidence of damping-off and loses in crop production is a function of environmental factors and how the plants can use from their genetic potential to resist against the pathogen (Ahmadi et al., 2008). Among the most important environmental factors which can favor the disease, temperature plays a key role. The optimum temperatures for infection of citrus fruits to brown rot that caused by Phytophthora palmivora were 27 to 30°C (Timmer et al., 2000). Ben-Yephet and Nelson (1999) studied the effects of 20, 24, 28 and 30°C on differential suppression of Pythium irregulare and showed that the pathogen caused damping-off in cucumber only at 20 and 24°C. Temperature had a profound impact on root rot development, plant growth and infection of carambola (Averrhoa carambola) roots by Pythium splendens (Ploetz, 2004). Infection of apples and pears with Phytophthora cactorum required 3 to 7 h of wetness at temperature 15 to 30°C (Grove et al., 1985). Pythium aphanidermatum and Pythium myriotylum are considered to be broad host range species favored by very warm conditions, whereas others such as P. ultimum and P. irreguare are considered to be broad host range species favored by cool conditions (Van der, 1981).

The incidence of damping off of safflower caused by P. splendens was reported to increase with temperature from 10 to 25°C (Thomas, 1970). Mundel et al. (1995) performed an experiment on infected soil with P. ultimum and indicated that temperature level affected emergence of safflower seedlings and incidence of damping-off. They showed that safflower should be seeded early when soil temperature is low, even though emergence may be slow. Also they concluded that if seeding is delayed until soil temperatures are higher than 10ºC, growers should consider not planting safflower if soil moisture levels are high (Mundel et al., 1995). On the other hand the optimum temperature for safflower seed germination is about 25°C (Gu and Xu, 1984). As noted, the effect of temperature on seed and seedling growth and incidence of Pythium damping-off in safflower has been investigated by researchers. But the best temperature for both pathogen and host has not been investigated. So, finding the optimum temperature in which favorable conditions provide to both fungal infection and expression of disease resistance in host is an important aim of safflower breeders. This study was undertaken to determine the temperature conditions decreasing seed rots and seedlings damping-off caused by P. ultimum in different safflower

genotypes. The objectives were to find temperature in which the lowest seed and seedling death takes place, recognize the most resistant genotype to the pathogen; and determine the effect of temperature × genotype interaction on the disease.

MATERIALS AND METHODS

This study was performed at Gorgan University of Agricultural Sciences and Natural Resources (GUASNR), Gogan, Iran in 2008. In this study germination of fourteen safflower genotypes was evaluated using a test by Pythium-infected and sterile paper towels at five temperatures in four replications.

Isolate of pathogen

Isolate of P. ultimum used in this study was recovered from rotted seeds and dead seedlings of safflower showing typical disease symptoms. The 3 to 5 mm pieces of diseased tissues were surface-sterilized by immersing in 0.5% hypochlorite sodium for 1 min. Then, the sterilized pieces were grown on a CMA selective medium that contained penicillin, streptomycin sulfate, pimaricin antibiotic and benomil (Singleton et al., 1992). After incubation on agar plates for 4 days at 25°C, some pieces of fungi cultures transferred to another plates containing 2% water agar. These plates keep for 24 h at room temperature and the pathogen purified by single hyphal tip isolation technique as describe by Singleton et al. (1992). Identification of Pythium ultimum was performed using previously published criteria (Dick, 1990; Van der Plaats-Niterink, 1981).

Zoospore production of P. ultimum was activated using a method described by Rahimian and Banihashemi (1979). To prepare zoospore suspension, 4×4 mm2 piece of fully grown agar plates were flooded in 500 ml flask containing sterilized distilled water and kept in light conditions for 72 h. These flasks were incubated for 10 min at 5°C and followed by keeping for 2 h at room temperature for releasing of zoospores. Zoospores concentrations were estimated with a hemacytometer, and the appropriate dilution was made with sterilized water to a final concentration of 10^5 zoospores ml-1.

Genotypes of safflower

Safflower genotypes tested in this experiment formed part of safflower collection held at the GUASNR, Gorgan, Iran. These genotypes were included cultivars, promising lines and plant introductions from Iran and other countries and to keep their genetic purity were grown along with controlling cross pollinations at least for three years in Research Farm of the GUASNR. The main reasons for choosing them to be in this study were their good performance, high seed production and considerable variability in seed size and oil content. The names and origins of selected genotypes are shown in Table 3.

Seeding, incubation, germination and disease assessments

50 seeds of each genotype were surface-sterilized in 2% sodium hypochlorite for 3 min, placed on a 50×50 cm2 paper towel. Then another paper towel was placed on seeds and former paper towel and all were rolled. All rolled towels were wetted with distilled water for control and with 10^5 zoospore suspension in Pythium-infected treatments. To keep the humidity on the paper towels (experimental units), all of them put in plastic bags. The experimental units were separately incubated at 10, 15, 20, 25 and 30°C. After 7 days keeping experimental units in incubator, number of germinated seeds (NGS), number of normal seedlings (NNS) and number of diseased

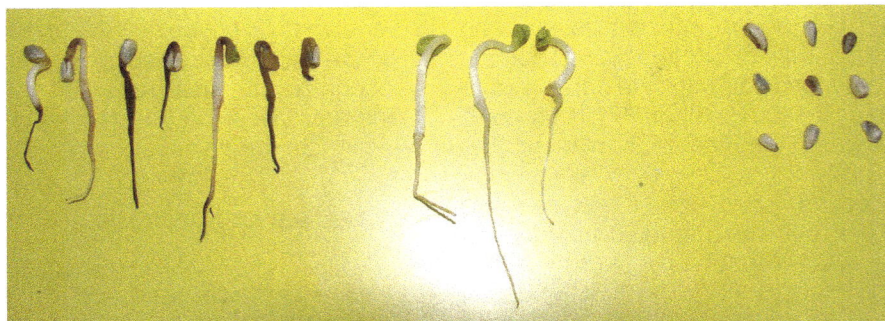

Figure 1. Occurrences of seed rot (right) and seedling death (left) in safflower due to *Pythium ultimum* infection. Normal seedlings are in the middle of the picture. The small gray or black spot on seeds, and dark-brown to black collapsed tissues on seedlings are obviously visible.

seedlings (NDS) were counted. Seedlings were considered germinated when the 3 mm of rootlets went out of the seed coat. The normal seedlings were all apparently healthy (symptomless) seedlings and diseased seedlings were those showed typical symptoms of Pythium damping-off such as brown discoloration (Figure 1).

To verify the cause of seed rots and seedlings damping-off, samples of germinated seed, diseased seedlings and ungerminated seeds were washed in sterile water, dried on paper towel, transferred to 2% water agar in petridishes. The petridishes were incubated at room temperature for 2 to 3 days, and examined for the presence of the pathogen, resembling those identified as *P. ultimum* which has been used for production of the zoospore suspension.

Experimental design and statistical analyses

The temperature levels (main plot; 10, 15, 20, 25 and 30 °C), media (sub plot; *Pythium*-infected and control) and genotypes (sub-sub plot, fourteen safflower genotypes) were run as a split-split plot design with four replications (Snedecor and Cochran, 1980). Each experimental unit was a two rolled paper towels containing 50 seeds. For each experimental unit NGS, NNS and NDS were recorded. Analysis of variance were carried out on NGS, NNS and NDS data to determine if temperatures, infection, genotypes and their interactions have significant effects on the recorded traits.

After examining data with a Kolmogorov-Smirnov (KS) test, data were put through a log transformation to stabilize the variance. Although data transformation decreased error in the coefficient of variation in analysis of variance table but had no significant effects on results, so the raw data were used in all analyses. Analysis of variance, least significant differences (LSD) test and KS test were carried out using the GLM procedure of SAS (SAS, 2004).

RESULTS

Some *Pythium*-infected seeds were rotted and small gray or black spots had been observed on their seed coat, so showed reduced germination (Figure. 1). The pathogen, *P. ultimum*, invades the hypocotyls, cotyledons or some parts of germinating seedlings and caused rotting and collapse of infected tissues and so showed damped-off seedlings. Temperature had a highly significant effect on number of germinated seeds (NGS) (Table 1). The highest NGS occurred at temperatures 10, 15, 20 and 25 °C and the lowest observed at 30 °C (Table 2). The diffe-

rence between *Pythium*-infected and control media for NGS was significant at 1% level and NGS was greater in control than *Pythium*-infected media (Tables 1 and 2). Genotypes differed significantly in their ability to germinate ($P < 0.01$) (Table 1). The results also showed that there is no significant interaction between genotype and media in this experiment (Table 1). The NGS of the genotypes had a variation between 39.1 and 47.8 (Table 3). In general, in every 50 seed plot, more than 45 seeds were germinated at all temperatures for all genotypes. Hartman, Dinger, LRV-51-51 and Arak-2811 with more than 45 germinated seeds were those genotypes that got letter 'a' in LSD test grouping at all temperatures (Table 3).

Because the interaction between genotype and temperature was highly significant for NGS (Table 1), the LSD test among genotypes was separately performed at each temperature (Table 3). At 10 °C, the greatest NGS belonged to a group of genotypes including all except genotypes CW-74 and 34074 (Table 3). At 15 °C, the highest NGS was observed in a group of genotypes including Acetria, LRV-55-295, Hartman, Dinger, LRV-51-51, Arak-2811 and Zarghan-259 (Table 3). At 20 °C, the greatest NGS belonged to a group of genotypes including Acetria, LRV-55-295, Hartman, Dinger, LRV-51-51, Arak-2811, Isfahan, Zarghan-259 and 34074 (Table 3). At 25 °C, the highest NGS were observed in a group of genotypes including Acetria, LRV-55-295, Hartman, Dinger, LRV-51-51, Arak-2811, Isfahan and Zarghan-259 (Table 3). And finally, at 30 °C the greatest NGS belonged to a group of genotypes including LRV-55-295, Hartman, Dinger, LRV-51-51, Arak-2811 and Zarghan-259. Temperature had a significant effect on number of normal seedlings (NNS) in *Pythium*-infected media (Table 1). Among the 5 different levels of treated temperature, the highest NNS were observed at 10, 15 and 20 °C (Table 2). Genotype had little effect on NNS, and all genotypes were similar and had a considerable NNS except genotypes CW-74 and 5-541 (Tables 1 and 3). The range of NNS for the genotypes was 14.90 to 22.20 (Table 3). Also the interaction bet-

Table 1. Analysis of variance of the effect of temperature and genotype on number of germinated seeds (NGS), number of normal seedlings (NNS) and number of diseased seedlings (NDS) in infected media with *Pythium ultimum* in safflower.

Sv	df	NGS	Sv	df	NNS	NDS
Temperature (T)	4	49.51**	Temperature (T)	4	4856.96**	3668.04**
Error 1	15	9.69	Genotype (G)	13	78.38	40.89
Media (M)	1	75.78**	G × T	52	39.20	48.02
M × T	4	5.54	Error	210	63.27	57.27
Error 2	15	6.82	Total	279	—	—
Genotype (G)	13	65.37**				
G × T	52	10.60**				
G × M	13	6.48				
G × M × T	52	4.17				
Error 3	390	6.01				
Total	559	—				

**; significant at % level.

Table 2. Effect of temperature and *Pythium*-infection on seed germination, number of normal and diseased seedlings in safflower.

Temperature (°C)	NGS	NNS	NDS	Media	NGS
10	45.4±2.59	25.7±4.91	19.7±4.81	Sterile	45.6±2.78
15	44.7±2.76	27.5±6.37	17.2±5.69	infected	44.9±2.91
20	45.8±2.37	25.2±11.31	20.5±10.32	LSD(0.05)	0.471
25	44.9±2.79	9.0±6.17	34.9±6.84		
30	43.6±3.43	9.5±8.14	33.3±8.05		
LSD (0.05)	0.888	2.963	2.819		

Means followed by the same letter within a column are not significantly different (P> 0.05) according to the least significant difference (LSD) test; NGS: number of germinated seeds; NNS: number of normal seedlings; NDS: number of diseased seedlings.

Table 3. Effect of safflower genotype on seed germination, number of normal and diseased seedlings in infected media with *Pythium ultimum* at temperatures 10, 15, 20, 25 and 30ºC.

Genotype	Origin	NGS					NNS	NDS
		10 º C	15 º C	20 º C	25 º C	30 º C		
Arak-2811	Iran	46.1±2.87	46.5±3.59	46.2±1.91	47.3±2.62	47.8±0.81	20.3±12.21	26.0±11.56
Isfahan	Iran	45.6±1.41	43.6±2.08	46.1±0.81	45.0±3.69	43.6±2.62	18.8±11.01	25.5±8.87
Zarghan-259	Iran	45.5±2.70	46.3±2.68	45.5±1.70	47.0±2.38	45.8±3.68	20.3±11.75	25.0±9.04
LRV-51-51	Iran	45.5±4.76	46.2±2.44	45.5±2.62	46.3±3.50	45.7±2.75	22.2±9.21	23.3±7.86
LRV-55-295	Iran	45.5±1.73	46.8±2.51	45.5±1.82	46.3±2.38	45.6±4.08	21.8±10.79	23.7±10.27
IL-111	Iran	45.8±2.50	44.1±0.95	44.7±1.91	43.1±3.86	41.8±1.73	18.8±11.39	25.6±9.00
Dinger	Turkey	46.6±2.62	45.3±2.06	45.8±1.29	45.3±1.70	46.0±2.62	19.8±11.50	25.4±11.94
Syrian	Syria	46.7±2.21	44.1±1.41	45.0±1.41	45.0±1.50	43.7±0.95	20.6±12.72	24.5±11.94
CW-74	USA	45.7±3.46	42.0±3.20	45.1±3.69	45.0±0.95	43.8±2.44	14.9±11.53	27.7±11.93
Hartman	USA	47.7±1.70	47.6±0.81	47.7±2.06	47.7±1.50	47.2±0.57	20.4±10.53	26.6±10.53
Aceteria	Canada	45.3±1.70	46.8±2.38	47.2±4.69	45.3±1.15	44.5±1.50	18.1±12.75	26.6±13.40
PI-250537	Unknown	46.8±2.51	44.7±2.61	47.2±2.36	44.8±2.94	42.6±2.50	19.5±10.80	24.0±10.42
5-541	Unknown	45.5±3.30	44.1±1.73	44.2±3.69	41.5±1.29	39.1±1.73	16.2±11.10	25.2±10.99
34074	Unknown	43.5±2.06	43.5±1.91	45.7±3.68	43.5±3.59	41.3±2.061	19.3±12.16	22.3±10.25
LSD (0.05)	—	2.283	2.390	2.534	2.377	2.653	—	—

Means followed by the same letter within a column are not significantly different (P > 0.05) according to the least significant difference (LSD) test; NGS: number of germinated seeds; NNS: number of normal seedlings; NDS: number of diseased seedling.

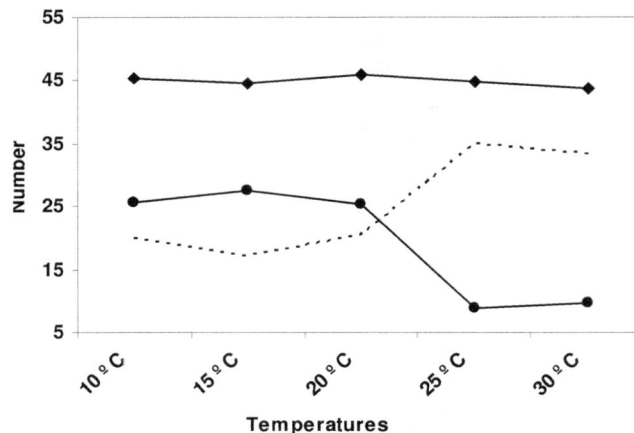

Figure 2. Effects of temperature on number of germinated seed (◆———◆), number of normal seedlings (●———●) and number of diseased seedlings (............) of safflower at infected media with *Pythium ultimum*.

ween genotype and temperature was not significant for NNS (Table 1).

There was a significant difference between used temperatures for their effects on number of diseased seedlings (NDS) in *Pythium*-infected media (Table 1). NDS was greatest at 25 and 30°C (Table 2). But, genotypes were similar to each other for NDS, because there were no significant difference among them (Table 1). NDS of the genotypes varied between 22.3 and 27.7 (Table 3). The lowest NDS belonged to genotype 34072 (Table 3). Also the effect of interaction between genotype and temperature was not significant on NDS (Table 1).

The effect of temperature on NGS, NNS and NDS are also shown in Figure 2. As shown in this figure more NNS was observed at 10, 15 and 20°C (always ≥ 25) than at 25 and 30°C. The best temperatures for development of damping-off were higher temperatures, because more NDS occurred at 25 and 30°C (always ≥ 33) than 10, 15 and 20°C (Figure 2).

For better understanding of variability in NGS, NNS and NDS of the studied genotypes over the temperatures, 3 graphs were created separately for each temperature (Figures 3, 4 and 5). The levels of temperature had a more differential effect on both NNS and NDS than NGS. In case of NGS, as shown in Figure 3, lines representing temperatures have many cutting off or overlapping. However, both NNS and NDS lines for temperatures 25 and 30°C were interestingly distinct in relation to lines of other temperatures (Figures 4 and 5). Also, there were considerably similar trend among temperature lines in both NNS and NDS than NGS (Figures 3, 4 and 5).

DISCUSSION

Both temperature and genotype affected number of ger-

minated seeds (NGS) in safflower. Effect of temperature on seed germination has been reported several times in safflower (Ayan et al., 2005) and other crop plants (Nyachiro et al., 2002; Riley, 1981). Maftoun and Sepaskhah (1978) studied effects of different temperatures on safflower seed germination and showed that the temperatures in which maximum germination took place were 10 and 20°C which were similar to the result of this study. Observation of the significant difference among genotypes along with non significant mean squares of media × genotypes for seed germination showed that the evaluated genotypes kept their germination potential whether at sterile or Pythium-infected media. It could be concluded that these genotypes have good potential for seed germination because almost all of them showed NGS over 45 per each 50 seed-plots (Table 3).

There was a significant difference among the 14 studied genotypes for number of normal seedlings (NNS) and number of diseased seedlings (NDS) in infected media under laboratory conditions. Genotypes LRV-55-259 and LRV-51-51 showed the higher NNS whereas CW-74 and 5-541 had the lowest NNS under *Pythium*-infected media. The lowest and highest NDS belonged to genotypes 34074 and CW-74, respectively. The variation in disease incidence among the genotypes could be attributed to differences in susceptibility to the pathogen. Presence of genotypic variation for response to infection with *Pythium* or other causal pathogens of damping-off have been reported by other researchers in safflower (Ahmadi et al., 2008; Heritage and Harrigan, 1984; Mundel et al., 1997). Also, it could be concluded that in *Pythium* free conditions, these genotypes should have good seedling establishment because all of them had a great percent of diseased seedlings under *Pythium*-infected conditions (Table 3). Breeding for resistance to damping-off in safflower is probably the most promising approach to minimizing

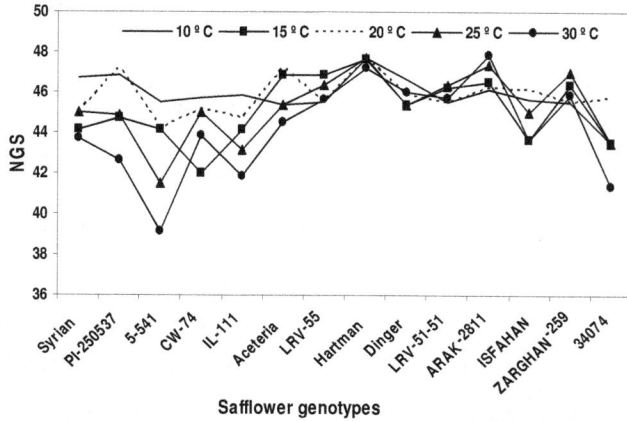

Figure 3. Effects of temperature on number of germinated seeds (NGS) in safflower at infected media with *Pythium ultimum*.

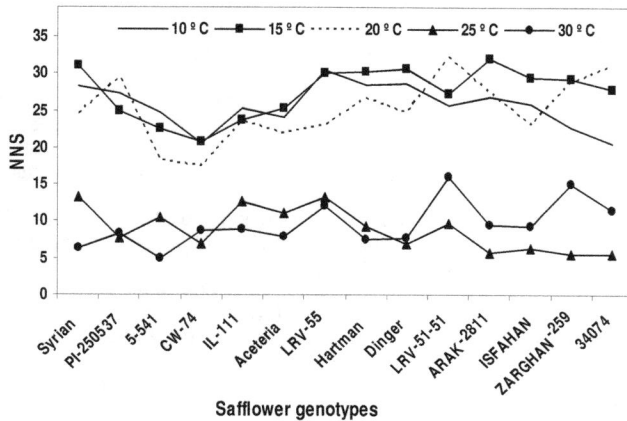

Figure 4. Effects of temperature on number of normal seedlings (NNS) in safflower at infected media with *Pythium ultimum*

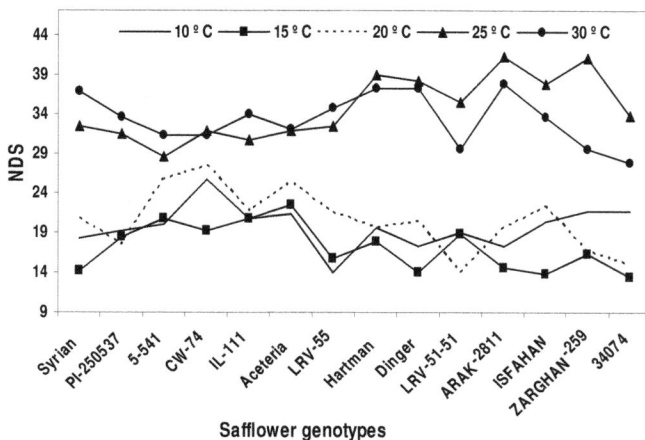

Figure 5. Effects of temperature on number of diseased seedlings (NDS) in safflower at infected media with *Pythium ultimum*.

mizing damages of the *P. ultimum*. Some of the studied genotypes had good potential of normal seedlings in *Pythium*-infected media and could be considered in breeding program for improvement of resistance to *Pythium* damping-off in safflower.

The number of diseased seedlings (NDS) caused by *P. ultimum* increased as incubation temperatures increased over 15 °C and reached a maximum at 25 and 30 °C (Figure 2). And, the maximum number of normal seedlings (NNS) were observed at 15 °C and decreased as temperature increased up to 30 °C (Figure 2). The increase in incidence of damping-off with temperature observed is in accordance with results of Thomas (1970) on *Pythium splendens*, and 24 °C is reported to be optimal for *P. irregulare* (Ben-Yephet and Nelson, 1999). Mundel et al. (1995) evaluated the effects of soil temperature, soil moisture and *P. ultimum* infection on the emergence of 12 safflower genotypes. They found that emergence in *Pythium*-infested soil was relatively high at 5, 10 and 15 °C, but was dramatically lower at 20 and 25 °C, particularly in soils with a moisture level of 30 kpa. They also observed post-emergence damping-off of seedlings in *Pythium*-infested soil, after maximum emergence at 15, 20 and 25 °C. The results of this study agree with the above findings. The favorability of higher temperatures on damping-off of cucumber caused by *P. irregulare* was also showed by Ben-Yephet and Nelson (1999). Contrary to the observations of Martin and Loper (1999) and Ploetz (2004) showing *Pythium* infection occurs mainly at lower temperatures, we observed an increase in disease incidence with temperature. This discrepancy may be due to differences between isolates, since isolates of the same *Pythium* species vary in their optimum temperature (Ben-Yephet and Nelson, 1999). The observation of Lifshitz and Hancock (1983) offer another possible explanation. They found that the optimum temperature for *Pythium* growth shifted to lower temperature when the fungus was added to nonsterile instead of sterile soil, indicating a difference between the physiological and ecological optima.

The results of this study indicate that to reduce the severity of damping-off in fields infested with the *Pythium* pathogen, safflower should be seeded early when soil temperature is low, even though germination may be slow or lower. In Golestan area, the mean temperature in March and April is 10.6 and 16.0 °C, respectively. In the absence of major seedling diseases, mid-March to mid-April has been identified as optimum planting period for safflower in this area. If seeding is delayed until soil temperatures are higher than 15 °C, safflower growers should consider not planting safflower if other conditions are favorable. Other studies with *Pythium* spp. have indicated that high soil temperature favours the development of seed and root rots (Ben-Yephet and Nelson, 1999; Thomas, 1970), and it may be easier to manipulate seeding date than to eliminate the pathogen. This study confirms these observations for *P. ultimum*, which is widespread in almost all fields and is the main inciter of damping-off of safflower in Iran (Ahmadi, 2008; Ahmadinejad and Okhovat, 1976).

Van der Plaats-Niterink (1981) in her monograph on *Pythium* noted that the role of *Pythium* spp. often depend on external factors. "When conditions are favorable for the fungus but less for the host, *Pythium* species can become very pathogenic". In the interaction that is described in this paper, *P. ultimum* appears to be an opportunist in that it causes its greatest damage on safflower not under temperature conditions that are most favorable for it, but when the host genotype is susceptible. From this study it can be concluded that temperature and genotype conditions play a major role in *P. ultimum* damping-off of safflower. Therefore, implementation of proper sowing time as mentioned above and selection of less susceptible safflower cultivars should be part of *Pythium* management practices to reduce damping-off incidence and severity under field cultivation.

ACKNOWLEDGMENT

In the memory of Miss Hasanpoor, one of our best students, who performed well in this study with us, but unfortunately she died in a motor accident 3 weeks after finishing this study.

REFERENCES

Ahmadi A, Pahlavani MH, Razavi SE, Maghsoudlo R (2008). Evaluation of safflower genotypes to find genetic sources of resistance to damping-off (*Pythium ultimum*). Elect. J.Crop Prcd. 1: 1-16. Istanbul, Turkey.

Ahmadinejad A, Okhovat M (1976). Pathogencity test of some soilborne fungi on some important field crops. Iranian Plant Pathol. 12: 13-16.

Ayan AK, Çırak C, Odabaş MS, Çamaş N (2005). Modeling the effect of temperature on the days to seed germination in safflower (*Carthamus tinctorius* L.). 6th International Safflower Conference. 6-10 June, 2005. *myriotylum* in compost at different temperatures. Plant Dis. 83: 356-360.

Ben-Yephet Y, Nelson EB (1999). Differential suppression of damping off caused by *Pythium aphanidermatum*, *Pythium irregulare* and *Pythium myriotylum* in compost at different temperatures. Plant Dis. 83: 356-360

Dick MW (1990). Keys to *Pythium*. Reading, UK: Published by the author. p. 64.

FAO, 2008. FAO database collection, www.fao.org.

Fortnum BA, Rideout J, Martin SB Gooden D (2000). Nutrient solution temperature affects *Pythium* root rot of tobacco in greenhouse float systems, Plant Dis. 84: 289-294.

Grove GG, Madden LV, Ellis MA (1985). Influence of temperature and wetness duration on sporulation of *Phytcphthora cactorum*. Phytopathology 75: 700-703.

Gu ZH, Xu BM (1984). Studies on the germination physiology and vigor of safflower seeds. Acta Phytophysiol. Sinica. 10: 305-314.

Heritage AD, Harrigan EKS (1984). Environmental factors influencing safflower screening for resistance to *Phytophthora cryptogea*. Plant Dis. 68: 767-769.

Huang HC, Morrison RJ, Mundel HH, Barr DJS (1992). *Pythium* sp. "group G", a form of *Pythium ultimum* causing damping-off of safflower. Can. J. Plant Pathology. 14: 229-232.

Kolte SJ (1985). Diseases of annual edible oilseed crops, Vol. III: Sunflower, safflower, and nigerseed diseases. CRC Press, Boca Raton, FL. 118 p.

Lifshitz R, Hancock JG (1983). Saprophytic development of *Pythium ultimum* in soil as a function of water matric potential and temperature. Phytopathology 73: 257-261.

Maftoun M, Sepaskhah AR (1978). Effects of temperature and osmotic potential on germination of sunflower and safflower and on hormone-treated sunflower seeds. Can. J. Plant Sci. 58: 295-301.

Martin FN, Loper JE (1999). Soilborne disease caused by *Pythium* spp.: Ecology, epidemiology and prospects for biological control. Crit. Rev. Plant Sci. 18: 111-181.

Mundel HH, Huang HC, Kozub GC, Barr DJS (1995). Effect of soil moisture and temperature on seedling emergence and incidence of *Pythium* damping-off in safflower (*Carthamus tinctorius* L.). Can J. Plant Sci. 75: 505-509.

Mundel HH, Huang HC, Kozub GC, Daniels CRG (1997). Effect of soil moisture and temperature on seed-borne *Alternaria carthami*, on emergence of safflower (*Carthamus tinctorius* L.). Bot. Bull. Acad. Sin. 38: 257-262.

Nyachiro JM, Clarke FR, DePauw RM, Knox RE, Armstrong KC (2002). Temperature effects on seed germination and expression of seed dormancy in wheat. Euphytica. 126: 123-127.

Ploetz RC (2004). Influence of temperature on *Pythium splendens*-induced root disease on carambola, *Averrhoa carambola*. Mycopathologia. 157: 225-231.

Rahimian MK, Banihashemi Z (1979). A method for obtaining zoospores of *Pythium aphanidermatum* and their use in determining cucurbit seedling resistance to damping-off. Plant Dis. Rep., 63:658-661.

Riley GJP (1981). Effects of high temperature on the germination of maize (*Zea mays* L.), Planta 151: 68-74.

SAS Institute (2004). SAS/STAT9.1. User's guide. SAS Inst., Cary, NC.

Singleton LL, Mihall JD, Rush CM (1992). Methods for research on soilborne phytopathogenic fungi, APS press, p. 265

Snedecor GW, Cochran WG (1980). Statistical methods. 7th edition, Iowa State University Press, Ames, Iowa.

Thomas CA (1970). Effect of seedling age on *Pythium* root rot of safflower. Plant Dis. Rep. 54: 1010-1011.

Thomas CA (1970). Effect of temperature on *Pythium* root rot of safflower. Plant Dis. Rep. 54: 300.

Timmer LW, Zitko SE, Gottwald TR, Graham JH (2000). *Phytophthora* brown rot of citrus: Temperature and moisture effects on infection, sporangium production, and dispersal. Plant Dis. 84: 157-163.

Van der Plaats-Niterink, AJ (1981). Monograph of the genus *Pythium*. Studies in mycology No 21, W. Gams RPWM. Jacobs, eds. Centraalgureau voor Schimmel-cultures, Baarn, Netherland.

Resistance in Kenyan bread wheat to recent eastern African isolate of stem rust, *Puccinia graminis* f. sp. *tritici,* Ug99

P. N. Njau[1]*, R. Wanyera[1], G. K. Macharia[1], J. Macharia[2], R. Singh[3] and B. Keller[4]

[1]Kenya Agricultural Research Institute (KARI) Njoro, Kenya.
[2]Egerton University Njoro. P. O. Box 536 Njoro, Kenya.
[3]CIMMYT Mexico. Apdo - Postal 6-641, 06000 Mexico, DF.
[4]Institute of Plant Biology, University of Zurich, Switzerland

Stem or black rust, caused by *Puccinia graminis*, has historically caused severe losses to wheat (*Triticum aestivum*) production worldwide. The causal race, commonly known as Ug99 and designated as TTKS based on the North American nomenclature, carries virulence for several genes commonly present in wheat germplasm. All Kenyan germplasm are known to be susceptible or partially susceptible to Ug99 although no proper documentation has been done. This study was aimed at evaluating the Kenyan bread wheat varieties on their response to Ug99. The varieties were screened for resistance at seedling stage and adult plant resistance stage. None of the varieties apart from Bonny were resistant at seedling stage. Some old Kenyan varieties were found to have adult plant resistance probably due to the presence of non-race specific gene Sr2 complex which among others can be exploited in breeding for resistance in Kenyan wheat.

Key words: Race specific resistance, adult plant resistance, genetic erosion.

INTRODUCTION

Stem or black rust, caused by *Puccinia graminis*, has historically caused severe losses to wheat (*Triticum aestivum*) production worldwide. Its control for over 30 years through the use of genetic resistance in wheat is a remarkable success story. In 1999, high susceptibility of International Maize and Wheat Improvement Center (CIMMYT) germplasm was noted in Uganda and an increase in stem rust incidence and severity was seen in Kenya.

McIntosh (2000) describes cereal rusts as 'social' diseases because the individual farmer is subject to airborne spores from areas beyond the farm and part of any inoculums produced on farm is dispersed to others. Control of rust diseases must therefore be addressed at the community level, preferably through the widespread use of resistant varieties.The causal race, commonly known as Ug99 and designated as TTKS based on the

North American nomenclature, carries virulence for several genes commonly present in wheat germplasm. A germplasm screening process was initiated in Kenya and Ethiopia to document the scope of virulence of the new race and also to identify any source of resistance. Over 80% of all the germplasm screened were susceptible (Wanyera et al., 2006; Singh et al., 2006; Jin et al., 2006).

The stem rust resistance gene Sr31, derived from Petkus rye (MacIntosh et al., 1995; Zeller et al., 1983) has been used worldwide in spring wheat through the widespread use of Russian and other East European wheat Kavkaz, Aurora and Loverin that originally carried the 1BL.1RS wheat–rye translocation (Zeller et al., 1973). This led to the release of numerous popular cultivars worldwide including several spring wheat cultivars derived from the spring wheat germplasm in CIMMYT. Gene Sr31 provided the main component for stem rust resistance in many wheat cultivars and continued to remain effective until recently, when isolates of *P. graminis* f. sp. *tritici* with virulence to Sr31 were detected from Uganda in 1999 (Pretorius et al., 2000). Similar virulence was observed in Kenya in 2003 and 2004 (Wanyera et al., 2006).

A few lines carrying Sr2 were found to have adult plant resistance. Gene Sr2, transferred to wheat from 'Yaroslav emmer' by McFadden (1930), is the only catalogued gene that is not race-specific. Sr2 can confer slow rusting (Sundrwirth and Roelfs, 1980) resistance of adult-plant nature. Resistance gene Sr2, in addition to other unknown minor genes derived from cultivar hope and commonly known as 'Sr2-Complex', provided the foundation for durable resistance to stem rust in germplasm from University of Minnesota in the USA, Sydney University in Australia, and the spring wheat germplasm developed by Dr N.E. Borlaug as part of a programme sponsored by the Mexican Government and the Rockefeller Foundation (McIntosh 1988, and Rajaram et al., 1988).

This study was aimed at evaluating the Kenyan germplasm and their response to Ug99. This will help the breeder in identifying the best parents to be used in the breeding program in fight against Ug99.

MATERIALS AND METHODS

Evaluation of seedling infection types

An isolate (04KEN156) of *P. graminis f. sp. tritici*, collected from Kenya in 2004 and identified as race TTKS (Wanyera el al., 2006) based on the 16 differentials in the *P. graminis f. sp. tritici* differential set of North America (Roelfs et al., 1990), in addition to six Pgt isolates of races QTCS, QTHS, RCRS, RKQQ, TPMK and TTTT representing broad virulence in the North American stem rust population, were included for comparisons of infection types for the 30 Kenyan varieties at St. Paul, MN, USA. The 30 Kenyan lines were planted in tray and allowed to grow to seedling stage as explained in yue Jin (2007). Urediniospores from long-term storage in a -80°C freezer were heat shocked at 40°C for 10 min and placed in a rehydration chamber for 2 - 4 h, where approximately 80% relative humidity was maintained by a KOH solution (Rowell, 1984).

The urediniospores were then suspended in a light mineral oil (Soltrol 170) and inoculated onto the fully expanded primary leaves of 7 - 9 day old seedlings of wheat lines. The inculated seedlings were then incubated in a dew chamber for 14 h at 18°C in the dark and then for an additional period of 3 - 4 h under fluorescent light. The plants were then placed on a greenhouse bench at 18 ± 2°C with a photoperiod of 16 h. Infection types (ITs), described by Stakman et al. (1962), were assessed 14 days post inoculation. ITs 0, 1, 2 or combinations thereof were considered low ITs indicating that the corresponding resistance gene is effective. ITs 3 - 4 were considered high ITs, indicating that the corresponding resistance gene is not effective against the race tested. In each test, 6 - 10 seedlings were evaluated and the test was repeated twice.

Field stem rust evaluations in Njoro

The 30 Kenyan bread wheat varieties were tested in 2006 and 2007 as part of a larger field stem rust screening nursery in Njoro, Kenya established by the Kenyan agricultural research institute in conjunction with CIMMYT and the global rust initiative (GRI).

The nursery site was located at 0°20'S, 35°56'E and 2,185 m in elevation with an average daily minimum temperature of 9.7°C (night) and an average daily maximum temperature of 23.5°C (noon). Variations of average daily temperatures are approximately ±2°C, occurring mostly during the day hours of the field evaluation period. Dew was formed nearly daily.

Entries were planted in single 1 m row plots on 30th June of 2006 and 15th June 2007. To facilitate inoculum build-up and uniform dissemination within the nursery, a continuous row of stem rust spreader (a mixture of susceptible cvs. Chozi and Duma carrying Sr31) was planted perpendicular to all entries. The spreader rows were inoculated once by dusting them with a mixture of urediniospores and talc powder. The source of inoculum was a bulk of urediniospores collected from experimental plots of Duma variety grown within the field where the experiments were carried out in Kenya. Plant response to rust infection at the adult plant stage was termed "infection response"

According primarily to the size of pustules and associated necrosis or chlorosis, infection responses were classified into four discrete categories; R = resistant, MR = moderately resistant, MS = moderately susceptible and S = susceptible (Roelfs et al., 1992). Infection responses overlapping between any particular two categories were denoted using a dash. For instance, "MR-MS" denoted an infection response class overlapped between the MR and MS categories. Stem rust severity was assessed following the modified Cobb scale (Peterson et al., 1948). Entries were evaluated for infection responses and stem rust severity two to three times between heading and plant maturity. The infection responses and stem rust severity at the soft-dough stage of plant growth were used to represent the final disease scores in this report.

RESULTS AND DISCUSSIONS

Seedling infection type of Kenyan wheat varieties

Kenyan varieties showed high level of resistance to the stem rust races found in the USA. All the varieties apart fro Duma was resistant to the QTCS race (Table 1). This trend was the same for all the other races but with race TTTT which is a more recent race the older Kenyan varieties showed more susceptibility at seedling stage. These included, K. Kudu, K. Nyangumi, K. Tama, K. Tembo among others as shown in Table 1. Such report had been reported earlier by Singh et al. (2006) while comparing various sources for durable or slow rusting type of resistance. When it comes to race TTKS which is popularly known as Ug99, all the Kenyan varieties become susceptible with the exception of Bonny which remained resistant whereas Tama showed two types of reactions. Given that the races have been evolving over time, it is clear that most of the Kenyan varieties have remained effective against the new races until the emergence of the new race Ug99. This can be explained from the point of view that most of our Kenyan varieties, especially those derived from CIMMYT carry the Sr31 stem rust resistance gene (Singh, 2006). The alien resistance gene Sr31 has been used in agriculture on the largest scale since 1980s in spring, facultative and winter wheat breeding programs worldwide except Australia. Its use in CIMMYT wheat improvement resulted in the release of several popular cultivars worldwide. The use of 1BL.1RS

Table 1. Seeding Infection type (IT) of Kenyan wheat varieties to various stem rust races including Ug99 tested at the at St. Paul, MN, USA.

Line	Test Race						
	QFCS	QTHJ	RCRS	RKQQ	TPMK	TTTT	TTKS
K-popo	0	23-	0;	0;	;1	;1	3
K-kudu	0	2/3	;	4	;	;1+/3(2 pl)	3
K-kulungu	0;	;2/3	4	;	;	;1	3
K-fahari	0;	2+/;	;1	0;/;2	;1	;	3-
Mbega	;/2	2-	1-	;1	2	2-	3+
K-nyangumi	0	;1--	;	0;	;1	3	3+
K-paka	0	;1	0;	0;	;1/2	;	3+
Tama	0	;	2=	;3	;	3-	;13
K-tembo	;	-	0;	-	-	3	3+;
Ngamia	1	23	;	22+	3	3	3+
Kwale	0	0/1	2-	2-	2	2-	3+;
Kipapu	0;/1	2	;	2+	;	2	2/4
K-chiriku	-	;1	2-	;	2/3	;/3/2	3+
Bonny	0/;1+	0	;	3	;	3; /;	-
Pasa	;	2	;1	2-	2	1	4
Swara	0;	1-	;/2-	0;/2+	;1	3	3+
Romany	0	;/3	0;	;/3	;1	;C	4
A-mayo	;1	2	2	0;	;1	3	3-;
Catcher	0;	2-	2-	2	2	2	3
Chozi	0;	;2-	2-	2-	2-	2-	3+
R-sabanero	;1	;1+	2+	2C	2	2	3+
Duma	2-	3-	;2-	2	3-	2/3	4
Bounty	0	-	-	-	0?	1	3+
K-yombi	0/;	2-	1;	2-	2-	1;	3+
K-heroe	0;	2	;1-	;1	2;	;2-	3+
K-mbweha	;2	2-	2-	2-	2-	2-;	3+
Regent	2	2	2+3	3	3+	3+	4
K-zabadi	2;	2	2	1;	;1	3-	3+
Gem	0;	;1	;/1	23-	3	;	3-
Ngiri	0;	;	0 esc?	0;	0;	0;	3

translocation was initially associated with increased grain yields and resistance to all three rusts and powdery mildew as it carried resistance genes for all these diseases on the same translocation. Large-scale deployment of Sr31 surprisingly did not result in its breakdown until the detection of race Ug99 in Uganda. In fact this gene probably further reduced the already low stem rust survival to almost nonexistent levels in most wheat growing regions to the extent that stem rust started to become a forgotten curse. This degree of stem rust susceptibility to a single race in Kenyan bread wheat has not been observed previously. It may indicate that the virulence combination of race TTKS is very unusual. More likely, however, it may indicate a serious erosion of the resistance package in the spring wheat germplasm that has provided stable stem rust resistance in the Kenya for over 50 years Jin and Singh (2006) compared seedling reac-

tions of US wheat cultivars and germplasm with highly virulent races present in the USA and race Ug99. Several wheat lines, especially spring wheat that were highly resistant to US races and did not carry the1BL.1RS translocation were also found to be susceptible to Ug99. This further supports the hypothesis that race Ug99 carries a unique combination of virulence to known and unknown resistance genes present in wheat germplasm. The major susceptibility is due to the specific nature of avirulence /virulence combination that Ug99 possesses which had led to the susceptibility of many wheat materials irrespective of where they were developed.

Adult plant resistance of Kenyan bread wheat toUg99

Among the 30 Kenyan wheat varieties evaluated for adult plant resistance in Kenya in 2006 and 2007, 20% were

Table 2. Adult plant response to infection and disease severity of the Kenyan bread wheat varieties.

Variety	Parentage	Disease sore		
		2006	2007	Mean (%)
K.kongoni	6410-2/6647-5=CI8154/2*Fr/2/3*ROM/3/Wis.245-II-50-17/CI8154/2/2*Fr	40S	40S	40
Mbuni	Za75/Ld357E-Tc3xGU⁻ 30520-1B-1B-3Y-0Y	40s	10MS	25
K.heroe	MBUNI/SRPC64//YRPC1	70s	40S	55
Chozi	F12.71/COC//GEN CM76689-3Y-08M-02Y-4B-2Y-OB	40S	30MS	35
Duma	AU/UP301//GLL/SX/3/PEW "S"/4/MAI "S"//PEW "S" CM67245	60S	60S	60
Mbega	Fink "S",CM41860-A-5M-2Y-3M-1Y-1M-1Y-OB-Opt2	30S	10MS	20
K.yombi	MBUNI/SRPC64//YRPC1	30S	20MS	25
Pasa	Buc "S"/Chat "S"	40S	30MS	35
K.chiriku	KTB/Carpintero "s"	30S	40MS	35
Kwale	Kinglet,CM33089-W	40MSS	10MSS	25
K.nyangumi	TZPP//SK^{E6}/LR64HDM/3/AFM/4/KSW/K4500-6	10M	5MSS	7
K.tembo	WIS.245/II-50-17//C.I 8154/2*Fr/3/2*Tob.66	30MSS	5MSS	18
K.fahari	TOBARI66/SRPC527//CI8154/3/2*FROCOR	60S	10MS	35
K.paka	(Wis245/II-50-17//CI.8154/2*Fr/3/2*Tob.66	40S	10M	25
K.popo	KL. Atl/Tob66//cfn/3/Bb	60S	10MS	35
Ngamia	SW 53=BUCKBUCK "S" CM31678	40S	10MS	25
Njbw I	KM14(PASA MUTANT)	40S	60MS	50
Njbwii	TNMU CM81812-12Y-06PZ-4Y-5M-0Y-2AL-0Y-2AL-0AL-OM	30S	40MS	35
K.kulungu	On/Tr. 207/3/cno//Son64/4/6661-53	50S	10MS	30
K.swara	(CI8254xFr2) x (T-K²xY.59.2.B)	15M	1MSS	8
K.kudu	K131xK184.P.2.A.I.F K1008.K.7.K.2	40S	5MSS	23
Bonny	YF3xBza² VI-116-2-4B-1T-2B-1T	5R	1MSS	3
Bounty	T-Kenya²x Bonza² VI-106-2t-3b-3t-1b-2t	10R	5RMR	8
Regent	H44xReward RL975.6	20M	5RMR	13
K.mamba	Africa Mayo48x/3/[(Wis.245xsup51)x(Fr.Fn/Y)² A]	4OS	5MSS	23
Gem	908-Frontana x Cajeme 54	20MS	5MS	13
K.grange	360.FxGranaderoklein	40MSS	MISSING	40
catcher	Thatcher-Santa Catalina x Frocor	60S	20MS	40
R.sabanero	Single Plant Selection	60S	5MSS	33
K.ngiri	CI8154/2*Fr//5*WRT.TC/3*MIT/3/2*Tob66	30S	10M	20
Tama	Yaktana54xLerma 52	20MS	5MS	13
K.zabadi	Son64/450^{5E}//Gto/3/Inia/4/K4500/Ksw/Tob66//CIANO	20S	10MS	15
K.mbweha	CI8154/2*F/3/2*GB54/36896//II-53-526	70S	10MS	40

moderately resistance (Table 2). These were mainly the old varieties which included K. Nyangumi, K. Swara, Bonny and Bounty which contain the Sr2 resistance gene complex usually associated with the Pseudo black chaff (PBC). All these varieties showing adult plant resistance were released during the early sixties and seventies when Sr2 complex was one of the main sources for stem rust resistance. The adult plant resistance gene Sr2 confers slow rusting (Sunderwirth and Roelfs, 1980).

Combination of Sr2 with other unknown slow rusting resistance genes possibly originating from Thatcher and Chris commonly known as the "Sr2-Complex" provided the foundation for durable resistance to stem rust in germplasm from the University of Minnesota in the United States, Sydney University in Australia and the spring wheat germplasm developed by Dr. N. E. Borlaug (McIntosh, 1988; Rajaram et al., 1988). Unfortunately, not much is known about the other genes in the Sr2 complex and their interactions. Knott (1988) has shown that adequate levels of multigenic resistance to stem rust can be achieved by accumulating approximately five minor genes. The varieties released in the eighties and nineties are the most susceptible with an average disease severity of 40% and above (Table 2). These include varieties like Kwale released in 1987, Chozi released in 1999 and Duma released in 1994. Again, this may indicate a serious erosion of the resistance package in the spring wheat germplasm over time especially when breeding was done in the absence of stem rust. The susceptible varieties also are suspected to have only single race specific type of resistance.

The decrease in incidence of stem rust to almost non significant levels by the mid-1990s throughout most of the wheat producing areas worldwide were coincident with a decline in research and breeding emphasis to such a level that in many countries including Kenya breeding was done in the absence of this disease. Most of the Kenyan wheat varieties released in the late 1980's and early 90's were selections from CIMMYT.

CIMMYT scientists continued to select for stem rust resistance in Mexico using artificial inoculation with six *P. graminis tritici* races of historical importance. New stem rust races have rarely occurred since the 'Green Revolution' in Mexico (Singh, 1991). Moreover, a majority of wheat lines selected in Mexico remained resistant at international sites either due to absence of disease, inadequate disease pressure, or presence of races that lacked necessary virulence for the resistance genes contained in CIMMYT wheat germplasm.

Almost 50 different stem resistance genes are now catalogued (McIntosh et al., 1998), several of which are incorporated in wheat from alien relatives of wheat. All but *Sr2* resistance genes are race-specific, and are expressed in both seedling and adult plants. Race specificity derives from the gene-for-gene relationship between the host plant resistance gene and corresponding virulence genes in the pathogen.

Conclusions and Recommendations

Most of the Kenyan lines are susceptible to the new race of stem rust TTKS (Ug99). This could be due to their common source for resistance Sr31 from CIMMYT germplasm. Some of the old Kenyan varieties like Bonny, Bounty, K. Swara and K. Nyangumi show adult plant resistance which could be associated to the presence of the adult plant resistance gene Sr2 complex. These varieties may be used in breeding for durable resistance to stem rust especially in combination with other major genes. More sources for resistance should be sought for to be incorporated in the Kenyan germplasm. Race specific type of resistance should be avoided to minimize epidemic as has been caused by Ug99 in future.

ACKNOWLEDGEMENT

The Global Rust Initiative (GRI) and Kenya Agricultural Research Institute (KARI) are acknowledged for the financial and logistic support given to the researchers. Yue Jin of St. Paul, MN, USA and Ravi Singh of CIMMYT helped in data collection and interplatation. The staff in Njoro did a wonderful job when it came to field maintenance and data collection. All are acknowledged.

REFERENCE

Knott DR (1988). Using polygenic resistance to breed for stem rust resistance in wheat. In "Breeding Strategies for Resistance to the Rusts of Wheat" (N. W, Simmonds S, Rajaram eds.), pp. 39-47.

CIMMYT, Mexico DF,Jin Y. Singh RP (2006). Resistance in US wheat to recent eastern African isolates of *Puccinia graminis* f. sp. *Tritici* with virulence to resistance gene *Sr31*. Plant Dis. 91: 1096-1099.

McFadden ES (1930). A successful transfer of emmer characteristics to vulgare wheat. J. Am. Soc. Agro. 22: 1020-1034.

McIntosh RA, Hart GE, Devos KM, Gale MD, Rogers WJ. (1998). Catalogue of gene symbols for wheat. In: Slinkard AE, editor. Proceedings of the 9th International Wheat Genetics Symposium, 2–7 August 1998, Saskatoon, Canada (5): 1- 235.

McIntosh RA. (1988). The role of specific genes in breeding for durable stem rust resistance in wheat and triticale. In: Simmonds NW, Rajaram S, editors. Breeding Strategies for Resistance to the Rust of Wheat. CIMMYT, Mexico, DF; pp. 1–9.

McIntosh RA, Wellings CR, Park RF (1995). Wheat Rusts: An Atlas of Resistance Genes. CSIRO, Australia.

McItosh RA (2000). Managing Cereal Rusts. National Perpective. Crop update.Cereals. Department of Agriculture and Food.

Peterson RF, Campbell AB, Hannah AE (1948). A diagrammatic scale for estimating rust severity on leaves and stems of cereals. Ca. J. Res. Sect. C. 26: 496-500.

Pretorius ZA, Singh RP, Wagoire WW, Payne TS (2000). Detection of virulence to wheat stem rust resistance gene *Sr31* in *Puccinia graminis* f. sp. *tritici* in Uganda. Plant Dis. 84: 203.

Rajaram S, Singh RP, Torres E. (1988). Current CIMMYT approaches in breeding wheat for rust resistance. In: Simmonds NW, Rajaram S, editors. Breeding Strategies for resistance to the Rust of Wheat. CIMMYT, Mexico, D. pp.101-118.

Roelfs AP, Singh RP, Saari EE (1992). Rust diseases of wheat: concepts and methods of disease management. CIMMYT, Mexico, D.F.

Roelfs AP, Martens JW (1988). An international system of nomenclature for Puccinia graminis f.sp. tritici. Phytopathology 78: 526-533.

Rowell JB (1984). Controlled infection by *Puccinia graminis* f. sp. *tritici* under artificial conditions. In: W. R. Bushnell and A. P. Roelfs (eds.). The Cereal Rusts, Origins, Specificity, Structure, and Physiology. Academic Press, Orlando. 1: 292-332

Singh RP (1991). Pathogenicity variations of Puccinia recondita f. sp. tritici and P. graminis f. sp. tritici in wheat-growing areas of Mexico during 1988 and 1989. Plant Dis. 75: 790-794.

Singh RP, Kinyua MG, Wanyera R, Njau P, Jin Y, Huerta-Espino J (2006). Spread of a Highly Virulent Race of *Puccinia graminis Tritici* in Eastern Africa: Challenges and Opportunities CAB Reviews: Perspectives in Agriculture, Veterinary Science, Nutr. Nat. Resour. (2006) 1: 054

Stakman EC, Steward DM, Loegering WQ (1962). Identification of physiologic races of *Puccinia graminis* var. *tritici*. U.S. Dep. Agric. Agric. Res. Serv. E-617.

Sunderwirth SD, Roelfs AP. (1980). Greenhouse characterization of the adult plant resistance of Sr2 to wheat stem rust. Phytopathology 70: 634-637.

Wanyera R, Kinyua MG, Jin Y, Singh R (2005). The spread of stem rust caused by *Puccinia graminis* f. sp. *tritici*, with virulence on *Sr31* in wheat in Eastern Africa. Plant Dis. 90: 113.

Zeller FJ (1973). 1B/1R wheat-rye chromosome substitutions and translocations In: Proc Int. Wheat Genet. Sympos. pp. 209-221

Zeller FJ, Hsam SLK (1983). Broadening the genetic variability of cultivated wheat by utilizing rye chromotin. In: Proc. Intl. Wheat Genet. Sympos. 6th. S. Sakamoto, ed. Kyoto, Japan. pp. 161-173

7

Genetic yield stability in some sunflower (*Helianthus annuus* L) hybrids under different environmental conditions of Sudan

Salah B. Mohamed Ahmed[1] and Abdel Wahab H. Abdella[2]

[1]Department of Crop Science, Faculty of Agriculture and Environmental Sciences, University of Gedaref, P.O. Box 449, Gedaref, Sudan
[2]Department of Agronomy, Faculty of Agriculture, University of Khartoum, P. O. Box 13314, Shambat, Sudan

Nineteen locally developed sunflower (*Helianthus annuus* L.) hybrids and an introduced one (Hysun 33) were evaluated at two irrigated locations, New Halfa and Rahad for two consecutive seasons (2003/04 and 2004/05) in order to estimate stability of performance for seed yield per ha (t). A randomized complete block design with four replicates was applied at each location. Data on seed yield was collected. The variance of genotype x environment interaction (GxE) was highly significant, suggesting that the yield of the hybrids was inconsistent in different environments. The average ranking of the 20 hybrids, according to stability parameters showed that Salih was the first ranked hybrid followed by Ka99x7, Ka99x29, Ka99x13 and Shambat 6, whereas Hysun33 was the last ranking one. The two released hybrids (Salih and Shambat 6) were adapted favorable environments, whereas Ka99x25-2 adapted unfavourable conditions. However, the graph for scatter points for yield of the locally developed hybrids revealed that Salih and Shambat 6 were stable, whereas that of Hysun 33 was unstable under these environments and its yield was only increased with improving the conditions.

Key words: Sunflower, hybrids, yield, GxE interaction, stability.

INTRODUCTION

In Sudan, commercial production of sunflower was initiated in the 1987/1988 season. The production of the crop dropped from 8 thousand tons in 1999/2000 to 4 thousand tons in 2000/2001 season with an average yield of 0.39 t/ha and 0.73 t/ha, respectively. In the following two seasons (2001/2002 and 2002/2003) the production increased to 18 thousand tons with an average yield of 1.5 t/ha. Khidir (1997) summarized the major problems facing sunflower production in the Sudan to be, lack of adequate information about the crop under Sudan condition, mal distribution and fluctuation of rains, a high percentage of empty seeds particularly in non-hybrid varieties, the difficulty in finding good seeds and high yielding cultivar, importing of hybrid seeds from over-seas lead to high cost of production and the damages caused by birds and termites. Plant breeders

Generally agree on the importance of high yield stabi-

lity, but there are fewer consensuses on the most appropriate definition of stability and on methods to measure and to improve yield stability. Stability in performance is one of the most desirable properties of a genotype to be released as a cultivar for wide range of application. The significance of genotype xenvironment (GxE) interaction in variety evaluation programme is well recognized (Miller et al., 1958). Finlay and Wilkinson (1963) pointed out that the slope of the linear regression (b_i) of the yield (Y_{ij}) of jth genotype and jth environment, on the mean yield (y.j) of all the genotypes in jth environment is helpful in testing the genotypic stability. They pointed that a genotype which has a $b_i = 1$ has average stability but genotypes which have slope greater than one and less than one are below and above average stability respectively. In addition to the b_i, the deviation mean square ($\sigma^2 d$) which describes the contribution of genotype to the GxE interaction is used (Eberhardt and Russell, 1966). Both statistics are used in different ways to assess the reaction of genotypes to the varying environments. The $\sigma^2 d$ is strongly related to the remaining unpredictable part of variability of

*Corresponding author. E-mail: salahballa72@yahoo.com.

Table 1. Pedigree of tested sunflower hybrids.

Entry No.	Parents		Hybrids
1	R 1	(Male)	Ka99x 1
2	R5	(Male)	Ka 99 x 5 (Salih)*
3	R 6	(Male)	Ka 99x 6 (Shambat 6)*
4	R 7	(Male)	Ka 99 x 7
5	R 11	(Male)	Ka 99 x 11
6	R 13	(Male)	Ka 99 x 13
7	R 15	(Male)	Ka 99 x 15
8	R 17	(Male)	Ka 99x 17
9	R 18	(Male)	Ka 99x 18
10	R 22	(Male)	Ka 99 x 22
11	R 25-1	(Male)	Ka 99 x 25-1
12	R 25-2	(Male)	Ka 99 x 25-2
13	R 29	(Male)	Ka 99 x 29
14	R 30	(Male)	Ka 99 x 30
15	R 32	(Male)	Ka 99 x 32
16	R 35	(Male)	Ka 99 x 35
17	R 37	(Male)	Ka 99 x 37
18	R 41	(Male)	Ka 99 x 41
19	R 42M	(Male)	Ka 99 x 42 M
20	Ka99	(Female)	
21	Hysun 33		Commercial hybrid

*Newly released hybrids

any genotype and is, therefore, considered a response of genotypes to environmental effects and may be regarded as a response parameter. According to Wricke (1962), a genotype with small values of ecovalance (Wi) and/or deviation from regression line ($\sigma^2 d$) was a stable one and hence it contributed least to GxE interaction. Bange et al. (1997) pointed out that the potential yield of sunflower is highly dependent on environmental conditions throughout the life of the crop.

Becker and Leon (1988) stated that successful new hybrids must show good performance for yield and other essential agronomic traits, their superiority should be reliable over a wide range of environmental conditions. Therefore, the objective of this study was to evaluate 19 locally developed sunflower hybrids and an introduced one (Hysun33) under different irrigated environments with aim of estimating GxE interaction and identifying stable hybrids for yield.

MATERIALS AND METHODS

Locations

This trial was carried out at two locations, Faculty of Agriculture and Natural Resources at New Halfa (Lat.15° 19' N, Long. 35° 36' E and Alt. 450m above the sea level), Eastern Sudan. The soil is calcareous, alkaline in reaction (pH ≈ 8.3), nonsaline, nonsodic and moderately fertile. The climate is semi arid with mean annual rainfall of about 200 mm and maximum temperature of about 42°C in summer and 21°C in winter, and at Rahad Agricultural Research Station

Farm (Lat. 13° 31' N, Long. 34° 32' E and Alt. 411 m above the sea level). The Farm lies in the central clay plains of the Sudan. The soil is Vertisol with about 70% clay of low water permeability, high water holding capacity and with very low nitrogen and organic matter content. The climate is poor Savanna with mean annual rainfall of about 300 mm and maximum temperature of about 40° C in summer and 20° C in winter. At both locations, irrigation water was provided at ten days interval. At each location, the trial was carried out in summer and winter seasons, that is, four environments.

The plant material

The plant materials used in the study (Table 1) consisted of 19 single cross (F_1) hybrids of sunflower (*Helianthus annuus* L.) derived from crossing nineteen locally generated restorer lines with one exotic male-sterile line (Ka99). Crossing was made by hand pollination at the University Farm, Faculty of Agriculture, University of Khartoum, at Shambat (Lat. 15° 46' N, Long. 32° 32' E and Alt. 380 m above the sea level) during winter season of 2003. Two of the resulting hybrids, recently released under the commercial names "Salih" and "Shambat 6" as well as the standard commercial hybrid "Hysun 33" were evaluated at the four mentioned environments.

Experimental procedures and data analysis

A randomized complete block design with four replicates was used for laying out the field experiment. Each block was divided into 20 plots, to which the hybrids were assigned randomly. The plot dimensions were six meters length and three meters width. Each accession was presented by four ridges, each six meters long and 0.70 m apart. Three seeds were sown in holes of 0.20 m distance along the ridge, and then thinned to one plant per hole, three

Table 2. Mean of squares from the combined analysis of Gx E interaction of seed yield (t/ha) for the 20 hybrids evaluated at the four environments.

Source of variation	df	Mean of squares
Genotypes (G)	19	2.382**
Environments (E)	3	129.320***
GxE	57	0.835**
Pooled error	228	0.517

** significant at P ≤ 0.01.   *** significant at P ≤ 0.001

weeks after sowing. At New Halfa, sowing date was on 9th July 2003 for summer season and 27th October 2004 for winter. At Rahad, the sowing was on 20th July 2003 for the summer season and 2nd December 2004 for the winter season. Data was collected on seed yield per area unit (t/ha) according to the following equation:

$$\text{Seed yield /ha (t)} = \frac{\text{seed weight (kg / plot)} \times 10000 \ m^2}{\text{Plot area } (m^2) \times 1000}$$

The data from the four environments were subjected to the combined analysis of variance to estimate the variance of genotype x environment interaction (Gomez and Gomez, 1984). Stability of performance for the tested hybrids was carried out following Eberhardt and Russell (1966) procedure. Moreover, the stability of performance for yield was estimated for the twenty hybrids over the four environments using the formula suggested by Wricke (1962) as follows:

$$W_i = \sum (Y_{ij} - Y_{i.} - Y.j + Y.)$$

Where;

Wi = Wricke ecovalance
Y_{ij} = mean yield of the ith genotype in the jth environment
Yi. = mean of the ith genotype
Y.j = mean of the jth environment
Y. = over all mean

Expected yield of the Ith genotype in j^{th} environment was calculated by the formula suggested by Eberhardt and Russell (1966) as:

$$\acute{Y}_{ij} = x_i + b_i l_j$$

Where:

$\acute{Y}_{ij}$ = expected yield of ith genotype in j^{th} environment.
Xi = the mean yield of a genotype over environments
b_i = coefficient of regression line and was calculated as:

$$bi = \sum_j Y_{ij} l_j / \sum_j l^2_j$$

Where:

$\sum_j Y_{ij} l_j$ = the sum of the products
$\sum_j l^2$ = the sum of squares.
l_j = environmental index.

The environmental index l_j for I^{th} environment is defined as the deviation of the mean of all the genotypes at a given location from the overall mean, and was calculated as:

$$l_j = [\{ \sum l \ Y_{ij} \} / g - \{ \sum_l \sum_j Y_{ij} \}_{/ gn]} \text{ with } \sum_i l_j = \text{zero}$$

Where:

g = number of genotypes and n = number of environments

Moreover, a scatter graph of the stability was also plotted to estimate the stability of performance for the yield. The abscissa was marked with yield levels (μ), and the ordinate with regression coefficients (bi). Observed yield of a given hybrid was then plotted against its bi value. The two vertical lines on the graph present the values $μ \pm σ$ and the horizontal ones denote the values $b \pm σ$.

RESULTS AND DISCUSSION

Seed yield (t/ha) showed highly significant relationship (P ≤ 0.01) of genotype x environment interaction (Table 2), suggesting an inconsistency in the performance of the hybrids across the four environments. Therefore, there is a need for assessing stability of performance for each of the twenty hybrids in order to identify hybrids with superior yield. These results are inconformity with those reported by Singh and Yadava (1986) and Bange et al. (1997).

Table 3 shows the means (μ), regression coefficients (bi), deviation from regression line ($σ^2d$) and Wricke ecovalance (Wi)) with ranking (R) for seed yield (t/ha). Ten hybrids had means exceeding the average (3.58 t/ha) by 0.6 - 31.8%. These were Ka99 x7, Ka99 x13, Ka99 x15, Ka99 x17, Ka99 x18, Ka99 x25-2, Ka99 x29, Salih, Shambat 6 and Hysun33. Eleven hybrids had bi around unity with probability level equal to one. These were Ka99 x 1, Ka99 x 7, Ka99x11, Ka99 x 15 , Ka 99 x 18, Ka99 x 25-1, Ka99 x 29, Ka99x30, Ka99 x 37, Salih and Shambat 6. However, none of the twenty hybrids had a bi significantly different from unity. Only two hybrids (Ka99 x 15 and Hysun33) had $σ^2d$ significantly different (P ≤ 0.05) from zero. More or less a pattern of ranking similar to the observed one based on the bi and $σ^2$d for the hybrids was followed by Wricke ecovalance (Wi).

With respect to stability parameters assessed, hybrids; Ka99 x 1, Ka99 x 7, Ka99 x 13, Ka99 x 17, Ka99 x 29, Salih and Shambat 6 had □ higher than the average, bi above unity and $σ^2d$ not significantly different from zero. They were considered to be below average in stability, so they were sensitive to environmental changes and hence they could be recommended for favorable environment. Hybrid Ka99 x 25-2 was above average in stability, in that it had mean seed yield above the average, bi below unity and $σ^2d$ not significantly different from zero. Hence, it

Table 3. Mean yield (μ), regression coefficient (bi), deviation from regression (σ^2d) and ecovalance (Wi) with ranking (R) on seed yield (t/ha) for the 20 sunflower hybrids evaluated at four environments.

Hybrid	Mean (μ)	R	bi	R	Probability	σ^2 d	R	Wi	R	Average(R)
Ka99X1	3.46	12	0.891	8	1.000	0.122	17	0.018	7	11
Ka99X7	3.60	10	1.128	12	1.000	0.042	5	0.000	1	2
Ka99X11	3.04	20	0.969	1	1.000	-0.116	15	0.299	19	18
Ka99X13	3.88	4	1.143	14	0.409	-0.037	3	0.085	12	4
Ka99X15	3.66	8	1.050	3	1.000	0.366*	19	0.005	4	7
Ka99X17	3.95	3	1.166	16	1.000	0.093	14	0.134	16	16
Ka99X18	3.74	5	1.111	9	0.337	-0.091	13	0.023	8	8
Ka99X22	3.41	13	0.468	19	0.114	0.059	6	0.032	9	14
Ka99X251	3.13	19	0.872	11	1.000	0.021	1	0.209	17	15
Ka99X25-2	3.64	9	0.676	18	0.288	0.119	16	0.002	3	12
Ka99X29	3.71	7	1.132	13	1.000	0.061	7	0.015	6	3
Ka99X30	3.25	17	0.913	6	1.000	-0.081	11	0.114	13	13
Ka99X32	3.23	18	1.104	7	0.055	-0.126	18	0.127	15	19
Ka99X35	3.33	14	0.840	15	0.365	-0.038	4	0.069	10	10
Ka99X37	3.31	15	1.078	5	1.000	-0.070	9	0.075	11	9
Ka99X41	3.56	11	0.879	10	0.333	-0.085	12	0.001	2	6
Ka99X42M	3.25	16	0.778	17	0.268	-0.026	2	0.114	14	17
Salih	3.71	6	1.036	2	1.000	-0.070	8	0.014	5	1
Shambat 6	4.05	2	1.063	4	1.000	0.080	10	0.213	18	5
Hysunn 33	4.72	1	1.704	20	0.161	0.376*	20	1.275	20	20
Overall mean	3.58									

* significant at $P \leq 0.05$

Table 4. Environmental indices (I_i) and expected means ($\acute{Y}$) for seed yield (t/ha) of 5 selected hybrids evaluated at four environments.

Hybrid	Envir. (E)	Index Expected (I) ($\acute{Y}$)	
Ka99 x 7	E_1	1.221	4.970
	E_2	-0.154	3.419
	E_3	0.644	4.319
	E_4	- 1.708	1.666
Ka99 X 18	E_1	1.221	5.098
	E_2	- 0.154	3.570
	E_3	0.644	4.465
	E_4	- 1.708	1.843
Salih	E_1	1.221	4.972
	E_2	-0.154	3.547
	E_3	0.644	4.374
	E_4	-1.708	1.937
Shambat.6	E_1	1.221	5.345
	E_2	-0.154	3.886
	E_3	0.644	4.733
	E_4	-1.708	2.250
Hysun33	E_1	1.221	6.800
	E_2	-0.154	4.456
	E_3	0.644	5.816
	E_4	-1.708	1.809

could be considered as adapted hybrids for unfavorable environmental conditions. However, hybrid Ka99x11, Ka99x15, Ka99x18, Ka99x22, Ka99x25-1, Ka99x30, Ka99x32, Ka99x35, Ka99x37, Ka99x41, Ka99x42M and

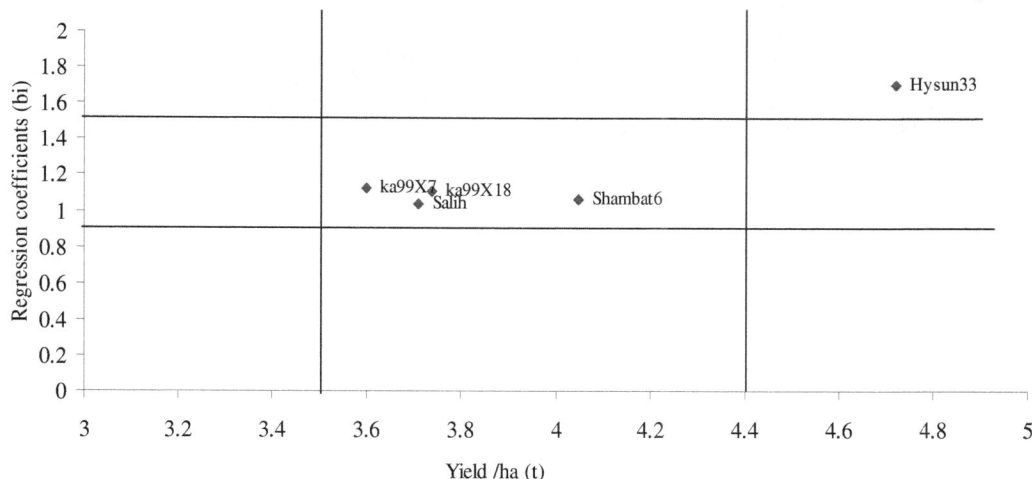

Figure 1. Stability of performance. Scatter graph for yeild (t/ha) on regression coeficients (bi), point for only five hybrids are shown

Hysun33 did not follow a particular pattern. They either had a mean yield below the average or a σ^2d significantly different from zero and therefore, they were not adapted. From the average ranking of the hybrids, Salih was the first ranked followed by Ka99 x 7, Ka99 x 29, Ka99 x 13 and Shambat 6. These five hybrids had the least contribution to G x E interaction as indicated by their low values of σ^2d and Wi. Therefore, they are considered as the most stable hybrids in terms of yield.For the sake of simplicity, the four environments, of summer season in New Halfain, winter season in New Halfa, summer season in Rahad and winter season in Rahad, are symbolized as E_1, E_2, E_3 and E_4 respectively. The environmental indices (I_i) and the expected means ($Ý$) for the seed yield (t /ha) of five arbitrary selected hybrids (Ka99 x 7, Ka99 x 18, Salih, Shambat 6 and Hysun33), evaluated at the four environments are depicted in Table 4. As for I_i, E_1, had the highest positive index of 1.221 and hence, it was the most favorable, followed by E_3, then E_2. On the other hand, E_4, recorded the highest negative index of -1.71, and thus it was the poorest one. The highest expected mean yield ($Ý$) for each hybrid was determined at E_1, whereas the lowest one was determined at E_4. Shambat 6 gave the highest seed yield (t/ha), under unfavorable environmental conditions followed by Salih, Ka99 x 18, and Hysun 33. El-Hity (1994) reported that the indigenous hybrids and the open pollinated crops are higher yielder and more adapted to the local environments than the introduced hybrids. The scatter graph of the stability for the five selected hybrids is shown in Fig.1. It is apparent that while the point for Hysun33 deviated from the zone of interception of the two vertical and the horizontal lines, the points for the locally developed hybrids were within the zone. Moreover, all the points for the locally developed hybrids fall near the lines for $Ý \pm \sigma$ and that for $b^- \pm \sigma$. Therefore, they were more stable in yield.

Dabholkar (2001) reported that, a point of stable genotype should fall within the zone of interception of the two vertical and horizontal lines of $Ý \pm \sigma$ and $b^- \pm \sigma$ in the scatter graph for estimating stability of performance of a genotype. Based on the results obtained in this study, the following conclusions could be drawn:-

1. The significant genotype x environment interactions for seed yield indicates the importance of evaluating the hybrids at different environments.
2. Salih and Ka99 x 17, are most likely adapted hybrids to favorable environmental conditions whereas, Ka99 x 25-2 is the most stable hybrid for yield under adverse conditions.
3. The yield of the locally developed hybrids is stable under different environmental conditions and less affectted by the poor environments, whereas the yield of Hysun33 fluctuated and only increased with improving the environmental conditions.
4. Emphasis should be made on Ka99 x 25-2, since it is stable for yield under adverse environmental conditions and had a wide range of adaptation.

REFERENCES

Bange MP, Hammer GL, Rickert KG (1997). Environmental control of potential yield of sunflower in the tropics. Austral. J. Agric Res. 48: 231-240.

Becker HC, Leon J (1988). Stability analysis in plant breeding. Plant Breed. J. 101, 1-23.

Dabholkar AR (2001). Elements of Biometrical Genetics, College of Agriculture. Concept Puplishing Company, New Delhi 110059, India, p. 427.

Eberhardt SA, Russell WA (1966). Stability parameters for comparing varieties. Crop Science 6 : 36 - 40.

El-Hity MA (1994). Evaluation of sunflower cultivars and their interaction with year. Alexandria J. Agric. Res. 39(2): 179 - 192.

Finlay KW, Wilkinson GM (1963). The analysis of adaptation in plant breeding programme Austral. J. Agric. Res. 14: 742-754.

Gomez KA, Gomez AA (1984). Statistical Procedures for Agricultural Research .2nd ed. John Wiley and Sons, Inc. New York. p. 680.

Khidir MO (1997). Oil Crops in Sudan. Khartoum University Press 1st ed. pp. 103-120. (in Arabic).

Miller PA, Williams JC, Robinson HP Comstock RE (1958). Estimation of the genotypic and environmental variances and covariances in upland cotton and their implications in selection. Agron. J. 60: 126-131.

Singh JV, Yadava TP (1986). Variability studies of some quantitative characters in sunflower. Harv. Agric. I Univ. J. Oilseeds Res. 3(1): 125-127.

Wricke G (1962). Über eine Methode Zur Erfassung der Ökologischen Streubreite in Feldversuchen .Z. PflanZenZüchtung, 47: 92-97.

8

Phenotypic variation of cacao (*Theobroma cacao* L.) on farms and in the gene bank in Cameroon

M. I. B. Efombagn[1]*, O. Sounigo[2], S. Nyassé[1], M. Manzanares-Dauleux[3], A. B. Eskes[2]

[1]Institute of Agricultural Research for Development (IRAD), P. O. Box 2067 or 2123, Yaoundé, Cameroon.
[2]CIRAD, UPR31, 34398 Montpellier Cedex 5, France.
[3]Agrocampus de Rennes, 65 Rue de St Brieuc, 35042, Rennes, France.

A survey was undertaken in the 2 major cocoa producing areas (Southern and Western) of Cameroon to study the morphological diversity existing in cacao farms in relation to genetic diversity in gene bank accessions. A total of 300 farm accessions (FA) were selected in the field which were compared to 77 gene bank accessions distributed into 4 groups (AGs) according to their origin. The 17 quantitative and qualitative descriptors used in this study were related to leaf (flush colour), flower (ligule colour), pod (weight, length, width, apex form, shape, rugosity, colour, husk hardness, basal constriction and pod index) and seed (number, length, width, dry weight and colour) characters. For the qualitative characters evaluated, considerable morphological variation was observed using the Shannon Weaver diversity index (SWDI) within FA and gene bank accessions. Among the FA, a differentiation between southern and western regions was only possible when using quantitative pod traits. Mean quantitative traits values of FA were not too different than those of most gene bank AGs, except for a few traits of agronomical interest (seed weight and pod index). No significant variation was observed for seed traits in all FA groups (southern/western). The morphological structure (quantitative traits) showed spatial differentiation between western and southern FA and a closer relationship between gene bank and some farm accessions. Furthermore, a molecular study done earlier using microsatellite profiles of the same FA did not show any genetic difference between FA of both regions, suggesting that the agromorphological performance of FA is rather due to non-genetic factors. In contrast, microsatellites have shown that most of the gene bank accessions were genetically distant from the FA, suggesting the low intake of some breeders' genotypes to farmers' fields. The level of diversity found in farmers' germplasm could enhance the gene bank and current breeding programs.

Key words: *Theobroma cacao* L., farm accessions, morphological diversity, breeding.

INTRODUCTION

Cacao (*Theobroma cacao* L.) is a perennial crop of significant economic importance in producing countries of West Africa, South America and Southeast Asia. The traditional classification of *T. cacao* assumes 3 horticultural races or main types: Criollo, Forastero and Trinitario. The conventional classification of cacao into Criollo and Forastero is based on distinct morphological, historical and commercial traits. Trinitario is an intermediate type between Criollo and Forastero, a hybrid group with traits that include the total variation of the species

(Motamayor et al., 2008).

In cacao, certain morphological characters of pods and seeds are used as the basis of classification into categories, which may be called varieties, cultivars, types or populations (Wood and Lass, 1985). Morphological descriptors are useful because they can help the breeders to select the best accessions for the breeding programme (Engels et al., 1980). Phenotypic characterization of the species, usually conducted by gene banks, involves leaf, flower, pod and seed descriptors (Engels et al., 1980; Bekele and Bekele, 1996). The phenotypic appearance of cacao fruits (pods) plays an important role in the definition of types and populations. Considerable variation is encountered at the level of seed (seed) size.

---

*Corresponding author. E-mail: efombagn@yahoo.fr.

Studies on morphological diversity have been carried out on flowers, fruits and leaves of accessions from cacao germplasm (Engels, 1986) which revealed the existence of 2 morphological groups: one composed of the Criollo and Trinitario accessions and the other composed of the Forastero accessions, with a continuous variation between the 2 groups due to several genetic admixtures that occurred in the species. This structure obtained by Engels (1986) was confirmed later by N'Goran (1994) using seed and pod characters. Flower traits used earlier by Enríquez and Soria (1967) and more recently by Lachenaud et al. (1999) allowed the detection of a great variability among cacao cultivars. Globally, all these results showed that morphological markers could allow the structuring of the diversity of different populations in germplasm collections in research stations.

Molecular diversity of cocoa found in Cameroon was previously analyzed by Efombagn et al. (2006, 2008). The need to complement this molecular work with a morphological diversity study became important. Therefore in the current study, there was phenotypic characterization of 300 cacao accessions collected in various farmers' fields, distributed over different cacao growing areas of Cameroon and 77 breeders' accessions available on-station in gene banks of the Institute of Agricultural Research for Development (IRAD), Cameroon. Simple sequences repeat markers were considered verifying if the molecular diversity of farmers and breeders material tallied with the results from the morphological study. The objective was to assess diversity in farmers' and breeders' germplasm material using morphological traits. The level of diversity in farmers' and breeders' popula-tions assessed with qualitative and quantitative phenoty-pic traits for leaves, pods and seeds, as well as the relation-ships between these 2 types of germplasm were measured. The paper also addressed the exploitation of the phenotypic variation found in gene bank and in farm cacao populations for the purpose of breeding.

MATERIALS AND METHODS

Study site

The study was conducted in the Southern and the Western agro-ecological areas where cacao is grown in the country. Within these 2 areas, the collecting sites ranged from latitude between N02°14.199' and N05°42.924' and longitude between E009°01.430' and E011°20.885'. The Southern part is characterised by heavy shaded cacao plantations with a low level of management and chemical inputs. The average yield varies between 100 and 500 Kg of fermented and dried cocoa per ha (Varlet and Berry, 1997). The soils are ferralitic and acidic and the rainfall pattern includes 2 wet and 2 dry seasons. In the Western area, the climatic conditions for cacao cultivation are relatively favourable and thus cacao plantations are relatively lightly shaded, with a systematic use of chemical inputs (Losch et al., 1992). Yield varies between 600 and 1200 Kg of fermented and dried cocoa per ha. The soils are volcanic and the rainfall pattern includes one wet and one dry season.

Plant material

Cacao accessions used in the study included 300 farm accessions (FA) collected in the Southern (145 FA) and the Western (155 FA) areas and 77 gene bank accessions (GA) as part of the cacao collections of IRAD at the Nkoemvone Research Station (southern Cameroon) (Table 1). Gene bank accessions belong to two genetic groups (GGs) of cacao, or hybrids between the 2 GGs. These GGs include the upper Amazon Forastero (UA) and Trinitario comprising a wide range of hybrids between the Criollo as defined by Cuatrecasas (1964) and Motamayor et al. (2002) and Amazon Forastero, both originating from South America. The cacao trees were randomly selected (by the farmers and the breeders) in the field and assigned accession numbers prior to their transfer and their vegetative propagation in the nurseries on-station. The data on the transferred material were collected in the field in September and October 2004.

Microsatellite profiles of the studied farm and gene bank accessions were used to generate the genetic structure of the cacao material under study. Different steps including PCR (polymerase chain reaction) and capillary electrophoresis were used (Efombagn et al., 2006; 2008).

The 17 morphological and agronomical traits that were recorded in the study are presented in Tables 2 and 3. Plant data including leaf, flower, fruit and seed traits were recorded following the identification of the so-called minimum descriptors from the Bioversity International (formerly called International Board for Plant Genetic Resources Institute (presently called) (Bekele and Butler, 2000; Eskes et al., 2000). These descriptors are reported of being the most discriminative and taxonomically useful ones which preclude redundancy (Bekele et al. (2006). They were also selected for ease of observation, reliability of scoring and for their relation to agronomical value, in the case of seed descriptors. The method used for pod and seed characterization is described by Bekele and Bekele (1996) and Bekele and Butler (2000).

Statistical analysis

Basic statistics and multivariate analysis for quantitative traits: As estimates of diversity study, data analysis for all quantitative morphological characters was done using MINITAB-15 software (Minitab Inc, 2007). Diversity statistics of quantitative traits consisted on descriptive statistics where means and coefficients of variation of different accession groups (AGs) were determined. Analysis of variance (ANOVA) was also performed among different AGs. All the morphological quantitative traits were subjected to Principal Components Analysis (PCA) using the correlation matrix to define the pattern of variation. PCA axes with Eigen values ≥0.8 were selected to define the variation among accessions for agronomic and morphological traits.

Estimates of diversity study with qualitative traits

For qualitative traits, the phenotypic frequency data were analysed by Shannon-Weaver Diversity Index (SWDI), H', using the formula:

$$H' = - \sum_{i=1}^{K} P_i \ln P_i$$

Where k is the number of phenotypic classes for a descriptor and P_i is the proportion of the total number of accessions (N) in the ith class. H' was estimated for each of the nine qualitative descriptors (Eskes et al., 2000).

Table 1. Farm and gene bank germplasm used in the study.

Accession status	Accession group (AG)	Genetic group (GG)	Country of origin	Acronym	Number of accessions studied
Gene bank (Nkoemvone Station)	SNK*	Trinitario(Tr)	Cameroon	SNK	36
	ICS		Trinidad	ICS	5
	SNKXICS		Cameroun	SNK600-Tr**	11
	IMC	Upper Amazon	Peru		
	UPA	Forastero(UA)	Ghana	UA	10
	UAT	UA X Tr	Ghana		
	SNKXUA		Cameroon	SNK600-UAxTr*	15
Total					77
Farm (FA)	FA-South		Southern Cameroon	FA-South	145
	FA-west		Western Cameroon	FA-West	155
Total					300
All				All	

*: SNK accessions were selected in farmers' fields in 1950s among the best farmers' trees in term of yield, while SNK600 series were selected on-station in Upper Amazon (UA) x Trinitario (Tr) and in TrxTr crosses, as well as in selfings of Tr (Efombagn, 2008).
**: These two were formerly considered as a common group usually called 'SNK600 series, despite the difference observed in their genetic origin.

RESULTS

Quantitative traits

Among the 300 FA of the study, pods were on average 14.8 cm in length, 7.0 cm in width and 510.6 g in weight (Table 2). Among gene bank accessions, the highest values were recorded in ICS for pod length (18.3 cm), width (8.3 cm) and weight (622.5 g). Farm accessions had a mean seed number of 40.5 for, mean seed length of 23.8 mm, for mean seed width of 13.1 mm and mean individual dry seed weight of 0.92 g. Seed number was lower in the gene bank compared to FA. The highest seed size and weight were recorded for ICS, with 26.6 mm for length, 14.5 mm for width and 1.34 g for weight. In the gene bank, the lowest seed length and weight values of 22.8 mm and 0.93 g, respectively, were found in UA. ICS and UA have registered respectively the lowest (21.2) and the highest (29.0) pod index. The average pod index of the FA was relatively high (26.3) compared to the average of all gene bank AGs. Differences between pod and seeds traits of all AGs of the study were significant, except for seed width (Table 2). Among FA, the accessions from the Western part of the country (FA-West) recorded the highest values for pod traits compared to those of the southern part (FA-South). However, the difference was not significant for seed width between both sets of AGs.

Diversity estimates for qualitative traits

Diversity for each of the nine qualitative descriptors was observed and high diversity values were obtained in both FA and gene bank accessions. Table 3 presents the estimates of SWDI (H') of farm and gene bank accessions. For all the FA, the minimum value of H' was 0.42 for pod basal constriction and the maximum value was 0.89 for pod apex form. For all gene bank accessions, the same descriptors recorded respectively the highest and the lowest values, with 0.56 for pod basal constriction and 0.89 for pod apex form. In FA, the H' values of flower ligule and leaf colours (not estimated in gene bank accessions) were respectively 0.67 and 0.61 (Table 3).

Principal component analysis

The four principal components (PC1 to PC4) explained 86.0% of the total phenotypic variance. The relationship of these principal components and quantitative variables is given in Table 4. In the first principal component (PC1), all the variables were prominent, except for seed number and width to a lesser extent. In the second principal component (PC2), seed size traits (length and width) were the most significant. A plot of PC1 and PC2 showed the relative grouping of FA and gene bank accessions (Figure 1). No clear separation was observed between FA and gene bank accessions. However, there was a spatial differentiation between the FA of the southern and the western origin.

Relationship between phenotypic and genetic structures

The phenotypic variance revealed by the principal

Table 2. Basic statistics for eight quantitative and agronomic traits in FA and GA.

Trait	Gene bank					Farm			F ratio
	SNK	SNK600-Tr	SNK600-UAXTr	ICS	UA	FA-South	FA-West	All FA	
Pod Length (cm)	15.8±2.1	14.1±2.3	13.6±1.6	18.3±1.5	10.8±2.4	13.5±2.8	16.2±2.52	14.8±2.9	2.7*
Pod Width (cm)	7.9±1.0	6.3±0.6	6.6±0.5	8.3±0.5	6.1±1.4	6.0±0.6	8.1±1.05	7.0±1.3	5.4****
Pod Weight (g)	561.2±178.3	515.1±150.9	538.8±101.9	622.5±188.7	471.3±163.9	453.4±134.3	568.1±129.6	510.6±143.7	2.0 *
Seed Number	36.7±4.6	36.7±4.6	40.0±7.6	35.1±1.6	38.8±5.3	39.7±4.5	41.2±5.4	40.4±5.0	2.8*
Seed Length (mm)	22.2±1.9	25.5±1.8	25.5±1.3	26.6±2.1	22.8±2.5	24.0±1.9	23.4±2.5	23.7±1.8	11.1****
Seed Width (mm)	12.3±1.1	13.6±2.3	13.5±1.6	14.5±0.8	13.0±1.6	13.2±1.2	12.9±2.7	13.1±2.1	1.7 ns
Dry weight (1 seed)	1.0±0.3	1.1±0.1	1.2±0.1	1.3±0.0	0.6±0.1	1.0±0.2	0.9±0.2	0.9±0.3	3.1**
Pod index[1]	23.6±3.8	21.8±3.7	23.7±3.3	21.2±2.1	38.0±6.8	26.9±8.6	25.5±6.5	26.3±7.7	2.7*

****p < 0.0001; ***p < 0.001; **p < 0.01; *p < 0.05
1: Pod index is expressed as the number of pod needed to produce 1 kg of dried cocoa.

component analysis (Figure 1) was compared with the genetic structure generated by microsatellite analysis (Figure 2). When the geographic distribution of FA was considered, the phenotypic variation showed a spatial differentiation between western and southern FA (Figure 1), while no genetic clustering could be observed with microsatellite for these 2 FA subgroups (Figure 2). In addition, the proximity between several gene bank accessions and the FA as revealed by phenotypic markers, were not confirmed with microsatellite markers.

DISCUSSION

The study revealed morphological variation within FA and gene bank groups. Pod weight and size, dry seed weight and pod index of the locally selected AGs (SNK and SNK600 series) of the gene bank were higher than those of the farm accessions. Except for seed width, all the quantitative morphological traits were found useful in differentiating farm from gene bank accessions.

There were no differences in seed size (length and width) among SNK accessions and SNK600 series. As observed in a previous study carried on wild cacao trees from French Guiana (Lachenaud and Olivier, 2005), morphological seed descriptors are not always able to discriminate among groups of cacao accessions.

Among AGs of Tr origin, the quantitative characters of the ICS group were higher than those of the locally-selected Tr. Tr differed from UA when all mean values were considered and previous morphological studies have already differentiated these 2 genetic groups (N'Goran, 1994; Bekele et al., 2006). Some AGs like ICS might be of particular interest to breeders because of their superior agronomic traits such as pod index and seed size. Pod characters differed between FA of Southern and Western origin.

According to Lachenaud (2007), variation based on pod traits might be associated with different morpho-geographic groups. Therefore, the difference between the 2 geographic FA groups of our study is rather due to the variation of ecological conditions under which the cacao is grown. Molecular characterization using SSRs markers did not reveal any genetic differences between FA of Southern and Western Cameroon (Efombagn et al., 2008), confirming the hypothesis of an environmental influence on field performances of cacao genotypes.

Based on SWDI (*H*), all the pod qualitative traits showed variation within farm and gene bank groups of accessions. These descriptors were among the traits found to be most useful for studying the variability of cacao populations (Engels, 1983; Raboin et al., 1993; Bekele et al., 1994; Lachenaud et al., 1999).

When mean values of all the 16 quantitative and qualitative morphological traits of the study were

Table 3. Comparison of Shannon-weaver Diversity Index (*H'*) values for qualitative traits studied in farms and gene bank.

Descriptor	Criterion of comparison			
	Origin		Type of cultivar	
	Farm	Gene bank	Traditional*	Hybrid**
Pod apex shape	0.89	0.89	0.86	0.91
Pod shape	0.69	0.74	0.19	0.86
Pod rugosity	0.65	0.58	0.50	0.70
Pod colour	0.59	0.69	0.56	0.65
Pod husk hardness	0.56	0.68	0.27	0.77
Pod basal constriction	0.42	0.56	0.19	0.56
Cotyledon colour	0.59	0.61	0.57	0.67
Flower ligule colour	0.67	-	0.59	0.73
Leaf flush colour	0.61	-	0.58	0.63

*: Primary germplasm.
**: material resulting from manual or natural pollination in seed gardens or cacao farms.

Table 4. Eigen values r and percentage of variation explained by the first 4 principal components for all the accessions included in the study.

Descriptor	PC1	PC2	PC3	PC4
Eigen values	3.15	1.92	1.08	0.82
Proportion variance (%)	0.39	0.24	0.13	0.10
Cumulative variance (%)	0.39	0.63	0.77	0.87
Pod weight	0.46	- 0.21	0.23	0.18
Pod width	0.38	- 0.34	0.22	0.03
Pod length	0.40	- 0.33	0.23	0.15
Seed number	0.16	- 0.19	- 0.78	0.44
Seed length	0.24	0.51	0.04	0.36
Seed width	0.15	0.56	0.17	0.38
Dry seed weight	0.39	0.30	- 0.03	- 0.58
Pod index	- 0.43	- 0.13	0.43	0.35

considered, the level of diversity between farm and gene bank materials did not vary considerably as shown in Table 2 (field measurements values) and Table 3 (*H'* values). In addition, the genetic diversity study carried out with microsatellite markers (Efombagn et al., 2006) revealed that the level of diversity in farmers' fields was genetically close to that of the accessions maintained in the gene bank. Therefore, the selection process conducted by farmers while growing cacao in Cameroon did not result in a reduction of the diversity existing in cacao farms over the time.

The morphological variation using quantitative traits (PCA) has shown a spatial differentiation between western and southern FA. However, no genetic difference was detected with microsatellite analysis between these two subgroups. The variation found was due to non-genetic factors such as the prevailing cacao growing conditions in both growing regions. According to the PCA, the distribution of FA in relation with gene bank accessions has confirmed that most of the accessions from farmers' fields were close to accessions of SNK and SNK600 series (breeders' material), except for some cultivars from Southern Cameroon. As implication to biodiversity and genetic conservation, the genetic material that harbour genes controlling agronomic traits such as resistance to *Phytophthora* pod rot, is conserved in farmers' fields over several decades through farmers practices. The farmers usually achieve this by selecting pods from their own or neighbouring plantations, prior to the production of seeds for next the plantations (Efombagn, 2006). Quantitative traits of other cacao organs such as leaf and plant growth habit were not used in this study because they were found to have no discriminative value (Ostendorf, 1957). For morphological

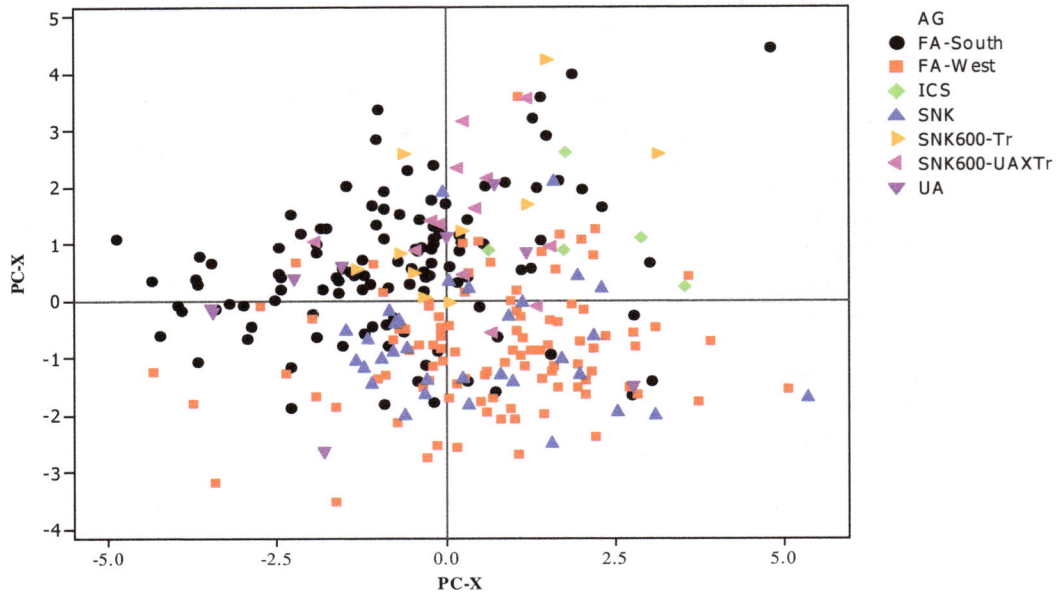

Figure 1. Phenotypic diversity revealed by principal component analysis of all the farm (FA) and gene bank accessions.

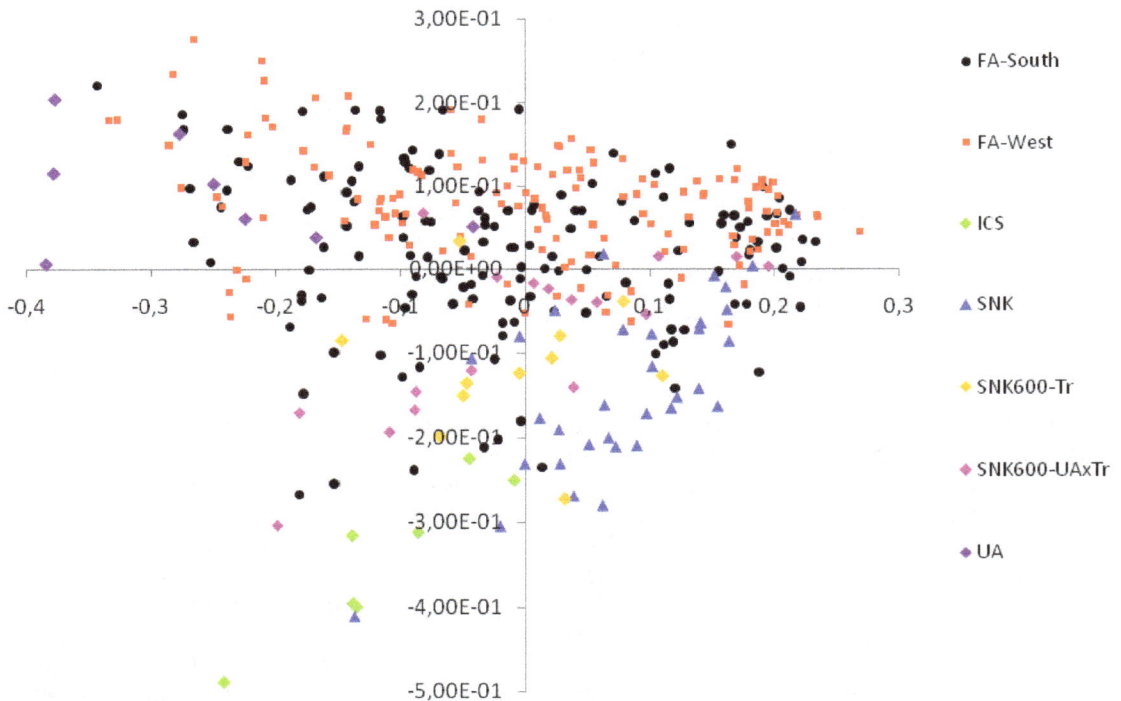

Figure 2. Genetic structure of all the farms and gene bank accessions studied based on microsatellite analysis.

traits of agronomic interest such as seed size, no significant difference was found among farm genotypes.

In the cacao selection process in Cameroon, it is therefore suggested to improve seed size by exploiting gene bank accessions with favourable seed characteristics. However, such traits which are usually quantitative and under polygenic control are under a strong environmental influence (Dias, 2001). Prior to their use in cacao breeding, stability of such quantitative traits should be studied to generate reliable, reproducible data in different ecological conditions (Simmonds, 1981; Engels, 1993). In future breeding programs, other phenotypic traits of agronomic interest may involve number of pods per tree, which was not investigated in this study, as well as resistance to pest and diseases, and the aromatic and flavour attributes (quality) of cacao.

ACKNOWLEDGEMENTS

The authors would like to thank the CFC/ICCO/Bioversity project titled 'Cacao Productivity and Quality Improvement, a Participatory Approach' for provision of funds and technical support. We also thank the Technicians of IRAD, F. Edoa, I. Badjeck and K.D. Vefonge for their assistance in field data collection.

REFERENCES

Bekele F, Butler DR (2000). Proposed of Cacao descriptors for characterisation. In: Working procedures for cacao germplasm evaluation and selection. In: Eskes A.B., Engels J.M.M. and Lass R.A. (eds), Proceedings of the CFC/ICCO/IPGRI Project Workshop, Montpellier, France, February 1-6, 1998. IPGRI, Montpellier pp. 41-48.

Bekele FL, Bekele I (1996). A sampling of the phenetic diversity of cocoa in the International Cocoa Gene Bank of Trinidad. Crop Sci. 36: 57-64.

Bekele FL, Bekele I, Butler RD, Gillian G Bidaisee (2006). Patterns of morphological variation in a sample of cacao (Theobroma cacao L.) germplasm from the International Cacao Gene bank, Trinidad. Genet. Resour. Crop Evol. 53: 933-948.

Bekele FL, Kennedy AJ, Mc David C, Lauckner B, Bekele I (1994). Numerical taxonomic studies on cacao (Theobroma cacao L.) in Trinidad. Euphytica 75: 231-240.

Cuatrecasas J (1964). Cacao and its allies: A taxonomic revision of the genus Theobroma. Contrib US Herbarium 35: 379-614.

Dias LAS (2001). Melhoramento Genético do Cacaueiro. L.A.S. Diaz (Ed), FUNAPE, UFG, p. 578. Efombagn MIB (2008). Diversité Génétique et Sélection du Cacaoyer au Cameroun: Approches participative, phénotypique et moléculaire. Thèse de Doctorat, Agrocampus Rennes, France p. 149.

Efombagn MIB, Motamayor JC, Sounigo O, Eskes AB, Nyassé S, Cilas C, Schnell R, Manzanares-Dauleux MJ, Kolesnikova-Allen M (2008). Genetic diversity and structure of farm and gene bank accessions of cacao (Theobroma cacao L.) in Cameroon revealed by microsatellite markers. Tree Genetics and Genome 4: 821-831.

Efombagn MIB, Sounigo O, Nyassé S, Manzanares-Dauleux M, Cilas C, Eskes AB, Kolesnikova-Allen M (2006). Genetic Diversity in cocoa germplasm of southern Cameroon revealed by simple sequences repeat (SSRs) markers. Afr. J. Biotechnol. 5(16): 1441-1449.

Engels JMM (1983). A systematic description of cacao clones 111. Relationships between clones, between characteristics and some consequences for the cacao breeding. Euphytica 32: 719-733.

Engels JMM (1986). The systematic description of cacao clones and its significance for taxonomy and plant breeding. PhD Thesis, Agricultural University, Wageningen, The Netherlands p. 125.

Engels JMM (1993). The use of botanical descriptors for cacao characterization: CATIE experiences. In: Proceedings of the Interna-

tional Workshop on Conservation, Characterisation, and Utilisation of Cacao Genetic Resources in the 21st Century. The Cocoa Research Unit, Trinidad pp. 69-76.

Engels JMM, Bartley BGD, Enríquez GA (1980). Cacao descriptors, their states and modus operandi. Turrialba 30: 209-218.

Enríquez S, Soria V (1967). Seleccion y studio de los caracteres utiles de la flor para la identificacion y descripcion de cultivares de cacao. Cacao, Turrialba, Costa Rica 12: 8-16.

Eskes AB, Engels JMM, Lass RA (2000). Working procedures for Cocoa Germplam Evaluation and Selection. Proceedings of the CFC/ICCO/IPGRI Project Workshop, 1-6 February 1998, Montpellier, France. International Plant Genetic Resources Institute, Rome, Italy.

Lachenaud P (2007). Fruit trait variability in wild cocoa trees (Theobroma cacao L.) from the Camopi and Tanpok basins in French Guiana. Acta Bot. Gallica. 154(1): 117-128.

Lachenaud P, Bonnot F, Olivier G (1999). Use of floral descriptors to study variability in wild cacao trees (Theobroma cacao L.) in French Guiana. Genet. Resour. Crop Evol. 46: 491-500.

Lachenaud P, Olivier G (2005). Variability and selection for morphological seed traits in wild cacao trees (Theobroma cacao L.) from French Guiana. Genet. Resour. Crop Evol. 52: 225-231.

Losch B, Fusillier JL, Dupraz P (1992). Stratégies des producteurs en zone caféière et cacaoyère au Cameroun. Quelles adaptations à la crise ? Montpellier, France, CIRAD-Dsa, Collection documents systèmes agraires n° 12, p. 252.

Minitab Inc. (2007). MINITAB User's Guide 2: Data analysis and Quality Tools. Release 15 for Windows. Minitab Inc. USA.

Motamayor JC, Risterucci AM, Lopez PA, Lanaud C (2002). Cacao domestication I: the origin of the cacao cultivated by the Mayas. Heredity 89: 380-386.

N'Goran J (1994). Contribution à l'étude génétique du cacaoyer par les marqueurs moléculaires: diversité et recherche de QTIs. Doctoral thesis, University of Montpellier II, Montpellier, France p. 105.

Ostendorf FW (1957). Identifying characters for cacao clones. In : Reuniao do comite tecnico interamericano de cacao, 6, Actas. Instituto de cacao da Bahia, Salvador pp. 89-110.

Raboin L-M, Paulin D, Cilas C, Eskes AB (1993). Analyse génétique de quelques caractères quantitatifs des fleurs de cacaoyers (Theobroma cacao L.). Leur intérêt pour l'évaluation de la diversité de l'espèce. Café Cacao Thé 37: 273-282.

Simmonds NW (1981). Principles of characterization and evaluation. In: International Conference on Crop Genetic Resources. Report of the FAO/UNEP/IBPGR, Roma pp. 33-35.

Varlet F, Berry D (1997). Réhabilitation de la protection phytosanitaire des cacaoyers et caféiers au Cameroun. Tome I : rapport principal ; tome II : annexes. Douala, Cameroun, Conseil interprofessionnel du cacao et du café pp. 202, 204.

Wood GAR, Lass RA (1985). Cocoa. Longman. 4th Ed. Tropical Agriculture series, London, UK p. 620.

9

Genetic polymorphism in exotic safflower (*Carthamus tinctorious* L.) using RAPD markers

Mahasi M. J.[1], Wachira F. N.[2], Pathak R. S.[2] and Riungu T. C.[3]

[1]Kenya Agricultural Research Institute (KARI), P. O. Private Bag, Njoro, Kenya.
[2]Egerton University, P. O. Box 536, Njoro, Kenya.
[3]KARI – Muguga South P.O. Box 30148 Nairobi - Kenya.

Safflower is a drought tolerant annual oil crop and this gives it an advantage over the other crops in the drier parts of Kenya. It is valued worldwide as a source of high quality vegetable oil. In the past, characterization of safflower using molecular markers has been limited. The objective of this study was to evaluate the degree of polymorphism in 36 safflower accessions using RAPDs. Sixty-one amplification products were scored using 14 random 10 mer primers and binary matrices subjected to statistical analyses using NTSYS. A resemblance matrix was developed using SMC, which was used with the UPGMA to compute cluster analysis and PCA. Eight groups were formed at a similarity coefficient of 0.79. Cluster two had 14 accessions originating from India, USA, Australia and Bangladesh while cluster three had 9 accessions from India, USA and Mexico. Proportionally accessions from India were highest in cluster one and two. The differences between pairs of accessions were basically related to the number of RAPD fragments shared. Four Indian accessions PI 214150, PI 199910, W6 16821 and PI 248359 clustered together. However, Girna also from India formed an independent cluster. SMC among accessions ranged from 0.37 (PI 248359 and PI 262419) to 0.98 (PI 560177 and T65). The last two accessions may be genetically related since they constituted the nearest to a complete match for all markers. Accessions from different countries tended to group together though random scattering often occurred. Using PCA the first three components explained 44% of the total variation. The results indicate genetic polymorphism between the safflower accessions under study.

Key words: Amplification, PCA, RAPDs, simple matching coefficients, electrophoresis.

INTRODUCTION

Safflower (*Carthamus tinctorius* L. 2n = 2 x = 24, family Asteraceae) has a strong central branched stem and varying number of branches. Typically the plants are, herbaceous and thistle-like usually with many long sharp spines on the leaves and bracts (Helm et al., 1991). Globally, over 60 countries grow safflower, on about 1.2 million hectares with annual production of 0.79 million tons (Singhal, 1999). India is the largest producer of safflower

flower (68%, 0.2 million tons) in the world with highest acreage (60%, 0.43 million hectares) and production is mainly for the domestic vegetable oil market (Johnson and Marter, 1993). Safflower has a wide range of related species. Within the genus of *Carthamus*, there are more than 20 species divided into 4 sections (Knowles, 1988). Section one (2n = 20) has (*oxyantha* and *palaestinu*), section two (2n = 24) has (*tinctorius, alexandrius, glaucus, syriacus* and *tenuis*) section three (*lanatus* 2n = 44) while section four (*baeticus* 2n = 64). The first two sections are diploids, the third is a tetraploid and the fourth section consists of hexaploid species (Khidir, 1969).

Safflower is a drought tolerant annual oil crop and this gives it an advantage over the other crops in the drier parts of Kenya. It is valued worldwide as a source of high quality vegetable oil. In the past, safflower germplasm was characterized entirely on the basis of morphological traits, abiotic stresses and (or) biochemical characters

---

*Corresponding author. E-mail: jendekamahasi@yahoo.com.

Abbreviations: UPGMA, Unweighted pair group method with arithmetic mean; **NSTYS**, numerical taxonomy and multivariate analysis system; **PCA**, principal component analysis; **SDW**, sterile distilled water; **SMCs**, simple matching coefficients; **Rpm**, revolutions per min. **UV**, utra violet.

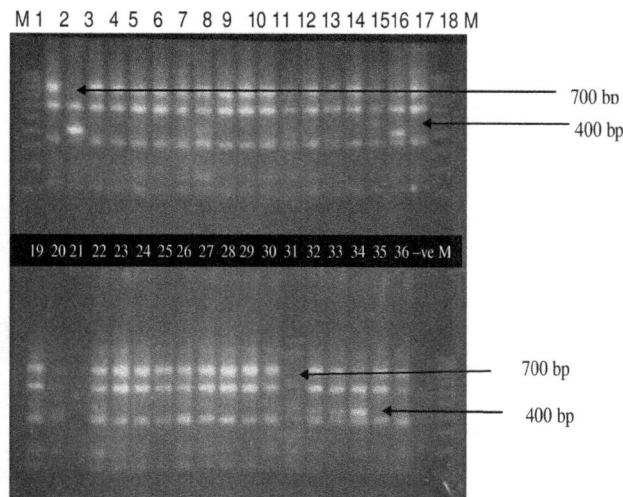

Figure 1. RAPD markers of 36 safflower accessions amplified with primer AB4 – 19 (sequence 5´-GGTGCACGTT- 3´) M = 100 bp ladder, lane numbers represent the accessions numbers while arrows indicate polymorphic markers

Which do not necessarily reflect genetic diversity (Fernandez-Martinez et al., 1993). The environmental has a strong influence on morphological traits (especially quantitative traits).Studies have also shown that there are not sufficient numbers of morphological markers to provide detailed coverage of most genomes (Shawla, 2002). Isozymes directly relate to genes, but their inherently low level of polymorphism among closely related cultivars, constrain their use in genetic linkage analyses and marker-assisted selection (Tanksley and Orton, 1983). Hence selection of genotypes based on molecular markers would be highly reliable and cost effective. Yazdi-Samadi et al. (2001) and Amini et al. (2007) used RAPD markers to detect genetic diversity in safflower accessions. Sehgal and Raina (2005) characterized 14 Indian safflower cultivars using RAPD, SSR and AFLP. AFLP markers were found to be the most efficient since two primer pairs were sufficient to genotype the cultivars. Using two marker systems RAPD and ISSR Yadla (2004) has reported that RAPD markers could be used to assess the intra varietal variation while ISSR markers could be useful tool in phylogetic analysis. Johnson et al. (2007) have found AFLP markers to be useful for distinguishing safflower variation within and among different geographical regions. This shows limited work has been reported in safflower on molecular methods. Hence the objective of this study was to evaluate genetic diversity in introduced safflower accessions using RAPDs and its application to germplasm identification and classification. Hypothesis for the study was that the "introduced safflower accessions are not genetically related".

MATERIALS AND METHODS

The study was conducted on 36 exotic safflower accessions, which were supplied from India, Bangladesh, China, Australia, USA, Mexico and FAO. DNA isolation was accomplished using a modified Cetyltrimethylammonium Bromide (CTAB) method (Hulbert and Bennetzen, 1991). The DNA was quantified on a spectro-photometer at a wavelength of 260 and 280 nm. While intactness and quantity was checked by running samples along some uncut unmethylated lambda (λ) DNA standards (20, 50 and 100 ng). The gel was stained in ethidium bromide (10 µg/ml) and visualized on an UV transilluminator. Out of 24 random 10-mer oligonucleotide primers obtained from Operon Technologies Inc. (USA), 14 primers that amplified clear and reproducible band profiles were selected. Each 10 µl PCR optimized reaction mixture contained 1X PCR buffer, 2.5 mM Mgcl₂, 100 nM dNTPs (-dATP, dCTP, dGTP and dTTP, Sigma Chemicals); 0.5 units of Taq DNA Polymerase (from Biolabs in UK), 200 nM Primer, 0.5 µl DNA template (10 ng), and 6.9 µl SDW. Amplification of DNA reactions were performed in a DNA thermocycler machine (Mastercycler) with a heated lid (93°C) programmed as follows; one hot start cycle of 93°C for 5 min (for strand separation), 40 cycles of 93°C for 1 min (DNA denaturation), 42°C for 1.5 min (annealing), 72°C for 1 min (DNA polymerization) and a final extension cycle of 72°C for 10 min. The samples were then maintained at 4°C. The PCR products were electrophoresed on 1.5% agarose gels stained in ethidium bromide and visualized on a UV light transilluminator. Band profiles were manually scored on two independent occasions and compiled into a rectangular binary matrix (not shown). Positive amplification were treated as separate characters and scored for the presence (1) or absence (0) of bands (Figure 1) Only intensely stained unambiguous bands were scored. The binary matrix file was transposed and the code "99" inserted for missing values before statistical analysis.

RESULTS AND DISCUSSION

A total of 61 amplification products were scored with an average frequency of 4.4 bands per primer. The binary matrices were subjected to statistical analyses using NTSYS (Rohlf, 1992). The calculation of SMCs was based on the presence or absence of RAPD fragments from paired samples given by $SM_{jk} = (a + d)/ (a + b + c + d)$. Where a, b, c, d are matches for each pair of accessions until a 36 x 61 similarity matrix was developed. The distance coefficient was the proportion of unmatched markers between a given pair of entries and this has been suggested as an appropriate estimator of relatedness under the assumption that the presence or absence of a discrete character in two or more genotypes results from the same genetic changes (Skrotch et al., 1992). The similarity matrix was then used to cluster the data using UPGMA algorithm and PCA. SMCs were generated using the NTSYS software (Table 2). The matrix values estimated the number of RAPD fragments shared (or not shared) between two accessions.

SMCs ranged from 0.37 (for accessions PI 262419 and PI 248359) and 0.39 (for accessions PI 537598 and PI 248359) that is, least genetic similarity to 0.98 (for accession PI T65), greatest genetic similarity. The latter two accessions (Oleic Leed and T65 originated from USA and India respectively). There is a strong indication that they may be genetically related since they constitute the nearest to a complete match for all markers (high degree of homology). However, similarities in accessions can also arise due to convergent evolution, selection, or shar-

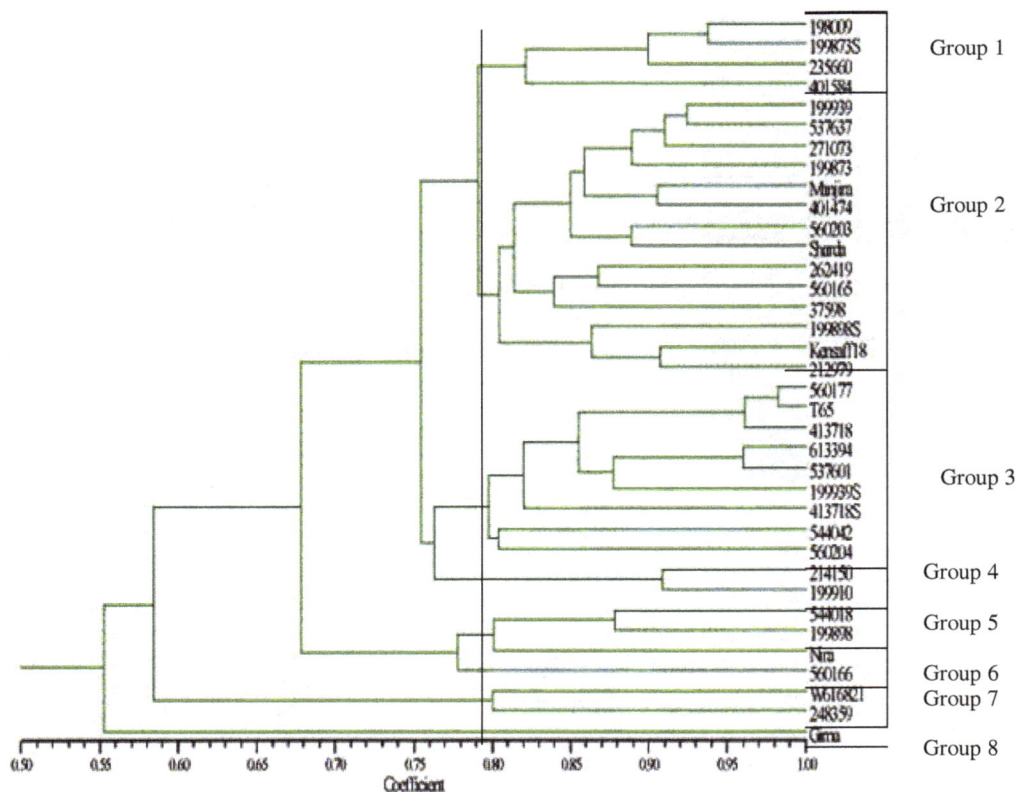

Figure 2. Dendrogram of 36 safflower accessions based on RAPD analysis of 61 markers using the simple matching coefficient and UPGMA clustering.

ing of a common parentage. This is consistent with earlier findings by Johnson et al. (2007) that molecular markers could be used for identification of duplicate accessions. Using RAPDs, breeders can select related or unrelated parental germplasm to maximize variability in the Kenyan breeding programme. Mating dissimilar accessions is likely to result into heterosis the magnitude of which deends on genetic distance. The RAPD results also suggest that some accessions from the same country were clustered together, other accessions were scattered with accessions obtained from different countries. Amini et al. (2007) have reported similar results in a genetic diversity study involving sixteen breeding lines and four safflower introductions. A small proportion of polymorphic primers would indicate that the germplasm is reasonably homogeneous. This observation among the safflower accessions is not unexpected due to movement of breeding material across the world into different programmes.

Cluster analysis

The UPGMA cluster analysis showed that the 36 accessions were grouped into eight marker-based groups (Figure 2). Cluster 2 had the largest number of accessions (14) originating from India, USA, Australia and Bangla-

desh. Cluster 3 had the second largest number of accessions from India, USA and Mexico. But proportionally accessions from India were represented at a higher level in both cluster 1 and 2. The dispersion into the various groups appeared to be at random though a few accessions formed distinct clusters (Figure 2). Since India is a secondary Centre of diversity for safflower, this helps to explain why accessions from there were found in almost every cluster due to movement of safflower from the Centre of origin.

The 4 Indian accessions PI 214150, PI 199910, W6 16821 and PI 248359 (group 4 and group 7) clustered together and diverged from the other accessions. Probably they have a common ancestor or could be an indicator of duplicates. However, similarities in accessions can also arise due to convergent evolution, selection, or sharing of a common parentage. Nicese et al. (1998) observed similar results in walnut and concluded that the RAPD assay can be useful in breeding programmes for identification of new cultivars as well as assessment of the genetic similarity among different genotypes. However, group 8 (Girna) formed an independent cluster and appeared to be most distantly related to all others (Figure 2). This accession originated from India and may have resulted from similar selection pressures at different pla-

Table 1. Eigen value, explained variance and cumulated variance in the PCoA using characters used to classify 36 safflower accessions using RAPDs.

Principal coordinate	Eigen value	Explained variance (%)	Cumulative variance (%)
1	1.93621729	20.6867	20.6867
2	1.26839267	13.5516	34.2383
3	0.95635288	10.2178	44.4561

Table 2. SMCs for some of the exotic safflower accessions studied.

PI No/Name	PI 235660	PI 99873S	PI 262419	PI 560177	PI 537598
PI 199939		0.90			
PI 271073		0.90			0.39
PI 401584		0.90			
PI 238359			0.37		
Kensaff 18	0.75				
PI 560203	0.75				
PI 212979	0.75				
T65				0.80	
PI 544042				0.80	
PI 199898				0.80	
Nira				0.80	

ces in an effort to develop uniform genotypes leading to similar forms with a different genetic background. Apparently the 5 accessions constitute an independent genetic pool, which could be of interest to safflower breeders. The results of this study agree with those reported by Amini et al. (2007) where cluster analysis based on RAPD markers and 54% coefficient of similarity divided the safflower genotypes into 5 distinct groups.

Principal component analysis (PCA)

The PCA indicated that the first 3 components accounted for 44% of the total variation (Table 1). Since molecular markers reveal neutral sites of variation at the DNA sequence level, this implies that most of the variation measured in the morphological study (Mahasi et al., 2006) was due to the environmental effects. The was similarity among markers (Table 2) for some accession pairs e.g. 99873S had a 0.90 genetic similarity with accession PI 199939, PI 211073 and PI 401584. While PI 413718 and PI 560177 and PI 413718 and T 65 had a genetic similarity of 0.96 between them. Several other accessions can be identified in the present study, which had genetic similarities e.g. 199898S with PI 401474, Sharda and PI 560165 at 0.75 genetic similarity (Table 2). This helps to explain the observation by Knowles (1989) that safflower has moved from the Centre of origin in the Middle East to other regions like India and accessions from the same areas would be expected to share at least some common genetic structure in the cause of germplasm movement across countries admixtures between accessions can occur.

Conclusions

It is evident from this study that the RAPD assay can be useful in safflower breeding programmes, for identification of new cultivars as well as assessment of the genetic similarity. Germplasm characterization from diverse world sources with large number of loci and more markers/different markers like AFLP and ISSR may provide guidance in enriching the present safflower collections in Kenya. During inter-mating accessions with greater genetic distance may provide unique genetic combination and useful variation for breeding. The results of the present study confirm that polymorphism based on RAPDs was useful in revealing genetic relatedness between the safflower accessions under study. Therefore any meaningful effort for construction of genetic relatedness trees can best be done using molecular markers as opposed to agronomic traits which are subject to environmental effects. Markers are useful for genotyping accessions and other factors but plant breeders still need the agronomic data to compliment molecular information in order to understand variation among accessions. Though some accessions from the same country were grouped together, some accessions grouped randomly with those introduced from different countries. The RAPD dendrogram revealed that the closer the geographical locations the closer the genetic relationships. The results of this study, have disapproved the hypothesis that "the introduced safflower accessions are not genetically related" as, reveal-ed by the RAPD groupings.

ACKNOWLEDGEMENTS

We wish to acknowledge the Director KARI for facilitating

this study, the Centre Director KARI Njoro for logistics and all the staff who participated and supported us.

REFERENCES

Amini F, Saeidi G, Arzani A (2007). Study of genetic diversity in safflower genotypes using agro-morphological traits and RAPD markers. Euphytica DOI 10 1007/sl0681-007-9556-6

Arid and Semi Arid Lands (Asals) of Kenya. Asian J. Plant Sci. 5: 1035-1038.

Bergman, Mundel HH (2001) (ed). V[th] International Safflower Conference Proceedings, Williston, ND, Sidney MT, USA.

Fernandez-Martinez J, Rio M and Haro A (1993). Survey of safflower (Carthamus tinctorius L.) germplasm for variants in fatty acid composition and other seed characters. Euphytica 19: 115-122.

Helm JL, Schneiter AA, Riveland N, Bergman J (1991). Safflower production. North Dakota State University, Fargo, ND 58105, USA, A-870 (revised): 14 AGR-6, 4

Hulbert SH, Bennetzen JL (1991). Recombination at the Rp1 locus of maize. Mol. Gen. Genet. 226: 377-382

Johnson A, Marter A (1993). Safflower Products: Utilization and Markets Marketing Series, Natural Resource Institute, No. 6: 26.

Johnson RC, Kisha T, Foiles C and Bradley V (2007). Characterizing safflower with AFLP Molecular Markers. Crop Sci. 47: 1728-1736

Khidir MO (1969). Evolution of the genetic system of safflower (Carthamus L.) Genetica 40: 84-88.

Knowles PF (1988). Carthamus species relationships. A lecture in Beijing Botanical Garden, Institute of Botany, Chinese Academy of Sciences.

Knowles PF (1989). Safflower. In: Oil crops of the World, G Robbelin RK Downey, Ashri A Eds. McGraw Hill Publishers, 361-384.

Mahasi MJ, Pathak RS, Wachira FN, Riungu TC, Kinyua MG and Kamundia JW (2006). Correlations and Path Coefficient Analysis in Exotic Safflower (Carthamus tinctorious L.) Genotypes Tested in the Arid and Semi Arid Lands (Asals) of Kenya. Asian J. Plant Sci. 5: 1035-1038

Shawla HS (2002). Introduction to plant biotechnology. Science Publishers Inc. pp. 329-330.

Nicese FP, Hormaza JI, McGranahan GH (1998). Molecular characterization and genetic relatedness among walnut (Juglans regia L.) genotypes based on RAPD Markers. Euphytica 101: 199-206.Rohlf FJ (1992). NTSYS – PC. Numerical Taxonomy and Multivariate Analysis System. Version 2.02k. State University of New York, Stony Brook N.Y.

Sehgal D, Raina SN (2005). Genotyping safflower (Carthamus tinctorius L.) cultivars by DNA fingerprints. Euphytica 146: 67-?7

Singhal V (1999). Indian Agriculture, F.A.O. report. pp. 300-301.

Skrotch P, Tivang J, Nienhuis J (1992). Analysis of genetic relationship using RAPD marker data. In: Application of RAPD Technology to Plant Breeding, pp. 26-30. Joint Plant Breeding Symposia Series.

Tanksley SD, Orton TJ (1983). Isozymes in plant genetic and breeding. Part B In: "Development in plant genetics and breeding" Elsevier Science Publishers BV, Amsterdam, Netherlands, pp. 401-421.

Yadla H (2004). Genetic diversity analysis in the genus Carthamus. MSc. in Agriclture. Acharya NG, Ranga Agricultural University.

Yazdi-Samadi B, Maali R, Amiri M, Ghannadha R, Abd-Mishan C (2001). Detection of DNA polymorphism in landrace populations of safflower in Iran using RAPD-PCR technique.. In J. Bergman and H. H. Mundel (ed). V[th] International Safflower Conference Proceedings, Williston, ND and Sidney MT, USA. p. 163.

10

Compatibility, production of interspecific F_1 and BC_1 between improved CMS *Brassica campestris* ssp. *pekinensis* and *B. oleracea* var. *acephala*

Peng-fang Zhu[1]*, Yu-tang Wei[2]

[1]Forestry Faculty, Shenyang Agricultural University, Shenyang, China, 110161, China.
[2]Horticulture Faculty, Shenyang Agricultural University, Shenyang, China, 110161, China.

The interspecific cross between improved cytoplasmic male sterility (CMS) *Brassica campestris* ssp.*pekinensis* (Lour.) Olsson and *B. oleracea* var. *acephala* DC. aims to obtain kale male sterile lines and enrich Chinese cabbage genetic resources. The results indicated that there were serious segregations between the 2 parents, so as to descendants could not be obtained by routine pollination methods. Otherwise, interspecific hybrids have been obtained by young embryos culture *in vitro* and the best time of embryo rescue was the 9 - 11 day after pollination. Interspecific embryo collapsed at sphere stage, with endosperm nuclei being surrounded by cell walls. The morphology, fertility, microspore growth, vernalization and numbers of chromosome were conducted to prove the truth of interspecific hybrids. Leaf morphology was intermediate of the 2 parents. Anther shape and plumpness, nectary volume and numbers were matroclinous. The male sterility showed 100% both in degree and ratio. Interspecific hybrids microspore aborted clearly from mononuclear to dinuclear stage. Vernalization was also intermediate of the 2 parents. Chromosome number of hybrids was confirmed to be 2n = 19, which was the total count of gametophyte chromosome of 2 parents. Also, we obtained BC_1 which was from F_1×kale, by pollinated several times, while the average compatibility index was only 0.0014. There were sluggish chromosomes in BC_1 PMCs anaphase.

Key words: CMS Chinese cabbage, kale embryo culture, chromosome, F_1, BC_1.

INTRODUCTION

Kale (*Brassica oleracea* var. *acephala* DC.), which is from Europe originally, cultivated as an ornamental plant in China recently. In our previous experiments, it can resist adverse conditions, especially cold temperature. So it has extensive perspectives in landscape North China. Random pollinations cause severe segregation in descendants because kale is cross-pollinated. Application of male sterility is one of the best measures in cabbage (Fang, 1984), as well as in kale. Early studies suggested that diploid *Brassica* species represent "balanced secondary polyploids" exhibiting internal chromosome homeology (Nagaharu, 1935; Olsson, 1960). Quite a few

interspecific and intergeneric hybrids have been synthesized subsequently in *Brassica* and *Cruciferae*, especially oil seeds and vegetables (Nobumichi, 1980; Mohapatra, 1986; Akbar, 1989; Sang et al., 1996; Huang BQ et al., 2001; Luo P et al., 2003). However, we have not found any report on interspecific crosses between improved CMS Chinese cabbage and kale. The present interspeific cross aims to transfer cytoplasmic male sterility (CMS) from Chinese cabbage (*Brassica campestris* ssp. *pekinensis* (Lour.) Olsson) to kale and further improve kale hybrid seed production. Meanwhile, the acquirement of interspecific hybrids can enrich genetic resources of Chinese cabbage, which is a significant vegetable in China for the importing of clubroot resistance (Bradshaw and Williamson, 1991; Nomura et al., 2005).

*Corresponding author. Email: pengfangzhu@yahoo.com.cn.

Table 1. Results of young embryos growth *in vitro*.

Days of embryos after pollination	Number of excised embryos	Number of plantlets get from excised embryos	Percentage of plantlets (%)
6 - 8	40	2	5.0
9 - 11	56	8	14.29
12 - 14	60	4	6.67
15 - 17	42	0	0
18 - 20	38	0	0

MATERIALS AND METHODS

MATERIALS

Improved CMS Chinese cabbages were supplied by Horticulture Faculty, Shenyang Agricultural University and the original CMS materials were introduced from America in 1996. Seeds of kale were bought on markets in Japan and Dalian, Liaoning Province. Both parents, F_1 and BC_1 were cultivated in the experimental plots in Shenyang Agricultural University.

METHODS

Manual crossing

Improved CMS Chinese cabbages were pistillate parents and kales were pollen parents. Manual pollinations were carried out during bloom.

Embryo growth

Ovaries, which were from female parent manual pollination, were made into wax slices after fixed in FAA (70% ethanol: glacial acetic acid:formaldehyde = 18:1:1) and dyed in Ehrlich's Haematoxylin. The thickness of slices was 8 μm.

Embryo rescue culture *in vitro*

Embryos of different development days since manual pollination (Table 1) were taken out from ovaries under stereoscope. Media was MS (Murashige and Skoog medium) with 2 mg·L^{-1} 6-benzyladenine, 0.1 mg·L^{-1} naphthyl acetic acid (NAA) and 30 g·L^{-1} cane sugar, were divided into 5 groups of embryos.

Morphology and vernalization

Typical morphological characters were observed during plantlet, lotus and bloom. Development of stages and vernalization characteristics were observed from sowing to bloom.

Fertility and microspore development

Numbers of nectary, shapes of anther, etc. were observed under a stereomicroscope. Vigor of pollen were evaluated by 3,3'-Dimethylbenzidine staining. Microspore was identified under Olympus-BH2 microscope through wax slices.

Cytology

F_1 root tips, which were from earlier plantlets *in vitro*, were pretreat-ed with saturated solution of 1,4-dichlorobenzene for 2 - 5 h at 20 then killed with Carnoy's fluid (3 alcholol + 1 acetic acid) for 5 - 16 h and hydrolyzed in equal volume raw HCl with 95% alcholol solution at 20°C for 10 - 15 min and stained in carbol fuchsin lastly. BC_1 buds' treatment were the same as root tips'.

RESULTS

Compatibility

We pollinated almost 900 flowers artificially, but none plump seeds were obtained *in vivo* under natural condition. So the reproductive isolation between the 2 parents is quite severe.

Embryos growth and rescue culture

The 5th day after manual pollination, multicellular spherical embryos were clear, while endosperm nuclei were surrounded slightly by cell wall (Figure 1). The 10th day after pollination, numbers of cells of spherical embryo increased a lot, while endosperm walls were more obvious (Figure 2). The 14th day after pollination, embryo was crushed extremely and the whole ovule tissue collapsed (Figure 3). The results shows that embryos collapsed at spherical stage, with endosperm nuclei being surrounded by walls gradually. In the end, ovule tissue inflated extremely and the young embryos crushed and aborted.

Table 1 shows percentage of plantlets got from 6th to 8th day after pollination is lower than that of 9th to 11th and the best period of embryos rescue was from the 9th to 11th day after pollination, when 14.29% embryos could become plantlets. From the 15th to 20th day after pollination, we have not got plantlets from young embryos *in vitro*.

F_1 morphology and bolting characteristics

Table 2 shows that the F_1 progenies were basically intermediate from the 2 parents, such as the color of embryonal axis, the volume of lotus leaf, etc. Some characters such as wax which located on the surface of the leaves were from pollen parents obviously.

Bolting characteristics were also intermediate, that is,

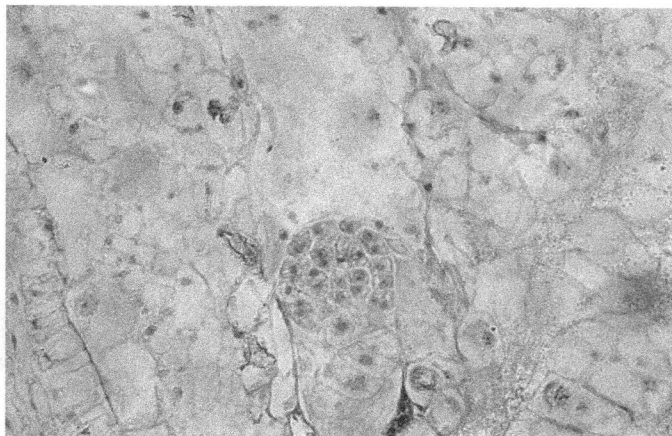

Figure 1. Multicellular spherical embryo, endosperm nuclei were surrounded by walls slightly (400×).

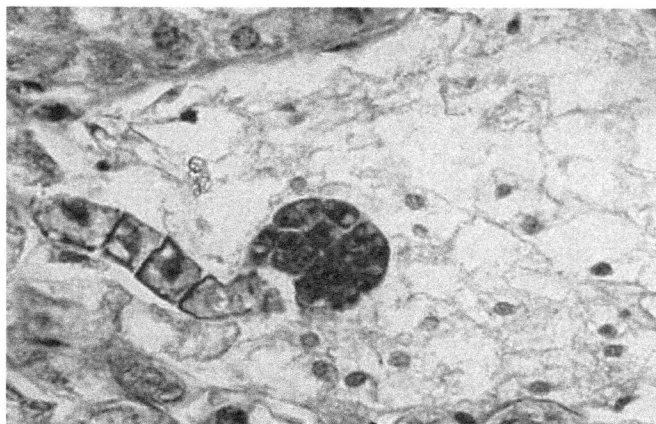

Figure 2. Multicellular spherical embryo, endosperm walls was obvious (400×).

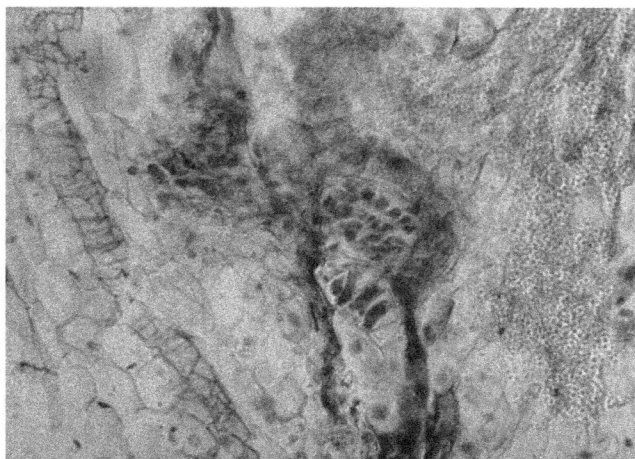

Figure 3. Embryo was crushed extremely and ovule tissue collapsed (400×).

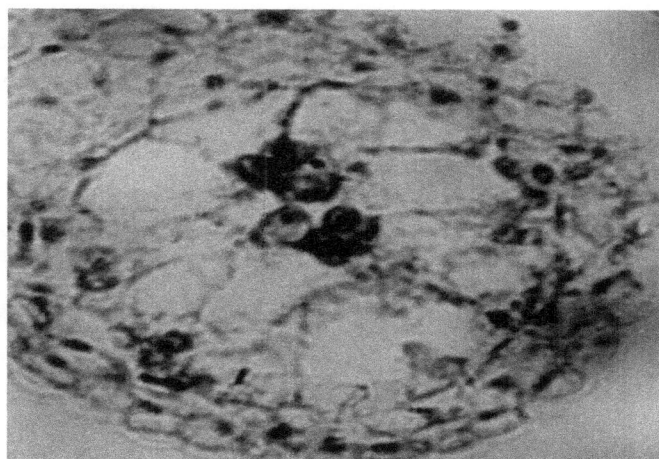

Figure 4. Tapetum cells vacuolation and inflation (400×).

the F_1 maintained rosette other than bolting when treated in the same conditions with female parent, which was bolting clearly. On the other hand, the F_1 bolting more easily and earlier than male parents, whose bolting was very strict to low temperature and the size of plants.

F_1 pollen fertility and microspore development

Table 2 also shows the F_1 hybrids anther shape, size and plumpness, nectary size and numbers, petal length/width were nearly matroclinous. There was no pollen under microscope so we can say that the F_1 progenies were 100% male sterile. In the meanwhile, the percentage of male sterility was also 100%.

Interspecific hybrids microspore aborted clearly from mononuclear to dinuclear stage. Most of tetrad micronucleus grew to uni-nucleate microspore stage and late uni-nucleate microspore stage. Nevertheless, since then, structures of anther wall appeared abnormality without tapetum cells disintegrate and uni-nucleate microspores were expanded and squeezed (Figure 4). In the end, most of uni-nucleate microspores collapsed and shrank and can not grow up to dinuclear microspores (Figure 5).

F_1 cytology

According to our observations, one of the typical F_1 progeny chromosome numbers were 2n = 19 (Figure 6) and the average cells ratio was 86.3%. Others were 2n = 29 (cells ratios 9.09%), which were presumably raised by unbalanced pairing of gametes from parents, for example, 20 gametes were from female parent and 9 ones from male parent, that is, double A group and one C group make a triploid of AAC and 2n = 38 (cells ratios 4.55%),

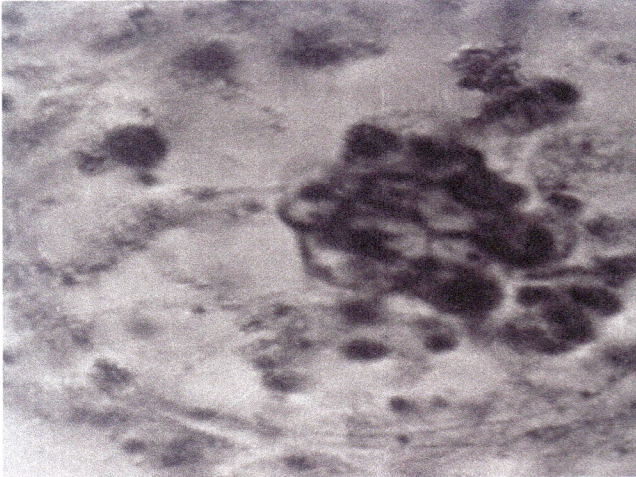

Figure 5. Microspore abortion (400×).

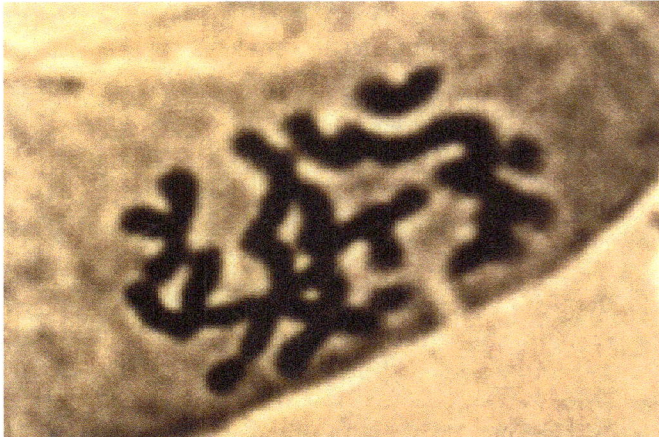

Figure 6. Number of root tip cells chromosome of F_1, 2n = 19.

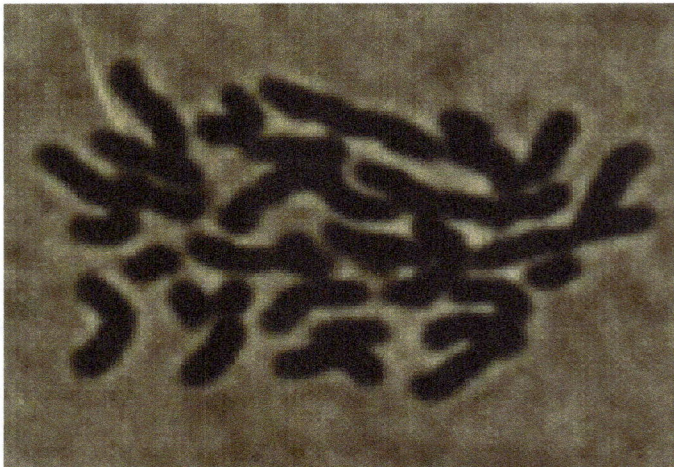

Figure 7. Number of root tip cells chromosome of F_1, 2n = 38.

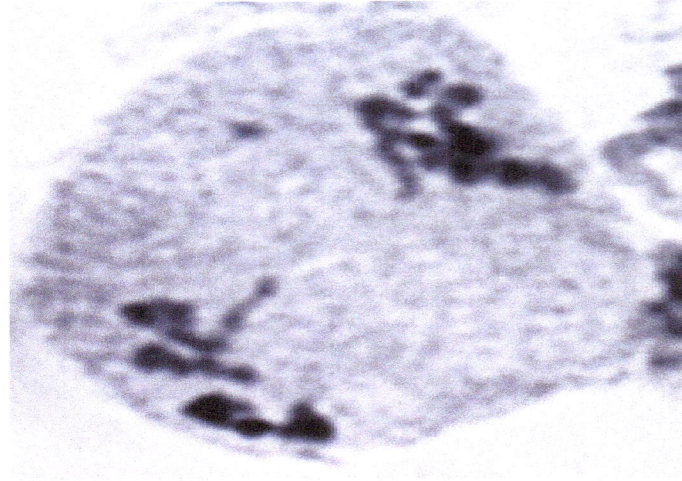

Figure 8. Meiophase anaphase, indicating sluggish chromosome (1000×).

which is an allotetraploid, were presumably raised by chromosome nondisjunction during meiosis or spontaneous chromosome doubling (Figure 7).

BC1 generation

The average compatibility index of BC_1, which was from F_1×kale by manual pollinating several times by cutting buds before blossoming, was only 0.0014. We observed sluggish chromosome in BC_1 PMCs anaphase (Figure 8) and the sluggish chromosome could be the main obstacle of chromosome mating (Liang ZhL et al., 1992). Meanwhile, a few PMCs chromosomes paired in metaphase normally (Figure 9).

DISCUSSION

We got none plump seed *in vivo* under natural condition by pollinating almost 900 flowers artificially. So the reproductive isolation is quite serious between *B. campestris* ssp.*pekinensis* (Lour.) Olsson and *B. oleracea* var. *acephala* DC. Production of interspecific hybrids between *B. campestris* and *B. oleracea* was also very difficult in previous work (Nagaharu, 1935; Hosoda et al., 1963; Hosoda et al., 1969). Crossability between *B. campestris* and *B. oleracea* was increased by excised rescue of ovaries *in vitro* (Inomata, 1983, 1996).

Morphological comparisons are traditional and reliable methods to analysis the truth of interspecific progenies. The F_1 progeny chromosome numbers of most of somatic cells (86.3%) were 2n = 19, that is, the total count of gametophyte chromosome of 2 parents (that is, 9 plus 10). Besides, we found some root tip cells with 29 chromosomes and some 38 with totally sum of 13.64% of

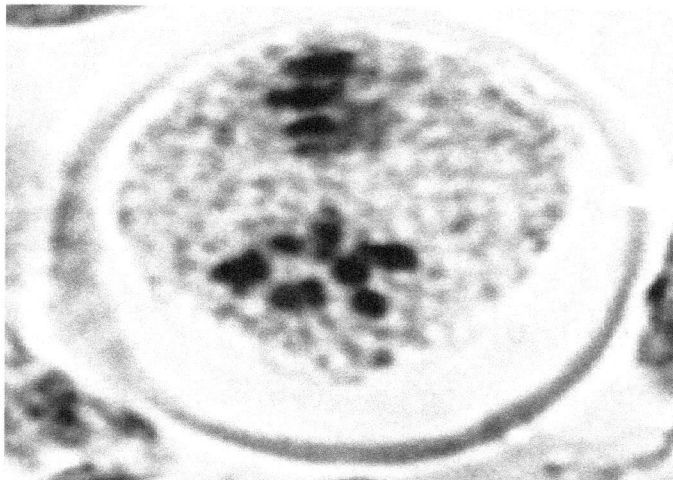

Figure 9. Metaphase, indicating chromosomes lay in equatorial plate (1000×).

the 2 latter types. According to Li and Zhang (1996), we can draw a conclusion from the number of the chromosomes directly that we got true interspecific hybrids.

We can say we have transferred cytoplasmic male sterility gene from Chinese cabbage to kale successfully because we got 100% male sterile F_1 progenies with normal nectaries, which are same to the female parent. Although during the interspecific cross, there is male sterility to some extent. For example, Olsson (1960) obtained hybrid plants from the cross between *B. campestris* and *B. oleracea*, which had 38 chromosomes in root tip cells. Pollen fertility ranged from 69 to 97%, with the mean of 86.8%. Hosoda et al. (1969) artificially synthesized *B. napus*, of which pollen fertility was from 45.0 to 77.5%.

The crushing of spherical embryos dues to the inflating excessively of ovule tissue. The insufficiency and obstacle of nutrient transportation may be the reasons of embryos abortion. Of course, the interspecific reproductive isolation was the key cause of imbalance between the embryos and endosperm. We have also observed embryos growth of *B. campestris* ssp. *Pekinensis* × *B. napus*, which is similar to diploid *B. campestris* ssp. *pekinensis* with separable endosperm nuclei without confining of walls. *B. napus* has genome C, which is from *B. campestris* originally and opposite to genome A and B in *Brassica* according to Nagaharu (1935), maybe the reason of coordination between embryo and endosperm development.

Though there were sluggish chromosomes in BC_1, some chromosomes mated normally. These would be keys of backcrossing successfully and obtaining expectant CMS kale for character transfer within the genus *Brassica* was possible by successive backcrossing (Inomata, 1996).

REFERENCES

Akbar MA (1989). Resynthesis of Brassica napus Aiming for Improved Earliness and Carried out by Different Approaches. Hereditas 111: 239-246.

Bradshaw JE, Williamson CJ (1991). Selection for Resistance to Clubroot (Plasmodiophora brassicae) in Marrowstem Kale (*Brassica oleracea* var. *acephala* L). Annals Appl. Biol. 119(3): 501-511.

Fang ZY (1984). Preliminary Results on Cabbage Cytoplasmic Male Sterility Lines Breeding. China Vegetables 4: 42-43.

Hosoda T, Sarashima M, Namai H (1969). Studies on the Breeding of Artificially Synthesized Napus Crops by means of Interspecific Crosses between n=10 Group and n=9 Group in Genus Brassica. Memo. Fac. Agri. Tokyo Univ. Educ. 15: 193-209.

Huang BQ, Chang L, Ju CM, Chen JG (2001). Production and Cytogenetics of Intergeneric Hybrids Between Ogura CMS *Brassica campestris* var. *purpuraria* and *Raphanus sativus*. Acta Genetica Sinica 28(6): 556-561.

Inomata N (1980). Hybrid Progenies of the Cross, *Brassica campestris* × *Brassica oleracea*. Cytogenetical Studies on F1 Hybrids. Japan. J. Genet. 55(3): 189-202.

Inomata N (1983). Hybrid Progenies of the Cross, *Brassica campestris*×*B. oleracea*. Crossing Ability of F1 Hybrids and Their Progenies. Jpn. J. Genet. 58: 433-449.

Inomata N (1996). Overcoming the Cross-incompatibility through Embryo Rescue and the Transfer of Characters within the Genus *Brassica* and between Wild Relatives and *Brassica* Crops. Academic Report Okayama University 85: 79-88.

Li MX, Zhang ZP (1996). Crop Chromosomes Technology. Beijing: China Agriculture Press p. 1.

Liang ZL, Jiang RQ, Zhong WN (1992). Studies on Chromosome Behaviour of F1 and Fertility Restoration in Hybrid of *Gossypium hirtutum* × *G. bickii*. J. Integrative Plant Biol. 34: 931-936.

Luo P, Fu HL, Lan ZQ, Zhou SD, Zhou HF, ALuo Q (2003). Phytogenetic Studies on Intergeneric Hybridization between *Brassica napus* and *Matthiola incana*. Acta Botanica Sinica 45(4): 432-436.

Mohapatra D (1986). Hybridization in *Brassica juncea* ×*B.campestris* through Ovule Culture. Euphytica 37(1): 83-88.

Nagaharu U (1935). Genome-analysis in Brassica with Special Reference to the Experimental Formation of *B. napus* and Peculiar Mode of Fertilization. Japan. J. Bot. 7: 389-452.

Nomura K, Minegishi Y, Kimizuka-Takagi C (2005). Evaluation of F2 and F3 plants introgressed with QTLs for Clubroot Resistance in Cabbage Developed by using SCAR Markers. Plant Breeding 124(4): 371-375.

Olsson G (1960). Species Crosses within the Genus *Brassica*. II. Artificial *napus* L. Hereditas 46: 315-386.

Sang WB, Yukio K, Yasuo M (1996). Production of Intergeneric Hybrids between Raphanus and Sinapis and the Cytogenetics of Their Progenies. Breeding Sci. 46: 45-51.

Zhang GQ, Tang GX, Song WJ, Zhou WJ (2004). Resynthesizing Brassica napus from Interspecific Hybridization between *Brassica rapa* and *B. oleracea* through Ovary Culture. Euphytica 140(3): 181-187.

Zhu PF, Wei YT (2004). Preliminary Studies on Interspecific Cross between *Brassica campestris* L.ssp. *pekinensis* and *B. oleracea* var.*acephala*. China Vegetables pp. 39-11.

11

Status of macadamia production in Kenya and the potential of biotechnology in enhancing its genetic improvement

L. N. Gitonga[1,3]*, A. W. T. Muigai[2], E. M. Kahangi[3], K. Ngamau[3] and S. T. Gichuki[4]

[1]Kenya Agricultural Research Institute, National Horticultural Research Center, P. O. Box 01000 - 220, Thika.
[2]Department of Botany, Jomo Kenyatta University of Agriculture and Technology, P. O. Box 00200-62000, Nairobi.
[3]Department of Horticulture, Jomo Kenyatta University of Agriculture and Technology, P. O. Box 00200-62000, Nairobi.
[4]Kenya Agricultural Research Institute, Biotechnology Center, P. O. Box 00200-57811, Nairobi.

Macadamia (*Macadamia* spp.) is considered the world's finest dessert nut because of its delicate taste and numerous health benefits. It is grown in Kenya both as a cash crop and foreign exchange earner with Kenya producing about 10% of the world's total production. Macadamia has great potential for poverty reduction due to the high value of its products and its low requirement for external inputs. Although the crop has been grown in the country for over 5 decades, the growth of the industry is not commensurate with the demand and market potential that exists. Some of the challenges facing the macadamia industry in Kenya include lack of cultivars adapted to various agro ecological zones, inadequate planting materials of high quality, high cost of the available good quality planting materials and pests and diseases that affect nuts thus lowering post harvest quality. This paper discusses the potential of agricultural biotechnology relevant to genetic improvement of macadamia to compliment other efforts for its improved productivity and value.

Key words: Macadamia, dessert nut, biotechnology, genetic improvement.

INTRODUCTION

Macadamia (family *Proteaceae*) is an ever-green tree growing up to 20 m (Duke, 1983). 2 species *Macadamia integrifolia* Maiden and Betche (smooth-shelled), and *Macadamia tetraphylla* L.A.S. Johnson (rough-shelled) are cultivated for their edible nuts (McHargue, 1996). The mature fruit consists of a cream to white seed (kernel, nut) enclosed in a hard brownish seed coat (shell) which is then enclosed in a green or grayish green pericarp (husk) (Bittenbender and Hirae, 1990; Yokoyama et al., 1990). The kernel can be eaten raw or fried (Duke, 1983) or as an ingredient into various confectionary products (Yokoyama et al., 1990; Sato and Waithaka, 1996). The oil extracted from macadamia is similar in composition to olive oil and is made up of 58.2% monounsaturated fatty acids (Cavaletto, 1980; Macfalane and Harris, 1981). It is

considered a healthy food product as it contains no cholesterol, thus maintains blood cholesterol levels in check (Onsongo, 2003). The seed cake that remains after oil extraction is used as a constituent of livestock feed (Woodroof, 1967). Rumsey (1927) also recommended the tree for timber and also as an ornamental.

Macadamia is mainly grown in Australia, Hawaii, South Africa, Kenya, Guatemala, Malawi, Brazil, Zimbabwe and Costa Rica in order of level of production (Table 1) (Wilkie, 2008). Other countries that cultivate the crop on a small scale include New Zealand, Mexico, Jamaica, Fiji, Argentina, Venezuela and Tanzania (Wasilwa et al., 2003.

Introduction in Kenya

Macadamia was introduced from New South Wales in Australia in to Kenya in 1946. 6 *M. tetraphylla* seeds were planted at Kalamaini estate in Thika district of central pro-

---

*Corresponding author. E-mail: lucygitonga2000@yahoo.com.

Table 1. Macadamia production (Tons in-shell) in major macadamia growing countries in the years 2006 and 2007.

| | Tons in-shell | |
Country	2006	2007
Australia	42000	39700
Hawaii	22000	20000
S. Africa	17230	19230
Kenya	11400	11100
Guatemala	7000	8300
Malawi	4230	7110
Brazil	3125	3750
Zimbabwe	800	770
Costa Rica	500	500
Others	2200	2500
Total	108285	110460

Adapted from Wilkie (2008).

vince (Harris, 2004). More seeds of *M. integrifolia*, *M. tetraphylla* and hybrids of the 2 were introduced in 1964 from Australia, Hawaii and California. In 1968, grafted seedlings were produced using scion material of superior *M. integrifolia* varieties which were imported from Hawaii. The grafted Hawaiian varieties were planted in different agro-ecological zones. These 3 sources were used by the Harries family (Bob Harris Ltd, (BHL)) to propagate and supply macadamia seedlings to farmers in central, eastern, rift and coast provinces, as an alternative cash crop to tea and coffee (Harries, 2004). By 1974, BHL had already supplied 800,000 ungrafted seedlings to farmers in central and eastern provinces (Waithaka, 2001; Harris, 2004). However, when the trees begun to bear, no marketing infrastructure had been organized and this discouraged farmers and some started cutting down the trees. Since most of the trees were seed-propagated, there was wide variation in yield and quality of nuts. Most trees produced 5 - 10 kg/tree/season and kernels had less than 70% oil content which was considered low quality (Ondabu et al., 1996). Varieties adapted to various agro-ecological zones were also lacking. As a conesquence, farmers were discouraged and some uprooted their trees.

History of Macadamia improvement in Kenya and current status

Improvement through selection breeding

Macadamia varietal development through selection breeding has extensively been done in other countries such as Hawaii (Bittenbender and Hirae, 1990), California (McHargue, 1996), Australia (Hardner et al., 2001) and South Africa (ARC, 2000) leading to varieties adaptable

to various AEZ. Introduction of such varieties in Kenya would broaden the genetic base. However, they would need to be evaluated for adaptability to Kenyan conditions. Hence, selection of varieties from already adapted germplasm in Kenya was necessary. Between 1971 and 1973, a feasibility study funded by the food and agriculture organization (FAO) was carried out by the Kenya government to determine the potential of revitalizing the macadamia industry and providing farmers with high quality planting materials (Hamilton, 1971; Waithaka, 2001). Results of the feasibility study indicated that macadamia had high potential as a cash crop and as foreign exchange earner for Kenya (Kiuru et al., 2004). The government of Kenya requested the government of Japan for financial support and technical expertise to assist Kenyan counterparts in rehabilitation of the macadamia industry. Hence, in 1977 agronomic surveys of macadamia trees planted in the later part of 1960s revealed 300 'superior' trees in terms of yield and quality of nuts. Out of these, scions were obtained from 30 most promising trees in farmers' fields and grafted clones were planted at the National Horticultural Research Centre (NHRC, Thika) of the Kenya Agricultural Research Institute, for detailed observations on yield, nut and kernel characteristics (Wasilwa et al., 2003).

Nut quality has been accepted as the first criterion for the selection of improved macadamia cultivars (Nissen and Williams, 1980). First grade kernels contain over 72% oil. Oil content is determined by specific gravity using dried kernels of below 2% moisture content. First grade kernels readily float in tap water. Kernels that contain 72% or below sink in tap water but float or sink in 1.025 specific gravity brine solution and are 2nd or 3rd grade respectively. Such nuts are usually immature and harder and they become over brown when roasted (Yokoyama et al., 1990). Kernel recovery ratio, expressed as a percentage of recovered kernels' weight to the total in-shell nut weight, should be over 32% while the weight of kernel should range between 1.55-3.14 g. First grade kernel ratio expressed as a percentage of nuts that float in tap water to the total weight of recovered kernels should be over 90% (Nissen and Williams, 1980). Yield, expressed as the number of kilograms per tree per year should be over 50 kg if the tree is growing under highly suitable conditions or 40 kg under moderate conditions (Ondabu et al., 1996). Based on the factors mentioned above, 7 promising cultivars were selected from the 30 trees with a yield potential of between 50-80 kg/tree/year, a 10 fold increase from average yield of 5 – 10 kg obtained previously by farmers. Nut and kernel characteristics were also improved to ideal ranges. These varieties were recommended for commercial planting in the late 1980s (Table 2).

Several other varieties including EMB-2, MRG- 2, MRG-25, TTW-2, Hawaiian varieties; HAES 246, HAES 294, HAES 333, HAES 508, HAES 660, HAES 741, HAES 788 (*M. integrifolia*), KMB-4, KMB-25, KMB-9,

Table 2. Yield, nut and kernel weights of the 7 varieties recommended for commercial planting in Kenya in the late 1980s. Data on yield was obtained from original mother trees in farmers' fields where they were best adapted while data on nut and kernel characteristics was taken from grafted clones at NHRC.

Variety	Species	Yield (in shell) at 15 years (Kg/tree)	Kernel recovery ratio (%)	Kernel weight (g)	First grade kernel ratio (%)
KRG-15	*M. integrifolia*	80.0	39.3	2.53	91.99
EMB-1	*M. integrifolia*	70.0	33.5	1.99	95.50
MRG-20	*M. integrifolia*	55.0	32.6	2.24	95.77
KMB-3	(*M. integrifolia* x *M. tetraphylla*) hybrid	60.0	34.8	1.90	94.30
KRG-1	*M. integrifolia*	50.0	29.6	2.20	72.70
KRG-3	*M. integrifolia*	50.0	31.7	2.30	90.00
KRG-4	*M. integrifolia*	65.0	32.8	2.10	56.70

Source: Hirama et al. (1987).

Table 3. Yield, Nut and Kernel weights of additional 10 varieties recommended for commercial planting in the late 1990s. Data on yield was obtained from original mother trees in farmers' fields where they were best adapted while data on nut and kernel characteristics was taken from grafted clones at NHRC.

Variety	species	Yield (in shell) at 15 years (Kg/tree)	Kernel recovery ratio (%)	Kernel weight (g)	First grade kernel ratio (%)
EMB-2	*M. integrifolia*	40 - 60	34.63	2.38	94.95
EMB-H	Hybrid	40 - 60	36.0	1.96	99.67
KMB-4	Hybrid	40 - 60	36.24	2.88	96.30
KMB-25	Hybrid	40 - 60	36.00	3.15	87.17
MRG-2	Hybrid	50 - 60	38.66	3.91	93.92
MRG-25	Hybrid	40 - 60	37.22	2.51	97.25
MRU-23	Hybrid	40 - 60	35.02	2.67	81.50
MRU-24	Hybrid	40 - 60	35.28	2.17	97.00
MRU-25	Hybrid	40 - 60	33.45	2.12	81.00
TTW-2	*M. integrifolia*	40 - 60	41.33	2.06	91.67

Modified from Tominaga and Nyaga, (1997)

EMB-H, MRU-23, MRU-24, MRU-25, MRG-1, MRG-2 and MRG-8 (*M. integrifolia* x *M. tetraphylla* hybrids) were planted in trials at different agro-ecological zones for continued observations on yield potential, tree and nut characteristics (Wasilwa et al., 2003). Subsequently, some addtional varieties, mostly natural (*M. integrifolia* x *M. tetraphylla*) hybrids were tentatively recommended for commercial planting in the late 1990s. These varieties have yield potential of 40 - 60 kg/tree/year, high kernel recovery ratio of between 33.45 - 41.33%, high kernel weight of between 1.96 - 3.15 g and over 80% first grade kernel ratio (Table 3) (Tominaga and Nyaga, 1997; Wasilwa et al., 2003)

Development of new varieties through cross breeding

Further agronomic observations revealed that some macadamia cultivars were adapted to different agro-ecological zones (AEZ) and were recommended accordingly

(Table 4). Only 4 varieties KRG-15, EMB-1, MRG-20 were permanently recommendation for 3 agro-ecological zones, sunflower-maize zone, marginal coffee and main coffee zones. To date, only one variety (KMB-3) adapted to high altitudes of 1750 - 1870 m above sea level (coffee tea zone), was identified and recommended (Tominaga and Nyaga, 1997). Hence, cross breeding work between 4 selected clones (EMB-1, KRG-3, MRG-20 and KMB-3) and 2 Hawaiian varieties (HEAS 508 and HAES 333) and further cross-pollination with clones MRG-1, MRG-2, MRG-8, KMB-9 and KRG-4 (Table 5) was initiated in 1991 with an aim of combining adaptability to high altitude and other nut quality characteristics. The progenies were planted at an altitude of 1850 m above sea level (Kiuru et al., 2004) for evaluation.

Multiplication of selected superior clones through vegetative propagation

Macadamia is preferentially (> 75%) out crossing (Sedg-

Table 4. Macadamia varieties recommended for 4 agro-ecological zones of Kenya.

Agro-ecological zone	Altitude (meters)	Mean Temperature (°C)	Rainfall in mm	Permanent recommendations	Tentative recommendations
Sunflower - Maize Zone	1280 - 1400 and below	19.5 - 20.7 and higher	800 - 900 and less	KRG-15, EMB-1 MRG-20	
Marginal Coffee Zone	1400 -1550	19.0 - 20.1	900 - 1200	KRG-15, EMB-1, MRG-20	KMB-4, MRG-25, TTW-2 (T/Taveta)
Main Coffee Zone	1550 - 1750	18.5 - 20.0	1200 - 1400	KRG-15, EMB-1, MRG-20, KMB-3	KMB-4, EMB-2, EMB-H, MRG-25 and TTW-2 (T/Taveta)
Coffee - Tea Zone	1750 - 1870	17.5 - 19.0	1400 - 1600	KMB-3	EMB-1, MRG-20, EMB-H, MRU-24 and MRU-25

Modified from Ondabu et al., (1996); Tominaga and Nyaga (1997); Wasilwa et al. (1999)

Table 5. Macadamia crosses performed in Kenya in 1991.

Cross	Number of crossed seed nuts	Cross	Number of crossed seed nuts
KMB-3 X KMB-9	474	MRG-20 X EMB-1	8
KMB-3 X KRG-3	33	MRG-20 X KRG-3	13
KMB-3 X EMB-1	3	KRG-3 X EMB-1	32
KMB-3 X MRG-20	26	EMB-1 X KRG-3	2
KMB-9 X KMB-3	27	EMB-1 X KRG-3	3
KRG- 3 X KMB-3	6	EMB-1 X MRG-20	2
MRG-20 X KMB-3	7		
Total			636

Source: JICA (1991)

ley et al., 1990). Production of true-to-type planting material is therefore by grafting scions from selected parents on to rootstocks raised from seed (Stephenson, 1983). In Kenya, grafting methods were evaluated between 1985 and 1997. 3 grafting methods namely (a) Top wedge (90 - 100% successful takes), (b) splice and (c) side wedge (70 - 100%) were fine-tuned and used for mass propagation of planting material of selected superior material. Using these methods seedlings can be produced within 1.5 - 2 years (Nyakundi and Gitonga, 1993; Gitonga et al., 2002). Grafting reduces the juvenile period before bearing from 7 - 10 years to 3 - 4 years. Further, grafting also contributed to dwarfing of trees from over 15 m to manageable heights of less than 10 m. Average yields were increased from between 5 and10 kg to between 50 and 80 kg/tree/season while nut quality was improved by increasing nut oil content to over 72% (Onadabu et al.,

1996) through propagation and distribution of superior cultivars. Selection of superior clones and development of vegetative propagation techniques and other agronomic packages has resulted into economic gain and importance of macadamia in Kenya. Currently, macadamia is a growing agro-processing industry that targets niche markets in Europe and the Orient (Rotich, 2004). The area under macadamia has increased from 469 ha in 1989 to an estimated 8000 ha in 2003 (Onsongo, 2003; Kiuru et al., 2004).

Macadamia has since become the most important nut crop in Kenya, with an annual production of about 10,000 metric tons. Being a low-input crop, macadamia is grown by over 100,000 small-scale farmers for income and livelihood (Waithaka, 2001). Further, macadamia can be grown as an intercrop with other cash and food crops (Onsongo, 2003; CABI, 2005). Small-scale farmers pro-

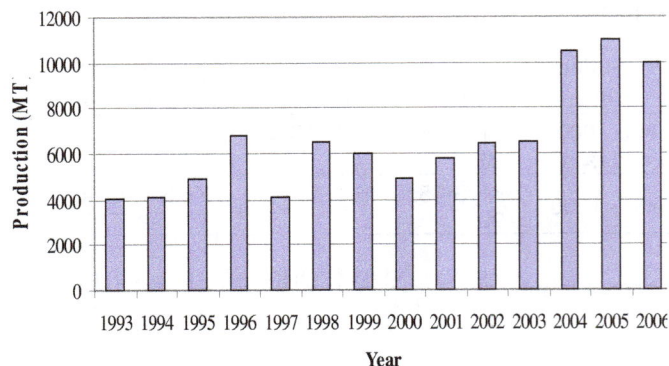

Figure 1. Macadamia nut production (in-shell MT) in Kenya from 1993-2006.
Source: USDA, 2008.

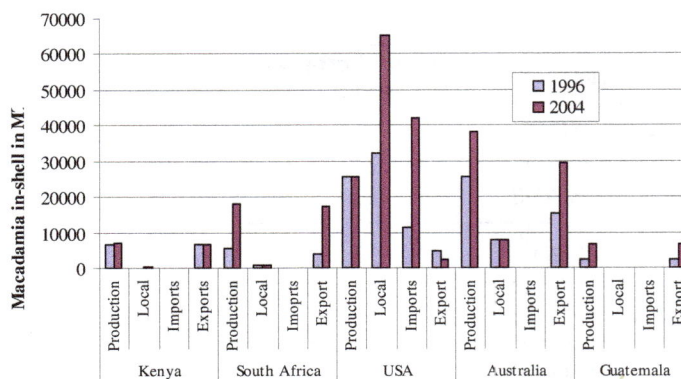

Figure 2. Macadamia in-shell basis - production, consumption, exports, and imports statistics for the years 1996 and 2004.
Source: USDA, 2008

duce about 70% of total production while 30% is produced by about 500 large-scale growers with at least 1000 trees each (Muhara, 2004). Kenya exports 93% of its total production to Japan, Hong Kong, Germany, USA, Canada and Switzerland (processed) and China, East Asia and India (unprocessed) while domestic market consumption is estimated at 50 tons of kernel annually (WHT and U.S.E.O, 2004; Rotich, 2004). The Kenya nut company (KNC), the major processor of macadamia nuts has branded its domestic products 'Nutfields' which include roasted nuts, chocolates and cookies sold in major supermarkets, hotels and airlines. International products are branded 'Out of Africa' (Onsongo, 2003).

Steady growth in production (Figure 1) and export of macadamia has been realized since the early 1990's. An increase from 4000 tons in 1993 to 7,300 tons in the year 2004 in export was registered (CABI, 2005). In 2003, the world production of processed kernel was 23,130 tons with Kenya producing 1520 tons (7%) from 1600 bearing trees (Onsongo, 2003; Lee, 2004).

However, the growth rate of the sub-sector is below the country's potential and is relatively slow when compared to the growth rate of competing countries such as South Africa to which Kenya has lost some of its market share (CABI, 2005). While the production and exports from Kenya barely changed between 1996 and 2004, those of South Africa tripled (Figure 2). Within the same period, USA doubled its local consumption and increased its imports more than 4 times without changing its production levels thus offering huge market potential for Kenyan macadamia.

Several factors have been cited as responsible for the slow production growth of which include among others, inadequate cultivars adapted to various agro ecological zones, inadequate planting materials of high quality, high cost of the available good quality planting materials and pests and diseases that affect nuts thus lowering post-harvest quality.

Prospects of biotechnology applications for Macadamia improvement

Macadamia breeders must explore as much genetic diversity as possible from which to select and recombine favorable traits through cross-breeding (McHargue, 1996) so as to develop varieties that are adapted to Kenyan conditions and those that can compete in the world market (Rotich, 2004). However, the genetic diversity of macadamia germplasm in Kenya is not known and this slows breeding efforts.

Currently, the genetic improvement of macadamia is based on introduction and selection from existing germplasm. Germplasm characterization is based on agro-morphological traits. This process has been slow and since 1977 only 4 varieties have been permanently recommended for commercial planting (Ondabu et al., 1996). Moreover, morphological traits alone represent only a small portion of the plant genome and are influenced by environmental factors, there by limiting their utility in describing the potentially complex genetic structures that may exist between and within species (Avise, 1994).

Application of DNA markers for macadamia germplasm characterization

Genetic markers are simply heritable characters with multiple states at each character. Typically, in a diploid genus such as Macadamia (2n = 28) (Hardner et al., 2005), each individual can have 1 or 2 different states (alleles) per character (locus). All genetic markers reflect differences in DNA sequences (Sunnucks, 2000). Basic DNA profiling techniques (DNA markers) include restriction fragment length polymorphisms (RFLPs), random amplified polymorphic DNA (RAPDs), amplified fragment length polymorphisms (AFLPs) and microsatellites. These techniques are continually being modified to improve resolution and information content. With the invention of the polymerase chain reaction (PCR) technique, DNA profiling took huge strides in both discriminating power and the ability to recover information from very small (or degraded) starting samples. With sequencing technologies associated with automated and/or semi automated large-scale screening systems, DNA-based polymorphisms are now the markers of choice for molecular-based surveys of genetic variation (Hannotte and Jianlin, 2006).

Among the DNA–based markers, microsatellites (Goldstein et al., 1995) are well known for genetic diversity studies due to their potentially high information content (Ferreira, 2006), the results obtained are reproducible, the data can be scored and analyzed using standardized methods (Lanteri and Barcaccia, 2006). Characterization of diversity of macadamia germplasm using molecular tools has been previously applied in other countries. As a tool towards quantitative breeding of macadamia in Hawaii, Vithanage et al. (1998) reported on the use of isozyme analysis to assess the genetic relatedness but found it limiting since several isozyme alleles were shared by most of the commercial cultivars. They developed random amplified polymorphic DNA (RAPD) and sequence tagged sites (STSs). Results showed that these markers followed Mendelian inheritance and these markers have been used to assay 76 individuals. Peace et al. (2002) used radio-labeled DNA amplification fingerprinting (RAF) technique for Macadamia chromosome mapping. They discovered 19 new RAF co-dominant markers which were used together with dominant markers to characterize 30 cultivars representative of the Hawaiian Macadamia industry. These markers were also used to assign cultivars to germplasm groups that reflected species status and breeding origins. Recently, Schmidt et al. (2006) isolated 33 microsatellite markers from *M. integrifolia* that were used to genotype 43 commercial cultivars of Australia generating an average polymorphic information content of 0.480. These 33 microsatellite loci represent a significant tool for genome mapping and population genetic studies and can therefore be immediately applied for genetic diversity studies of macadamia selections in Kenya.

Other approaches for molecular characterization with increased robustness include single nucleotide polymorphisms (SNPs), allele mining and diversity array technology (DArT). SNPs are DNA sequence variations that occur when a single nucleotide (A, T, C, or G) in the genome sequence is altered. For example the change of the DNA sequence AAGGCTAA to ATGGCTAA (Kahl et al., 2005). For a variation to be considered a SNP, it must occur in at least 1% of the population (allele frequency ≥1%) (Rapley and Habron, 2004).Since SNPs are located throughout the genome, whole genome scans may help identify regions associated with agronomically important traits and hence can be applied for high-throughput marker-assisted breeding, expressed sequence tags (EST) mapping and the construction of genetic linkage maps (Rafalski, 2002).

SNPs have been identified and utilized several crops including barley (Soleimani et al., 2003) and *Solanum caripense*, a wild relative of potato and tomato (Nakitandwe et al., 2007). In nut crops, SNPs have been detected in coconut and are being used for studying genetic population structure (Mauro-Hennera et al., 2006) while in almond (Shu-Biao et al., 2008) validated 100 SNPs based on the predicted SNP information derived from the almond and peach EST database. Detection and validation of SNPs in macadamia would greatly enhance diversity analysis and genome mapping.

Allele mining exploits the DNA sequence of one genotype to isolate useful alleles from related genotypes (Lathar et al., 2004). This can be done using DNA chip technology whereby the basic DNA sequence of a gene is spotted on a chip in the form of large series of sequence-overlapping probes consisting of 15 - 20 bases. Each base position in a fluorescently labelled sample is then interrogated for the presence of point mutations by monitoring hybridization signals with the spotted probes. Allele mining therefore helps to detect new point mutations, in relatively large DNA fragments. Once allelic variants of interest have been identified, the approach can be optimized by focusing on target sets of polymorphisms, for example by using SNP detection methods (Lathar et al., 2004; Upadhyaya et al., 2006). This method can therefore be applied to differentiate between genotypes or to detect new evolving macadamia varieties.

DArT was developed to provide a practical and cost effective whole genome fingerprinting tool with high through put (Jaccoud et al., 2001). The technology has the ability to generate reproducible molecular markers with no prior DNA sequence information (Kilian et al., 2005). Development of DArT starts with assembling a pool of DNA samples that encompass the diversity of the species (Wenzl et al., 2004). This may include the primary gene pool of a crop species, 2 parents of a cross (if the goal is creation of a genetic linkage map), or secondary gene pools. The DNA mixture representing the gene pool is then processed to produce a defined fraction of genomic fragments referred to as 'representation' which are then used to create a library in *Escherichia coli* (Kilian et al., 2005). DArT operates on the principle that the genomic 'representation' contains 2 types of fragments. Constant fragments, found in any 'representation' prepared from a DNA sample from an individual belonging to a given species and variable (polymorphic) fragments (molecular markers), only found in some but not all of the 'representations'. The variable fragments, referred to as the DArT markers are informative because they reflect sequence variation that determines the fraction of the original DNA sample that is included in the 'representation' (Kilian et al., 2005). The presence or absence of DArT markers in a genomic 'representation' is assayed by hybridising the 'representation' to the library of that species (Wenzl et al., 2004). Application of DArT for germplasm characterization has been applied in a number of crops. Xia et al. (2004) developed DArT for cassava and typed a group of 38 accessions. The average call rate was 98.1% with scoring reproducibility of 99.8%. The DArT markers displayed fairly high polymorphism information content (PIC) values and revealed genetic relationships among the samples consistent with the information available on the accessions. Yang et al. (2006) developed DArT for pigeon pea and its wild relatives and typed 96 accessions representing nearly 20 species of *Cajanus*. A total of nearly 700 markers were identified with the average call rate and scoring reproducibility of 96.0 and 99.7% respectively. DArT markers revealed genetic relationships among the accessions consistent with the available information and systematic classification. Since DArT has high throughput due to a high level of multiplexing and requires no prior sequence information (Kilian et al., 2005), it can immediately be applied for diversity studies in macadamia and identification of superior varieties.

Marker assisted selection for enhanced breeding of macadamia

Considerable developments in biotechnology have led plant breeders to develop more efficient selection systems to replace traditional phenotypic-pedigree-based selection systems (Ribaut and Hoisington, 1998). Marker assisted selection (MAS) is the indirect selection process where a trait of interest such as disease resistance, abiotic stress tolerance, and/or quality is selected based not on the trait itself but rather on a marker linked to it. The marker may be morphological, biochemical or one based on DNA/RNA variation (Semagn et al., 2006a). For example if MAS is being used to select individuals with disease, the level of disease is not quantified but rather a marker allele which is linked with disease is used to determine disease presence (Mingyao et al., 2005). Once genetic linkage has been identified for a disease, the next step is often association analysis, in which the markers within the linkage region are genotyped and tested for association with the disease (Mingyao et al., 2005). A review on the opportunities and constraints for marker assisted selection in macadamia breeding was done by Hardner et al. (2005). They suggested that detection of associations between marker-Quantitative trait loci (QTL's) using multiple families linked through pedigree was the most attractive strategy. Large scale association studies can be done using high throughput inexpensive genome-wide markers such as DArT (Baundouin et al., 2006) or SNP typing with DNA chips (Twyman, 2003; Ganal et al., 2009). DNA chips contain thousands of short DNA sequences immobilized at different positions and are used to discriminate between alternative bases at the site of a SNP. 2 chip-based typing methods are widely used. One method relies on allele-specific hybridization. Short DNA sequences on the chip represent all possible variations at a polymorphic site and a labeled DNA will

only hybridize if there is an exact match. The base is identified by the location of the fluorescent signal (Ganal et al., 2009).

In the second method, the oligonucleotide on the chip may stop one base before the variable site and typing relies on allele-specific primer extension. A DNA sample stuck onto the chip is used as a template for DNA synthesis, with the immobilized oligonucleotide as a primer. The four nucleotides, containing different fluorescent labels, are added along with DNA polymerase. The incorporated base, which is inserted opposite to the polymorphic site on the template, is identified by the nature of its fluorescent signal and the variation of the added nucleotide is identified by mass spectrometry (Tang et al., 1999). A recent advance for the detection of SNPs known as the high resolution melting curve (HRM) has been developed. The technique measures temperature induced strand separation of short PCR amplicons and is able to detect variation as small as one base difference between samples. It has been applied to the analysis of almond SNP discovery and genotyping (Shu-Biao et al., 2008).

MAS will be useful in macadamia where selected characters like yield are expressed late in plant development due to long juvenile period. Selection of genotypes for such traits will not wait until fruiting time. Selection for high nut yield and quality of macadamia from an array of cross-bred population would also benefit from MAS such that useful crosses can be selected and advanced early enough. MAS can also be used to select for disease and pest resistance. However, limited molecular mapping work on macadamia has been reported. The first molecular linkage map of macadamia (*M. integrifolia* and *M. tetraphylla*) based on 56 F_1 progeny of cultivars `Keauhou' and `A16' was reported by Peace et al. (2003). The map comprised of 24 linkage groups with 265 framework markers: 259 markers from randomly amplified DNA fingerprinting (RAF), 5 random amplified polymorphic DNA (RAPD) and one sequence-tagged microsatellite site (STMS). This molecular study is the most comprehensive examination to date of genetic loci of macadamia and is a major step towards developing marker-assisted selection for this crop. DArT is a powerful tool towards achieving faster marker assisted selection and construction of genetic linkage maps. Semagn et al. (2006b) constructed a genetic linkage map of 93 doubled-haploid lines derived from a cross between *Triticum aestivum* L. and a Norwegian spring wheat breeding line, NK93604, using DArT amplified fragment length polymorphism (AFLP), and simple sequence repeat (SSR) markers. The map has been successfully used to identify novel QTLs for resistance to *Fusarium* head blight and powdery mildew. In barley, Wenzl et al. (2006) developed a high density consensus map of barley that resulted in identification of 14 ± 9 DArT loci within 5 cm on either side of SSR, RFLP or STS loci previously identified as linked to agricultural traits. Validation of DArTs for whole genome profiling was initiated in coconut and over 350 markers were isolated from 120 lines from CIRAD and SriLanka (Wongtiem et al., 2005). It is proposed that, isolation of DArTs in macadamia be done to speed up genome mapping for agronomically important traits since a single DArT assay can simultaneously type hundreds to thousands of SNPs and insertion/deletion polymorphisms spread across the genome (Wenzyl et al., 2006).

Application of DNA markers for macadamia germplasm conservation

Most of the macadamia germplasm that is included in breeding programs is conserved on-farm. The existence of clonally propagated crops conserved on-farm is endangered by several factors including the introduction of alternative improved varieties (de Vicente et al., 2006). With multiplication of the few selected varieties of macadamia farmers have continued to cut or uproot original plantings and replacing them with the new cultivars. A recent survey (Gitonga et al., 2008a), revealed that out of 11 breeding selections 4 had already been cut down by the farmers despite a formal agreement with KARI to preserve the trees on-farm. 2 other trees had also died from natural causes between the years 2006 and 2007. Hence, there is risk of losing valuable germplasm leading to genetic erosion.

For *ex situ* conservation, genetic markers may contribute to the sampling, management and development of "core" collections as well as utilization of genetic diversity. For *in situ* and on-farm conservation strategies, genetic markers might help in recognizing the most representative populations within the gene pool (Lanteri and Barcaccia, 2006; de Vicente et al., 2006). Several criteria have been suggested for sampling the maximum possible genetic variation including, the number of alleles per locus, evaluated as those that are common in one to several populations but not in the species as a whole and those that are identified as unique (private) alleles (Lanteri and Barcaccia, 2006). Hence, information derived from genetic markers offers a good basis for better conservation decision making (de Vicente et al., 2006).

Genetic improvement of macadamia through tissue culture techniques

Mass multiplication of superior varieties

The conventional method of propagating macadamia in Kenya by grafting is constrained by requirement for long nursery period of 18-24 months (Nyakundi and Gitonga, 1993; Gitonga et al, 2002). Other methods of propagation have been found successful on macadamia including chip budding (Leigh, 1973), air layering (Kadman, 1982) and cuttings (Gitonga et al., 1997) but with little comercial viability.

Tissue culture techniques allow propagation of plant material with high multiplication rates on sterile artificial nutrient medium in an aseptic environment (George, 1993), a term usually referred to as cloning (Pfeiffer, 2003). Mass multiplication of clonal plant material through tissue culture techniques has now been widely used in nut crops such as cashew nut (Thimmappaiah et al., 2007), chestnut (Vieitez et al., 2007) and walnuts (Leal et al., 2007). Mulwa and Bhalla (2000) reported successful regeneration and rooting of shoots from nodal segments of *M. tetraphylla*. Regeneration of shoots from somatic embryos of *M. tetraphylla* has also been reported (Mulwa and Bhalla, 2006). Gitonga et al. (2004, 2008b) reported successful shoot regeneration from *M. integrifolia*, but with low multiplication rates and rooting problems. With the development of a viable tissue culture system for *M. integrifolia* and (*M. integrifolia* x *M. tetraphylla*) hybrids which form the main commercial cultivars in Kenya superior varieties can more efficiently be multiplied and distributed. Plantlets can also be enriched *in vitro* using endophytes to induce resistance of the resultant macammia trees to disease phytotoxins and other biotic stresses (Saikkonen et al., 2004). *In vitro* enrichment with endophytes has been reported in *Theobroma cacao* (Arnold et al., 2003) and banana (Kavino et al., 2007).

Development of new varieties through somaclonal variation and genetic transformation

Somaclonal variation is the term used to describe the variation seen in plants that have been produced by tissue culture (Skirvin, 1993). The variations can be phenotypic or genotypic, caused by changes in chromosome numbers (polyploidy and aneuploidy), chromosome structure (translocations, deletions, insertions and duplications) and DNA sequence (base mutations) (Peschke and Phillips, 1992). Hence, somaclonal variation leads to the creation of additional genetic variability. Somaclonal mutants can be enriched during *in vitro* culture to induce resistance to disease phytotoxins, tolerance to environmental or chemical stress (Bajaj, 1990; Haines, 1993). An example of such enrichment is with carrot somaclones resistant to *Alternaria dauci* (Dugdale et al., 2000) and *Pythium violae* (Cooper et al., 2006).

Plant genetic transformation involves the stable introduction of foreign DNA sequences, usually into the genome of the target plant. Specific new or modified genetic traits are added to existing plant varieties (Gayser and Fraley, 1989). Transgenic plants are then produced by regenerating shoots and roots from cells containing the foreign DNA through tissue culture techniques (McClean, 1998). Though macadamia is affected by relatively few pathogens that include *Phytophthora* spp., *Pythium* spp., *Botrytis* spp., *Rhizoctonia* spp., and *Pestalotia* spp. (Kiuru et al., 2004), the crop is highly affected by other stresses such iron chlorosis which affects both young and old plantations and can cause up to 50% losses (Wallace, 1957; Handreck, 1992). Such stresses can benefit from somaclonal variation or genetic transformation by developing resistant varieties. Pena and Séguin (2001) reported on the successful incorporation of transgenes for shortening the juvenile phase into forest trees and this can be of great benefit to macadamia.

In vitro conservation of macadamia germplasm

Macadamia is currently conserved on-farm and in field gene banks. Field gene banks are costly to maintain and they require considerable inputs in the form of land, labour, management and material inputs (Uyoh et al., 2003) and this also limits the extent to which replication of accessions can be done and how much diversity can be maintained.

Tissue culture techniques (*in vitro* conservation techniques) offer 2 efficient complementary options to field gene bank *ex-situ* conservation. Slow growth storage (the medium-term conservation of stock cultures at few degrees above zero, with or without the addition of compounds known to retard plant growth) and cryopreservation (freeze-preservation at ultra-low temperatures of 196°C in liquid nitrogen) (Benson, 1999). Cryopreservation is a sound alternative for long term conservation of plant genetic resources because at this temperature, all cellular divisions and metabolic processes are stopped (Panis and Lambardi, 2006). In the recent years, the 'vitrification/one step freezing' technique has been continuously improved and applied to woody species after the invention of the Plant Vitrification Solution 2 (PVS2) (Sakai et al., 1990; Panis and Lambardi, 2006). However, for successful *in vitro* preservation an adequate protocol for micropropagation has to be worked out first for regeneration of the preserved tissues and hence this calls for concerted efforts to develop a tissue culture protocol for macadamia. *In vitro* cultures also simplify quarantine procedures for local and international exchange of germplasm since there is no risk of disease transfer, once the material is screened of known diseases while in the source country (Frison and Putter, 1993).

Conclusions

Macadamia genetic improvement in Kenya has been slow owing to the use of conventional breeding methods. It is clear that sustainable use and conservation of macadamia genetic resources would be enhanced greatly through biotechnology applications to complement conventional breeding and conservation efforts. However, use of biotechnology tools applications should be prioritized to yield maximum practical benefits in the shortest possible time. Use of molecular techniques for germplasm characterization and marker assisted selection should be a high priority.

ACKNOWLEDGEMENTS

The authors are grateful to Dr. Lusike Wasilwa of KARI, Kenya for providing valuable information on macadamia. The valuable input from the reviewers of this manuscript is also highly appreciated.

REFERENCES

Agricultural Research Council (ARC) (2000). ARC- Institute for Tropical and Subtropical crops, Pretoria, South Africa.

Arnold AE, Mejia LC, Kyllo D, Rojas E, Maynard Z, Robbins N, Herre EA (2003). Fungal endophytes limit pathogen damage in a tropical tree. Proceedings of the National Academy of Sciences of the United States of America. 100(26):15649-15654.

Avise JC (1994). Molecular markers, natural history and evolution. Chapman and Hall, Recent research and development in Forest Genetic Resources. Proceedings of the training workshop on the conservation and sustainable use of forest genetic resources in Eastern and Southern Africa. 6th-11th December 1999, Nairobi, Kenya.

Bajaj YPS (1990). Somaclonal variation - Origin, induction, cryopreservation and implication in plant breeding. In. Bajaj YPS (ed) Biotechnology in Agriculture and Forestry II – Somaclonal variation in crop improvement I. Springer-Verlag Berlin, Heidelberg, New York, Tokyo, pp. 1-48.

Baundouin L, Berger A, Lebrun P (2006). Assessing linkage disequilibrium in coconut. In: de Vicente MC, Glaszmann J-C (eds) Molecular markers for allele mining. Bioversity International, p. 36.

Benson EE (1999). Cryopreservation. In: Benson EE (ed) Plant conservation biotechnology. London, England, Taylor & Francis, pp. 83-95.

Bittenbender HC, Hirae HH (1990). Common problems of Macadamia Nut. In: Hawaii Research Extension Series 112. College of Tropical Agriculture and Human Resources, HITAHR, University of Hawaii.

CABI (2005). Report on the identification of income diversification options and development of small to medium enterprise for high quality coffee and macadamia in Kenya, CAB International Africa Regional Center, Nairobi, Kenya.

Cavaletto CG (1980). Macadamia nuts. In: Nagy SS, Shaw PE (eds) Tropical and subtropical Fruits, composition, properties and uses, AVI Publications Company pp. 542-559.

Cheplick GP, Faeth SH (2009). Ecology and Evolution of the Grass-Endophyte Symbiosis. Oxford University Press, Oxford.

Cooper C, Crowther T, Smith BM, Isaac S, Collin HA (2006). Assessment of the response of carrot somaclones to *Pythium violae*, causal agent of cavity spot. Pathology 55 (3) 427-432.

de Vicente MC, Guzman FA, Engels J, Rao VR (2006). Genetic characterization and its use in decision-making for the conservation of crop germplasm. In: Ruane J, Sonnino A (eds) The role of biotechnology in exploring and protecting agricultural genetic resources. FAO, Rome, Italy. pp. 129-138.

Dugdale LJ, Mortimer AM, Isaac S, Collin HA (2000). Disease response of carrot and carrot somaclones to *Alternaria dauci*. Plant Path. 49(1):57-67.

Duke JA (1983). Handbook of energy crops.

Ferreira ME (2006). Molecular analysis of gene banks for sustainable conservation and increased use of crop genetic resources. In: Ruane, J, Sonnino, A (eds) The role of biotechnology in exploring and protecting agricultural genetic resources FAO, Rome, Italy pp. 121-127.

Frison EA, Putter CAJ (eds) (1993). FAO/IBPGR Technical guidelines for the safe movement of coconut germplasm. Food and Agriculture Organization of the United Nations. Rome/International Board for Plant Genetic Resources, Rome.

Ganal MW, Altmann T, Röder MS (2009). SNP identification in crop plants. Current Opinion in Plant Biol. 12(2):211-217.

Gayser CS, Fraley RT (1989). Genetically engineered plants for crop improvement. Science 244: 1293-1299.

George PS, Ravishanker GA, Ventakaraman LV (1993). Clonal multiplication of *Gardenia jasminoides* Ellis through axillary bud culture. Plant Cell Rep. 13:59-62.

Gitonga LN, Kahangi EM, Muigai AWT, Ngamau K, Gichuki ST, Mutuma E, Cheluget W, Watiki BG (2008a). Farmers' knowledge on Macadamia genetic diversity in Kenya as a means for *in situ* conservation. CATRINA 3(1): 55-60.

Gitonga LN, Kahangi EM, Gichuki ST, Ngamau K, Muigai AWT, Njeru ES, Njogu N, Wepukhulu S (2008b). Factors influencing the *in vitro* shoot regeneration in *Macadamia integrifolia*. Afr. J. Biotechnol. 7(22): 4202-4207.

Gitonga LN, Njeru ES, Kuria S, Muli S (2004). *In vitro* shoot proliferation of *Macadamia integrifolia*. Paper presented at the 9th KARI Scientific Conference. Nairobi.

Gitonga LN, Nyakundi W, Ruto ST, Watiki B, Balozi F, Takayama E (2002). Vegetative propagation of macadamia nut (*Macadamia integrifolia*, (*M. integrifolia* * *M. tetraphylla*) hybrids). In: Wesonga, JM, Losenge, T, Ndungu, CK, Obwara, FK, Agong, SG, Fricke, A, Hau, B, Stutzel, H (eds) Proceedings of the horticultural seminar on sustainable horticultural production in the tropics. 3rd-6th October 2001. JKUAT, Kenya.

Gitonga LN, Nyakund, W, Takayama E (1997). Effects of varietal differences on ability to root macadamia cuttings. Proceedings of the 1st National Horticultural seminar on sustainable horticultural production in the tropics. JKUAT, Kenya.

Goldstein DB, Linares AR, Cavalli-Sforza LL, Feldman MW (1995). An Evaluation of genetic distances for use with microsatellite loci. Genetics 139: 463-471.

Hamilton RA (1971). Macadamia nut growing, marketing, processing and research problems-Report to the Government of Kenya. FAO Rome No. TA 2996.

Hardner C, Peace C, Henshall J, Manners J (2005). Opportunities and constraints for marker-assisted selection in macadamia breeding. ISHS, Acta Hort. 694: 85-90.

Hardner C, Winks C, Stephenson R, Gallagher E (2001). Genetic parameters for nut and kernel traits in macadamia. Euphytica 117:151-61.

Haines RJ (1993). *In vitro* selection in breeding. In: Imrie BC, Hacker JB (eds) Focused plant improvement: towards responsible and sustainable agriculture pp. 210-219.

Handreck KA (1992). Iron-phosphorus interactions in the nutrition of seedling macadamia in organic potting media. Aust. J. Exp. Agri. 32(6): 773 - 779.

Hannotte O, Jianlin H (2006). Genetic characterization of livestock populations and its use in conservation decision-making. In: Ruane, J, Sonnino, A (eds) The role of biotechnology in exploring and protecting agricultural genetic resources. FAO, Rome, Italy pp. 89-96.

Harris M (2004). Involvement of the Harries family in the development of macadamia in Kenya, 1946-1974). Proceedings of the Macadamia Stakeholders meeting. 15th June, 2004. KARI HQTs. Nairobi, Kenya.

Hirama S, Ondabu N, Wasilwa L (1987). Observations of the differences on the characteristics of tree growth and bearing characteristics of five macadamia clones at different agro-ecology zones. Horticultural Development Project, Thika, Kenya.

Jaccoud D, Peng K, Feinstein D, Kilian K (2001). Diversity arrays: A solid state technology for sequence information independent genotyping. Nuc. Acid Res. 29: 25.

JICA (1991). Joint Review Report on the Horticultural Development Project in Kenya by Government of Japan, Japan International Cooperation Agency.

Kadman A (1982). Experiments with propagation of macadamia by air layering. Reprint from California Macadamia Society (CMS) Yearbook.

Kahl G, Mast A, Tooke N, Shen R, van den Boom D (2005). Single nucleotide polymorphisms. In: Meksem K, Kahl G (eds). The handbook of plant genome mapping: genetic and physical mapping. Wiley-VCH. pp. 75-104.

Kavino M, Harish, S, Kumar N, Saravanakumar D, Damodaran T, Soorianathasundaram K, Samiyappan R (2007). Rhizosphere and endophytic bacteria for induction of systemic resistance of banana plantlets against bunchy top virus. Soil biol. and Biochem. 39(5):1087-1098.

Kilian A, Huttner E, Wenzi P, Jaccoud D, Carling J, Caig V, Evers M, Heller-Uszynska, Uszynski G, Cayla C, Patarapuwadol S, Xia L, Yang S, Thompson B (2005). The fast and the cheap: SNP and DArTs-based whole genome profiling for crop improvement. In: Tuberosa R, Phillips RL, Gale M (eds). Proceedings of the international congress 'In the Wake of the Double Helix: From the Green Revolution to Gene Revolution, 27-42 May, 2003, Bologna, Italy pp. 443-461.

Kiuru P, Nyaga AN, Wasilwa L (2004). A Review of macadamia research in Kenya. Proceedings of the Macadamia Stakeholders meeting. KARI HQTs, Nairobi, Kenya.

Lanteri S, Barcaccia G (2006). Molecular marker based analysis for crop germplasm preservation. In: (eds) Ruane J, Sonnino A. The Role of Biotechnology in exploring and protecting agricultural genetic resources. FAO, Rome, Italy pp. 105-120.

Latha R, Rubia L, Bennett J, Swaminathan MS (2004). Allele mining for stress tolerance genes in Oryza species and related germplasm. Mol. Biotechnol. 27(2):101-108.

Leal DR, Sánchez-Olate M, Avilés F, Materan ME, Uribe M, Hasbún R, Rodríguez R (2007). Micropropagation of Juglans regia L. In: (Jain SM, Häggman H, (eds.) Protocols for micropropagation of woody trees and fruits. Springer, Heidelberg pp. 381-390.

Lee P (2004). Macadamia Working Group. Paper presented at the IX World Tree nut Convention, Las Vegas.

Leigh DS (1973). Notes on macadamia propagation in New South Wales. Reprint from California Macadamia Society Yearbook, 1973. http://www.coopersnuthouse.com/NotesOnMacadamiaPropagationInNewSouthWales20.ht.

Mauro-Herrera M, Meerow AW, Borrone JW, Kuhn DN, Schnell RJ (2006). Ten Informative markers developed from WRKY sequences in coconut (*Cocos nucifera*). Mol. Eco. Notes. 6(3):904-906.

McClean PE (1998). Analyzing plant gene expression with transgenic crops. Phillip McClean. http://www.ndsu.nodak.edu/instruct/mcclean/pls c731/transgenic/transgenic4.htm.

Macfalane N and Harris RV (1981). Macadamia nuts as an edible oil source. In: Pryde, EH, Princen, LH, MukNerjee, XD (ed) American Oil Chemiss Society.

McHargue LT (1996). Macadamia production in southern California. In. Janick J (ed), Progress in new crops. ASHS Press, Arlington, VA. pp. 458-462.

Mingyao L, Boehnke M, Abecasis GR (2005). Joint Modeling of Linkage and Association: Identifying SNPs Responsible for a Linkage Signal. Am. J. Hum. Genet. 76(6): 934–949.

Muhara J (2004). Role of private sector in processing and marketing of Macadamia in Kenya: Kenya Farm Nut experience. Proceedings of the Macadamia Stakeholders meeting. KARI HQTs, Nairobi, Kenya.

Mulwa RMS, Bhalla PL (2000). *In vitro* shoot multiplication of *Macadamia tetraphylla* L. Johnson. J. Hort. Sci. 75(1): 1-5.

Mulwa RMS and Bhalla PL (2006). *In vitro* plant regeneration from imamture cotyledon explants of macadamia (*Macadamia tetraphylla* L. Johnson). Plant Cell Rep. 25: 1281-1286.

Nakitandwe J, Trognitz F, Trognitz B (2007). Reliable allele detection using SNP-based PCR primers containing Locked Nucleic Acid: application in genetic mapping. Plant Methods 3:2.

Nissen RJ, Williams RR (1980). Assessment of Macadamia selections based on nut quality. Reprint from CMS Yearbook, 1980.

Nyakundi W, Gitonga L (1993). Macadamia Propagation handbook. JICA, Kenya.

Ondabu N, Nyaga AN, Tominaga K (1996). Macadamia clonal selection in Kenya. 5th Biennial KARI Scientific Conference, Nairobi, Kenya.

Onsongo M (2003). Kenya Trees Annual. USDA Global Agricultural Information Network.

Panis B, Lambardi M (2006). Status of cryopreservation technologies in plants (crops and forest trees). In: Ruane, J, Sonnino, A (eds) The role of biotechnology in exploring and protecting agricultural genetic resources. FAO, Rome, Italy pp. 61-78.

Peace C Vinathage V, Turnbull C, Carroll BJ (2002). Characterizing Macadamia germplasm with codominant radio-labeled DNA amplification fingerprinting (RAF) markers. International Symposium on Tropical and Subtropical fruits. Acta Hort. 575: 381-387.

Peace CP, Vithanage V, Turnbull CGN. Carroll BJ. (2003). A genetic map of macadamia based on randomly amplified DNA fingerprinting (RAF) markers. Euphytica, 134 (1): 17-26.

Pena L, Seguin A (2001). Recent advances in the genetic tra-nsformation of trees. Trends in Biotechnol. 19(12):500-506

Peschke V, Phillips RL (1992). Genetic implication of somaclonal va-riation in plants: Adv. Genet. 30: 42-75.

Pfeiffer TW (2003). From classical plant breeding to modern crop improvement. In. Chrispeels MJ, Sadava DE (eds). Plants, Genes and Crop Biotechnology. 2nd Edition. Jones and Bartlett Publishers, Inc. pp. 360-389.

Rafalski A (2002). Applications of single nucleotide polymorphisms in crop genetics. Current Opinion in Plant Biology 5(2):94-100.

Rapley R, Harbron S (Eds.) (2004). Molecular Analysis and Genome Discovery. Chichester. John Wiley & Sons Ltd.

Ribaut JM, Hoisington DA (1998). Marker assisted selection: New tools and Strategies. Trends Plant Sci. 3: 236-239.

Rotich M (2004). Development of macadamia industry: global markets of macadamia. Proceedings of the Macadamia Stakeholders meeting. KARI HQTs, Nairobi, Kenya.

Rumsey HJ (1927). Australian nuts and nut growing in Australia. Part 1. The Australian nut. Sidney.

Sakai A, Kobayashi S, Oiyama I (1990). Cryopreservation of nucellar cells of navel orange (Citrus sinensis Osb. var. brasiliencis Tanaka) by vitrification. Plant Cell Reports 9: 30-33

Sato Y, Waithaka JHG (1996). Integrating tree cash crops in agricultural production systems. Kenya.

Saikkonen K, Wäli P, Helander M, Faeth SH (2004).Evolution of endophyte-plant symbioses. Trends in Plant Science 9(6):275-280.

Schmidt AL, Scott L, Lowe AJ (2006). Isolation and characterization of microsatellite loci from Macadamia. Molecular Ecology Notes 6(4):1060-1063.

Sedgley M, Bell FDH, Bell D, Winks CW, Pattison SJ, Hancook TW (1990). Self-and cross compatibility of Macadamia cultivars. J. of Hort. Sci. 65: 205-218.

Semagn K, Bjørnstads Á, Ndjiondjop MN (2006a). Progress and prospects of marker assisted backcrossing as a tool in crop breeding programs. Afr. J. Biotechnol. 5:2588-2603.

Semagn K, Bjørnstad A, Skinnes H, Marøy AG, Tarkegne Y, William M (2006b) Distribution of DArT, AFLP, and SSR markers in a genetic linkage map of a doubled-haploid hexaploid wheat population. Genome 49(5): 545–555.

Shu-Biao-Wu SB, Wirthensohn MG, Hunt P, Gibson JP, Sedgley M (2008). High resolution melting analysis of almond SNPs derived from ESTs. Theor. Appl. Genet. 118:1-14.

Skrvin RM, Nortom M, McPheeters KD (1992). Somaclonal variation: Has it proved useful for plant improvement. Acta Hort. 336: 333-340.

Soleimanni VD, Baum BR, Johnson DA (2003). Efficient validation of Single Nucleotide Polymorphisms in plants by allele-specific PCR, with an example from barley. Plant Mol. Biol. Rep. 21: 281-288.

Stephenson RA (1983). Farm Characteristics and Management Practices in the Australian Macadamia Industry. Queensland Department of Primary Industries, Brisbane.

Sunnucks P (2000). Efficient genetic markers for population biology. Tree 15:199-203.

Tang K, Fu D-J, Julien D, Braun A, Cantor CR, Koster H (1999). Chip-based genotyping by mass spectrometry. Proceedings of the National Academy of Sciences of the United States of America 96(18):10016-10020.

Thimmappaiah, Shirly RA, Iyer RD (2007). Protocol for micropropagation of Castanea sativa Mill. In: (Jain SM, Häggman H, (eds.) Protocols for micropropagation of woody trees and fruits. Springer, Heidelberg pp. 313-320

Tominaga K, Nyaga AJN (1997). Breeding of Macadamia nuts. Evaluation report, 1994-1997.

Twyman R (2003). SNP typing with DNA chips. http://genome.wellcom.ac.uk/ doc_WTD021043.html.

United States Department of Agriculture (2008). Macadamia in-shell basis production by country in metric tons. http://www.indexmundi.com/agriculture/commodity=Macadamia&graph=production.

Upadhyaya HD, Furman BJ, Dwivedi SL, Udupa SM Gowda CLL, Baum M, Crouch JH, Buhariwalla HK, Singh S (2006). Development of a composite collection for mining germplasm possessing allelic variation for beneficial traits in chickpea. Plant Genet. Resour.: Characterization and Utilization, 4: 13-19.

Uyoh EA, Nkang AE and Eneobong EE (2003). Biotechnology, genetic conservation and sustainable use of bioresources. Afr. J. Biotechnol.2 (12): 704-709.

Vieitez AM, Sánchez MC, García-Nimo ML, Ballester A. (2007). Protocol for micropropagation of Castanea sativa Mill. In: (Jain SM, Häggman H, (eds.) Protocols for micropropagation of woody trees and fruits Springer, Heidelberg pp. 299-312.

Vithanage V, Hardner C, Aderson KL, Meyers N, McConachie C, Peace C (1998). Progress made with molecular markers for genetic improvement of macadamia. International Symposium on Biotechnology of Tropical and Subtropical Species, Part 2. Acta Horti. 461:199-207.

Waithaka JHG (2001). Sustainable commercial tree crop farming: A case for Macadamia nuts. Paper presented at the USAID African Sustainable Tree Crops Programme Conference, Nairobi, Kenya.

Wallace A (1957). Chlorosis in Macadamia. California Macadamia Society Year Book.

Wasilwa LA, Ondabu N, Watani G (2003). Adaptability of Kenyan macadamia selections and Hawaiian varieties in four agro ecological zones. A paper presented at the 1st KARI Adaptive Research Conference, Nairobi, Kenya.

Wasilwa LA, Watani GW, Ondabu N, Nyaga A, Muli H, Kagiri B, Kiiru S (1999). Performance of Macadamia varieties in three agro-ecology zones. Hortscience 35: 477.

Wenzl P, Li H, Carling J, Zhou M, Raman H, Paul E, Hearnden P, Maier C, Xia L, Caig V, Ovesná J, Cakir M, Poulsen D, Wang J, Raman R, Smith KP, Muehlbauer GJ, Chalmers KJ, Kleinhofs A, Huttner E, Kilian A (2006). A high-density consensus map of barley linking DArT markers to SSR, RFLP and STS loci and agricultural traits. BMC Genomics 7:206.

Wilkie JH (2008). Macadamia industry update. Paper presented at the 5th China International Foodstuff Exposition, Liuhua, China.

Wongtiem P, Perera C, Risterucci AG, Xia L, Patarapuwadol S, Carling J, Caig V, Evers M Uszynski G, Huttner E, Wenzi P, Fregene M, Vicente C, Glaszmann JC, Kilian A, (2005). Validation of Diverty Array Technology as a platform for whole-genome profiling in orphan crops. GCP meeting, Rome.

Woodroof JG (1967). Tree nuts, production, processing, products. AVI publications Company 1:313-337.

World Horticultural Trade & U.S. Export Opportunities (2004). Macadamia situation and outlook in selected countries.

Xia L, Peng K, Yang S, Wenzl P, de Vicente P, Fregene M, Kilian A (2005). DArT for high-throughput genotyping of cassava (Manihot esculenta) and its wild relatives. Theor. Appl. Genet. 110: 1092-1098.

Yang S, Pang W, Ash G, Harper J, Carling J, Wenzyl P, Huttner E, Kilian A (2006). Low level of genetic diversity in cultivated pigeon pea compared to its wild relatives is revealed by Diversity Arrays Technology (DArT) Theor. Appl. Genet. 113: 585-595.

Yokoyama KM, Wanitprapha K, Nakamoto ST and Bittenbender HC (1990). Macadamia Nut Economic Fact Sheet. 9. Dept Agric. Res. Econ., CTAHR, University of Hawaii.

12

Genotype × environment interactions and heritability of quantitative resistance to net blotch in Tunisian barley

M. Cherif[1]*, S. Rezgui [1], P. Devaux[2] and M. Harrabi[1]

[1] Institut National Agronomique de Tunisie. 43 Avenue Charles Nicolle. 1082 Cité Mahrajène, Tunis, Tunisia.
[2]Laboratoire de Biotechnologie, Florimond Desprez, PB 41, 59242 Capelle en Pévèle, France.

A doubled-haploid barley population derived from a cross between the Tunisian cultivar 'Roho' and the local line '90' was used to assess the genotype x environment interaction, heritability estimates and correlations between disease parameters of net blotch resistance at adult growth stage in three environments. Net blotch reaction was evaluated using the mass disease index, the area under the disease progress curve and the apparent infection rate. The DH lines showed different levels of quantitative resistance to *Pyrenophora teres* under low, moderate and high epidemic conditions. Large variation of the mass disease index, the area under the disease progress curve and the apparent infection rate was obtained under high-pressure conditions that facilitate identifying tolerant lines. Significant genotypic differences were noted, however preponderant genotype x environment interactions were the major sources of variation. Broad sense heritability values were large for all parameters using data from single environment and low for the mass disease index, the area under disease progress curve and the apparent infection rate using estimates from different growing conditions. In a single environment, resistance related parameters were found to be genetically associated. Excepting mass disease index, lack of associations between area under the disease progress curve and apparent infection rate measured on the three field conditions were obtained. The results obtained suggest that loci for mass disease index may be pleiotropic or linked and loci for areas under the disease progress curve and apparent infection rate may be independent. Multi-location screening for quantitative resistance to *P. teres* should be considered in a breeding program.

Key words: Heritability, genotype x environment interactions, genetic correlation, disease parameter, *Hordeum vulgare*, *Pyrenophora teres*.

INTRODUCTION

Net blotch of barley (*Hordeum vulgare* L.) caused by *Pyrenophora teres* (Died), is a major disease in most barley growing areas around the world. In Tunisia, Cherif et al. (1994) reported that net blotch is the most common disease of barley that was associated with high severity levels (70 - 80%) in some regions. Both net and spot forms are prevailing in Tunisia. Management practices used, the intensification of barley cultivation and the absence of resistant varieties are among the factors favouring net blotch incidence. Thus, selecting for quantitative resistance to *P. teres* could be efficient. Quantitative or partial resistance as a reduction in the percentage of leaf tissue affected was reported in this pathosystem (Douglas and Gordon, 1985; Robinson and Jalli, 1997; Steffenson and Webster, 1992; Steffenson et al., 1996). It was termed adult or field resistance since it reduces the rate of disease development in the field. Screening for quantitative resistance in the field is often

---

*Corresponding author. E-mail: cherif_majda@yahoo.fr.

Abbreviations: AUDPC, Area under the disease progress curve; **AUDPC$_{MDI}$,** AUDPC calculated for MDI in 2003-2004; **AUDPC$_S$,** AUDPC calculated for S; **DH,** doubled-haploid, **ESA-Mograne,** ecole supérieur agricole de mograne, **INAT,** institut national agronomique de Tunisie; **MDI,** mass disease index; MDI$_1$, MDI calculated in 2002 – 2003; **MDI$_2$,** final MDI calculated in 2003-2004; **r,** apparent infection rate; **r$_{MDI}$,** apparent infection rate calculated for MDI in 2003 – 2004; **r$_S$,** apparent infection rate calculated for S; **S,** final disease severity.

difficult. For net blotch resistance, Cakir et al. (2003) and Spaner et al. (1998) used the scale of Tekauz (1985). Others as El Yousfi and Ezzahiri (2001), Steffenson and Webster (1992) and Steffenson et al. (1996) used the percent disease severity developed by Burleigh and Loubane (1984). Area under the disease progress curve (AUDPC) describes disease development in the field and can differentiate, at a critical point, between two epidemics with equal final disease severities. This parameter was used by Burleigh and Loubane (1984), El Yousfi and Ezzahiri (2001), Robinson and Jalli (1997) and Steffenson and Webster (1992) to estimate resistance to net blotch. Apparent infection rate (r) was first proposed by Vanderplank (1963) and was later applied by many authors for several pathosystems (Burleigh and Loubane, 1984; El Yousfi and Ezzahiri, 2001; Robinson and Jalli, 1997; Steffenson and Webster, 1992). Information given by the AUDPC and the r is different when disease increases rapidly within a short period. The environment may alter the expression of quantitative resistance (El Yousfi and Ezzahiri, 2001). Little attention has, however, been directed toward the analysis of the genotype x environment interactions and the heritability estimates of these parameters. Doubled-haploid (DH) lines are suited for this analysis since these lines present an unlimited number of individuals facilitating the estimation of the genotype x environment interactions by multiplying replicates (Choo et al., 1985) and thus estimating heritability (Marwede et al., 2004; Spaner et al., 1998). The objectives of this study were:

(1) To investigate quantitative expression of resistance to *P. teres* in different environments using a DH population of barley.
(2) To evaluate genotype x environment components.
(3) To estimate heritability.
(4) To determine potential associations between disease related parameters.

MATERIALS AND METHODS

Plant material

A total of fifty-nine DH barley lines were obtained at Florimond Desprez, using anther culture and the *Hordeum bulbosum* method (Devaux and Pickering, 2005). These DH lines were developed from F1 plants of the cross between the Tunisian cultivar 'Roho' and the local 'line 90' which was carried out at INAT (Institut National Agronomique de Tunisie). 'Roho' is a widely grown two-row barley cultivar susceptible to net blotch. 'Line 90' is a six-row fixed line selected from the Tunisian national breeding program for its resistance to net blotch in the field. This line was selected from a cross between Local Cap-Bon and Jérusalem à barbes lisses/CI 10836 ICB 77-319-1AP-0SH-2AP-1AP-0AP.

Field experiments

The experiments were conducted during two consecutive growing seasons 2002 - 2003 and 2003 - 2004 at the experimental station of

ESA-Mograne (Ecole Supérieur Agricole de Mograne) associate which is, with, considered a particularly hot spot area for net blotch disease and during 2004 at the experimental station of INAT (Tunis) corresponding to a moderate site for net blotch infection.

Mograne trials 2002-2003 and 2003-2004

Parents and DH lines were sown in single rows on 5 December, 2002 and in two rows on 17 November, 2003 using a randomised block design with three replications. Row-lines were 1 m long and spaced 0.5 m apart. The Tunisian commercial cultivar 'Martin' was planted as a susceptible check every 14 entries. In 2002 - 2003, plants were inoculated at the mid-tillering stage of growth (GS 22 - 26) and again at the early stem elongation (GS 30 - 33) stage (Zadoks et al., 1974) using infected barley seeds with a mixture of local isolates including both net and spot forms of *P. teres* according to Onfroy (1997). In 2003 - 2004, inoculations were made twice: at early tillering stage of growth (GS 20 - 23) with barley straw collected from infected volunteer barley plants in neighbouring fields and at the end of tillering stage (GS 28 - 29) with infected barley seeds prepared as in 2002 - 2003. Disease incidence (percentage of plants having at least one lesion) and severity (average percent of leaf area affected by the disease) were assessed on 10 randomly selected plants per line according to Yahyaoui et al. (2003). Both disease incidence and severity were estimated only once in 2002 - 2003 at the mid-dough growth stage (GS 85 - 87) and four times in 2003 – 2004 starting at stem elongation growth stage (GS 35 - 37) to the dough development stage (GS 85 - 87).

Tunis trial 2004

Five plants of each parent and DH lines were sown on 15 January, 2004 in plastic pots (0.25 m diameter) filled with loamy-clay soil and grown at INAT station. A completely randomised design with three replications was adopted. Plants were inoculated three times every 15 days from mid-tillering growth stage (GS 22 - 26) by spraying a mono-conidial suspension of 'Bir Mcharga' isolate of *P. teres* f. *teres* adjusted to 10^4 conidia/ml for which parents present differential reaction (unpublished data). Pathogen culture and inoculum preparation were done as described by Steffenson et al. (1996). Inoculated plants were then covered with plastic sheets to insure a high level of humidity. Disease reactions were recorded three times after symptoms apparition from ear emergence stage (GS 53 - 58) to milk development stage (GS 72 - 76) using the percent net blotch severity (including both chlorotic and necrotic areas) according to the scale devised by Burleigh and Loubane (1984).

Evaluation criteria and statistical analysis

The data collected from Mograne trials were transformed in to mass disease index (MDI) following the method of Ding et al. (1993):

$$MDI = (DI \times DS) / 100$$

Where: DI is the disease incidence and DS is the disease severity. The area under the disease progress curve (AUDPC) and the apparent infection rate (r) were then estimated. The AUDPC was calculated using the equation:

$$AUDPC = \Sigma^n_{i=1} [(Y_{i+1} + Y_i) \times 0.5] [T_{i+1} - T_i]$$

Where: Y_i is the MDI at the ith observation, T_i the time (in day) at the ith observation and n the total number of observations. The r was estimated by a linear regression on time (T) of the natural logarithm

Table 1. Variation of disease parameters[a] for net blotch reaction in 59 doubled-haploid lines of barley under three environments.

	MDI$_1$	MDI$_2$	AUDPC$_{MDI}$	r$_{MDI}$	S	AUDPC$_S$	r$_S$
Mean	3.150	22.091	947.522	0.094	13.852	260.864	0.404
Min	0.080	1.500	33.700	0.068	5.000	82.670	0.381
Max	10.000	58.333	2441.000	0.109	28.333	542.080	0.427
LSD 0.05	1.990	9.306	452.720	0.007	4.905	63.787	0.007
'Martin'	5.003	59.000	2133.760	0.111	-	-	-

[a] **MDI$_1$,** mass disease index calculated in Mograne during 2002 - 2003; **MDI$_2$,** mass disease index calculated on the basis of data obtained from the last rating date in Mograne during 2003-2004; **AUDPC$_{MDI}$,** area under the disease progress curve calculated from mass disease index estimated in Mograne during 2003 - 2004, **r$_{MDI}$,** apparent infection rate calculated from mass disease index estimated in Mograne during 2003 – 2004; **S,** final disease severity estimated in Tunis 2004; **AUDPC$_S$,** area under the disease progress curve calculated from disease severity estimated in Tunis 2004; **r$_S$,** apparent infection rate calculated from disease severity estimated in Tunis 2004.

(ln) of the diseased tissue proportion (x) divided by the proportion of non-diseased tissue (1-x). The slope of the regression line is then taken as an estimate of r.

For Tunis trial, AUDPC and r were calculated using disease severity. In this trial, the severity was considered equivalent to the mass disease index because controlled conditions have induced a disease incidence of 100%.

Components of variance were estimated for each of the seven evaluation parameters in a single environment, while the analysis for the MDI, the AUDPC and the r was carried out across environments. These analyses were achieved using REML of PROC MIXED of SAS (SAS Institute, 1988).

The models I and II were used for parameters estimated at Mograne (2002 - 2003 and 2003 - 2004) and at Tunis 2004, respectively.
Model I:

$$Y_{ij} = \mu + B_i + G_j + \varepsilon_{ij}$$

with Y_{ij} = observation of genotype j in block i, μ = general mean, B_i = effect of block i, G_j = effect of genotype j and ε_{ij} = residual error.
Model II:

$$Y_{ij} = \mu + G_i + \varepsilon_{i(j)}$$

with Y_{ij} = observation of genotype i in replication j, μ = general mean, G_i = effect of genotype i and $\varepsilon_{i(j)}$ = residual error.

The model III was used for parameters estimated across environments (MDI, AUDPC and r)
Model III:

$$Y_{ijk} = \mu + E_i + R_j(E_i) + G_k + E_iG_k + \varepsilon_{ijk}$$

with Y_{ijk} = observation of genotype k in environment i in replication j, μ = general mean, E_i = effect of environment i, $R_j(E_i)$ = effect of replication j in environment i, G_k = effect of genotype k, E_iG_k = genotype x environment interaction of genotype k with environment i and ε_{ijk} = residual error. .

Broad sense heritability (H^2) estimates were calculated from variance components. For a single environment, heritability on a plot level was estimated from the following equation:

$$H^2 = \sigma^2g / (\sigma^2g + \sigma^2\varepsilon),$$

and for MDI, AUDPC and r, heritability for mean values across environments was estimated according to the formula proposed by Marwede et al. (2004):

$$H^2 = \sigma^2g / (\sigma^2g + \sigma^2ge/E + \sigma^2\varepsilon/ER)$$

where σ^2g, σ^2ge and $\sigma^2\varepsilon$ are the genotypic variance, the genotype x environment interaction variance and the environmental variance respectively. E and R are number of environments and replicates, respectively.

Genetic correlation coefficients were estimated between parameters investigated in each environment and between the same parameters evaluated in different environments. Variance and covariance components were estimated by the REML method of SAS PROC MIXED (SAS Institute, 1988).

RESULTS

Table 1 showed that MDI (or S for Tunis 2004) ranged from 0.08 to 10.00 at Mograne 2002 - 2003, from 5.00 to 28.33 at Tunis 2004 and from 1.50 to 58.33 at Mograne 2003 - 2004, with mean values of 3.15, 13.85 and 22.09, respectively. The magnitude of variations and the mean values of MDI (or S) would suggest that the prevailing growing conditions represent a major component of the disease inducing factors. Thus, lower mean with a limited range values indicate a low epidemic of net blotch, while intermediate mean and range and superior mean with higher range represent moderate and high epidemic of net blotch respectively. Moreover, for the same growing season (2003 - 2004), AUDPC estimates were higher at Mograne (ranging from 33.70 to 2441.00 with a mean of 947.52) than at Tunis (ranging from 82.67 to 542.08 with a mean of 260.86). However, r were lower at Mograne (ranging from 0.068 to 0.109 with a mean of 0.094) than at Tunis (ranging from 0.381 to 0.427 with a mean of 0.404). At Mograne, infection responses of most of the DH lines were significantly lower than that of the susceptible check 'Martin' suggesting quantitative resistance to P. teres of these lines. At Tunis, the susceptible check was not used; however, the DH lines showed a relatively moderate degree of resistance, as measured by final net blotch severities (S). Selection, for quantitative resistance to P. teres could be more appropriate under high epidemic conditions as those prevailed at Mograne 2003 - 2004 because the DH lines exhibited a large variation in their reaction as described by MDI$_2$, AUDPC$_{MDI}$ and r$_{MDI}$ (Table 1).

Table 2. Variance components and heritability estimates of disease parameters for net blotch reaction in 59 doubled-haploid lines of barley in single environment.

Disease parameter [a]	MDI_1	MDI_2	$AUDPC_{MDI}$	r_{MDI}	S	$AUDPC_S$	r_S
σ^2_g [b]	8.89**	214.94**	44.89 x 10^4**	7.49 x 10^{-5}**	26.54**	1.11 x 10^4**	9.94 x 10^{-5}**
σ^2_ε [c]	1.82	40.15	9.43 x 10^4	2.40 x 10^{-5}	10.44	0.19 x 10^4	2.19 x 10^{-5}
H^2 [d]	0.83	0.84	0.82	0.76	0.72	0.85	0.82

[a] MDI_1, mass disease index calculated in Mograne during 2002-2003; MDI_2, mass disease index calculated on the basis of data obtained from the last rating date in Mograne during 2003-2004; $AUDPC_{MDI}$, area under the disease progress curve calculated from mass disease index estimated in Mograne during 2003-2004; r_{MDI}, apparent infection rate calculated from mass disease index estimated in Mograne during 2003-2004; S, final disease severity estimated in Tunis 2004; $AUDPC_S$, area under the disease progress curve calculated from disease severity estimated in Tunis 2004 and r_S, apparent infection rate calculated from disease severity estimated in Tunis 2004.

[b] σ^2_g, genetic variance.

[c] σ^2_ε, error variance.

[d] H^2, broad sense heritability estimates.

The variance components of the seven evaluation parameters in each environment were estimated from the analysis of variance (Table 2). All the parameters studied showed highly significant differences (P < 0.01) among the DH lines, indicating the presence of genetic variation (σ^2_G). Thus, broad-sense heritability (H^2) estimates on a plot level were high since it ranged from 0.72 to 0.85 for the different disease parameters investigated (Table 2). Similar values of heritability calculated in each of the three environments suggest that residual error variance represent almost 25%.

Pooled analysis of variance across environments indicated the presence of highly significant differences among genotype, environment and genotype x environment interactions for MDI, AUDPC and r (Table 3). Although the genotypic effects (σ^2_G) were highly significant, genotype x environment interaction effects (σ^2_{GE}) were larger than the genotypic variance for the three studied parameters. In fact, superiority of the interaction variation (σ^2_{GE}) relative to the genotypic variation

(σ^2_G) was 513, 1803 and 435% for MDI, AUDPC and r, respectively. Moreover, a relatively high Experimental error was also observed for all parameters. These results affected the broad sense heritability (H^2) estimates for mean values across environments. These estimates were 0.35, 0.09 and 0.29 for MDI, AUDPC and r respectively (Table 3).

Genetic correlation coefficients were calculated between the parameters studied in the three environments (Table 4). Genetic correlation between parameters evaluated within each environment ranged from 0.75 to 0.95 and were all highly significant (P < 0.01). Mass disease index evaluated at Mograne 2002 - 2003 (MDI_1) was genetically associated with MDI evaluated at Mograne 2003 - 2004 (MDI_2) and with disease severity evaluated at Tunis 2004 (S). A lack of significant correlation between MDI_2 and S was noted. Similarly, no significant correlation was observed between AUDPC estimated at Mograne 2003 - 2004 ($AUDPC_{MDI}$) and at Tunis 2004 ($AUDPC_S$), as well as between r calculated in these

two environments (r_{MDI} and r_S).

DISCUSSION

The DH population was screened under low, moderate and high net blotch pressure conditions respectively at Mograne 2002-2003, Tunis 2004 and Mograne 2003 - 2004. These differences noted on the epidemiological levels were mostly attributed to variable sowing dates and the effects of rainfall and temperature that prevailed at the three growing conditions. In the three environments, the disease level can be compared only on the basis of MDI because time between inoculation and the final observation was longer for Mograne trial 2003 - 2004 than for Tunis trial 2004. Therefore, AUDPC will be larger in the first trial even though the terminal MDI values were equal. Similar trend is noted for the apparent infection rate represented by the linear regression coefficient of disease proportion logit on time. For similar terminal disease proportions, the apparent

Table 3. Variance components and heritability estimates for the mass disease index (MDI), the area under the disease progress curve (AUDPC) and the apparent infection rate (r) in 59 doubled-haploid lines of barley evaluated under three environments.

	MDI	AUDPC	r
σ^2_g [a]	13.61^{**}	12084.00^{**}	$16.46 \times 10^{-6**}$
σ^2_e [b]	89.03^{**}	225687.90^{**}	$4.87 \times 10^{-2**}$
σ^2_{ge} [c]	69.85^{**}	217926.80^{**}	$71.57 \times 10^{-6**}$
σ^2_ε [d]	17.47	48098.90	22.48×10^{-6}
H^2 [e]	0.35	0.09	0.29

$**P<0.01.$ [a]σ^2_g: genetic variance, [b]σ^2_e: environmental variance, [c]σ^2_{ge}: variance of genotype × environment interaction, [d]σ^2_ε: error variance, [e]H^2: broad sense heritability estimates.

Table 4. Genetic correlation coefficients between disease parameters[a] for net blotch reaction in a doubled-haploid barley population under three environments.

		MDI_2	$AUDPC_{MDI}$	r_{MDI}	S	$AUDPC_S$	r_S
Mograne 2002 - 2003	MDI_1	0.48^{**}			0.28^*		
Mograne 2003 - 2004	MDI_2		0.95^{**}	0.91^{**}	0.20^{NS}		
	$AUDPC_{MDI}$			0.85^{**}		0.17^{NS}	
	r_{MDI}						0.18^{NS}
Tunis 2004	S					0.75^{**}	0.94^{**}
	$AUDPC_S$						0.86^{**}

$*$ $P<0.05$, $**$ $P<0.01$, ^{NS} not significant at $P<0.05$.
[a] MDI_1, mass disease index calculated in Mograne during 2002 – 2003; MDI_2, mass disease index calculated on the basis of data obtained from the last rating date in Mograne during 2003 – 2004; $AUDPC_{MDI}$, area under the disease progress curve calculated from mass disease index estimated in Mograne during 2003 – 2004; r_{MDI}, apparent infection rate calculated from mass disease index estimated in Mograne during 2003 – 2004; S, final disease severity estimated in Tunis 2004; $AUDPC_S$, area under the disease progress curve calculated from disease severity estimated in Tunis 2004 and r_S, apparent infection rate calculated from disease severity estimated in Tunis 2004.

infection rate is superior when disease evaluation period is shorter explaining superior apparent infection rates found at Tunis than that noted at Mograne although the MDI is more significant in this latter environment.

DH lines produced from the cross: 'Roho' x 'Line 90' showed a better level of adult resistance in natural conditions (Mograne) especially in 2003 - 2004 than the susceptible check 'Martin'. Thus, these lines possess different levels of quantitative resistance to net blotch in the field. In controlled conditions (Tunis), the DH lines exhibited a relatively low degree of resistance with different reaction pattern. This result would assume that these DH lines at Tunis express partial resistance to *P. teres*. Quantitative resistance recognized by a reduction of percentage of tissue affected by net blotch was reported by Gupta et al. (2003), Steffenson and Webster (1992) and Tuohy et al. (2006). They consider the term quantitative resistance as a practical term to designate

incomplete resistance which have no relation with quantitative genetic. Slow-scalding resistance was considered synonymous with quantitative resistance (Sorkhilalehloo et al., 2002). It exhibited a compatible reaction with *Rhynchosporium secalis* coupled with low to intermediate levels of disease incidence and severity. In this investigation, MDI is used to describe quantitative resistance since it combines disease severity and incidence. Moreover, the quantitative resistance observed in the three environments was characterized by the three following evaluation parameters: MDI, AUDPC and r. They were used to assess the relative reduction on the final infection level, the late symptoms apparition and slow development of the disease.

Broad sense heritability values were large for all parameters using data from single environment. This result supports the finding of Grewal et al. (2008b), Robinson (1999) and Spaner et al. (1998) in the field conditions and those of Grewal et al. (2008a,b; 2010) in the growth chamber conditions. However, heritability estimates from different growing conditions were low to very low because of genotype x environment interactions as noted for the AUDPC. It is expected that the genotype x environment interactions were significant since the three trial growing conditions differed greatly from each other. Tunis experiment is carried out using a specific monoconidial isolate of *P. teres* to inoculate plants grown in pots; whereas trials at Mograne were conducted under natural growing conditions and inoculated with mixture of local isolate of *P. teres*. Moreover, sowing dates, climatic conditions and variable concentrations of inoculum may result in different levels of net blotch epidemics in the three trials associated with a significant genotype x environment interactions. The AUDPC is the most influenced parameter by the genotype x environment interactions. These results are expected because for each genotype, the disease development depends on the age of the plant, the environmental factors and the aggressivity of the inoculum. Similar genotypes x environment interactions for quantitative resistance to net blotch were obtained for terminal severity, AUDPC and r (Pinnschmidt and Hovmøller, 2002; Robinson and Jalli, 1999; Steffenson and Webster, 1992). These results were explained by the presence of multiple resistance loci in the two parents with environment-specific genetic resistance (Grewal et al., 2008b; Naz.et al., 2008). Earlier study demonstrated that interactions of QTL with environments were significant in the region on chromosome 6H contributing an additional proportion of phenotypic variance for net blotch resistance (Spaner et al., 1998). In addition, the advent of molecular markers has shown that quantitative resistance is due to the segregation of QTLs for pathotype-specific resistance of major effect, along with QTLs of minor effects (Jones et al., 1995).

In a single environment resistance related parameters were found to be associated. These results suggest that quantitative resistance to net blotch is equally described by any criteria. However, results of this investigation indicated that AUDPC would be a valuable tool to select for quantitative resistance to *P. teres* since it describe the pattern of disease across varying growth stages and its magnitude was usually greater than these of the two other parameters. Nevertheless, Steffenson and Webster (1992) found that relationship between the final disease severity and the AUDPC is highly influenced by the environment. In addition, they noted that high values of apparent infection rate could occur sometimes on genotypes with reduced disease severities when there is a rapid increase of net blotch attacks from a low to a moderate level within a short period. Thus, they proposed the AUDPC as the most reliable statistic for assessing quantitative resistance to *P. teres*. However, El Yousfi and Ezzahiri (2001), found that the environmental effect remained significant until the cultivars reached the flowering stage at which the environment have no effect on the linear relationship between disease severity and the AUDPC. It

was thus suggested that, discrimination between genotypes for quantitative resistance is best expressed at this growth stage.

In this investigation, genetic correlations were detected only between MDI evaluated at Mograne 2002-2003 with those evaluated at Mograne 2003 - 2004 and at Tunis 2004. Li et al. (2009) and Mahmoud et al. (2006) explained the significant genetic correlations by genetic linkage and/or pleiotropic effects of loci affecting the considered traits. Thus, the results obtained suggest that loci for MDI may be pleiotropic or linked and loci for AUDPC and r may be independent. Further QTL-based analyses are required to elucidate the genetic correlation between net blotch resistance parameters and to well know whether these correlations were due to linkage among genetic factors or pleiotropy or both.

Results obtained from this research show that net blotch resistance is quantitative and those genotypes x environment interactions were the major sources of variation. In a single environment, resistance related parameters were found to be genetically associated. Excepting mass disease index, lack of associations between area under the disease progress curve and apparent infection rate measured on the three field conditions were observed.

REFERENCES

Burleigh JR, Loubane M (1984). Plot size effects on disease progress and yield of wheat infected by *Mycosphaerella graminicola* and barley infected by *Pyrenophora teres*. Phytopathol. 74: 545-549.

Cakir M, Gupta S, Platz GJ, Ablett GA, Loughman R, Emebiri LC, Poulsen D, Li CD, Lance RCM, Galwey NW, Jones MGK, Appels R (2003). Mapping and validation of the genes for resistance to *Pyrenophora teres* f *teres*. in barley (*Hordeum vulgare* L.). Aust. J. Agric. Res. 54: 1369-1377.

Cherif M, Harrabi M, Morjane H (1994). Distribution and importance of wheat and barley diseases in Tunisia, 1989 to 1991. Rachis 13: 25-34.

Choo TM, Reinbergs E, Kasha KJ (1985). Use of haploids in breeding barley. Plant Breed. Rev. 3: 219-252.

Devaux P, Pickering R (2005). Haploids in the improvement of Graminaceous species. In: Palmer D, Keller W, Kasha K (eds.): Haploids in crop improvement II. Springer, Heidelberg pp. 214-242.

Ding G, Xung L, Oifang G, Pingxi L, Dazaho Y, Ronghai H (1993). Evaluation and screening of faba bean germoplasm in China. Fabis Newsletter 32: 8-10.

Douglas GB, Gordon IL (1985). Quantitative genetics of net blotch resistance in barley. New Z. J. Agric. Res. 28: 157-164.

El Yousfi B, Ezzahiri B (2001). Net blotch in semi arid regions of Morocco. I Epidemiology. Field Crops Res. 73: 35-46.

Grewal TS, Rossnagel BG, Scoles GJ (2008a). The utility of molecular markers for barley net blotch resistance across geographic regions. Crop Sci. 48: 2321-2333.

Grewal TS, Rossnagel BG, Pozniak CJ, Scoles GJ (2008b). Mapping quantitative trait loci associated with barley net blotch resistance. Theor. Appl. Genet. 116: 529-539.

Grewal TS, Rossnagel BG, Scoles GJ (2010). Validation of molecular markers associated with net Blotch resistance and their utilization in barley breeding. Crop Sci. 50: 177-184.

Gupta S, Loughman R, Plantz GJ, Lance RCM (2003). Resistance in cultivated barleys to *Pyrenophora teres* f. *teres*. and prospects of its utilisation in marker identification and breeding. Aust. J. Agric. Res. 54: 1379-1386.

Jones ES, Liu CJ, Gale MD, Hash CT, Witcombe JR (1995). Mapping quantitative trait loci for downy mildew resistance in pearl millet. Theor. Appl. Genet. 91: 448-456.

Li Y, Wang Y, Wei M, Li X, Fu J (2009). QTL identification of grain protein concentration and its genetic correlation with starch concentration and grain weight using two populations in maize (*Zea mays* L.). J. Genet. 88: 61-67.

Mahmoud T, Rahman MH, Stringam GR, Yeh F, Good AG (2006). Identification of quantitative trait loci (QTL) for oil and protein contents and their relationships with other seed quality traits in *Brassica juncea*. Theor. Appl. Genet. 113: 1211-1220.

Marwede V, Schierholt A, Möllers C, Becker HC (2004). Genotype x environment interactions and heritability of tocopherol contents in canola. Crop Sci. 44: 728-731.

Naz AA, Kunert A, Lind V, Pillen K, Léon J (2008). AB-QTL analysis in winter wheat: II. Genetic analysis of seedling and field resistance against leaf rust in a wheat advanced backcross population. Theor. Appl. Genet. 116: 1095-1104.

Onfroy C (1997). Maladies fongiques aériennes des légumineuses alimentaires. Identification, Diagnostic et Techniques de laboratoire. Document technique, INRA, Station de Pathologie Végétale p. 68.

Pinnschmidt HO, Hovmøller MS (2002). Genotype x environment interactions in the expression of net blotch resistance in spring and winter barley varieties. Euphytica 125: 227-243.

Robinson J (1999). Diallel analysis of net blotch resistance in doubled haploid lines of Nordic spring barleys. Euphytica 110: 175-180.

Robinson J, Jalli M (1997). Quantitative resistance to *Pyrenophora teres* in six Nordic spring barley accessions. Euphytica 94: 201-208.

Robinson J, Jalli M (1999). Sensitivity of resistance to net blotch in barley. J. Phytopathol. 147: 235-241.

SAS Institute (1988). SAS / STAT user's guide. Release 6.03 ed. SAS Inst., Cary, NC p. 1028.

Sorkhilalehloo B, Tewari JP, Turkington TK, Capettini F, Briggs KG, Rossnagel B, Singh RP (2002). Genetics of slow-scalding resistance in barley. Abstracts, Alberta Regional Meeting, The Canadian Phytopathological Society, 2002. Can. J. Plant Pathol. 24: 504-507.

Spaner D, Shugar LP, Choo TM, Falak I, Briggs KG, Legge WG, Falk DE, Ullrich SE, Tinker NA, Steffenson BJ, Mather DE (1998). Mapping of disease resistance loci in barley on the basis of visual assessment of naturally occurring symptoms. Crop Sci. 38: 843-850.

Steffenson BJ, Hayes PM, Kleinhofs A (1996). Genetics of seedling and adult plant resistance to net blotch (*Pyrenophora teres* f. *teres*) and spot blotch (*Cochliobolus sativus*) in barley. Theor. Appl. Genet. 92: 552-558.

Steffenson BJ, Webster RK (1992). Quantitative resistance to *Pyrenophora teres* f. *teres* in barley. Phytopathol. 82: 407-411.

Tekauz A (1985). A numerical scale to classify reactions of barley to *Pyrenophora teres*. Can. J. Plant Pathol. 7: 181-183.

Tuohy JM, Jalli M, Cooke BM, Sullivan EO (2006). Pathogenic variation in population of *Drechslera teres* f. *teres* and *D. teres* f. *maculate* and differences in host cultivar responses. Eur. J. Plant Pathol. 116: 177-185.

Vanderplank JE (1963). Plant diseases: Epidemics and control. Academic Press, New York p. 349.

Yahyaoui AH, Ezzahiri B, Hovmoller M, Jahoor A, Wolday A (2003). Field guide for barley and wheat diseases and cereal disease management in Eritrea. ICARDA Aleppo, Syria p. 84.

Zadoks JC, Chang TT, Konzak CF (1974). A decimal code for the growth stages of cereals. Weed Res. 14: 415-421.

13

Genetic and environmental correlations between bean yield and agronomic traits in *Coffea canephora*

Anim-Kwapong Esther* and Boamah Adomako

Cocoa Research Institute of Ghana, P. O. Box 8, New Tafo-Akim, Ghana.

Early identification and selection of genotypes with high yielding potential is a main breeding objective of *Coffea canephora*. Eighteen genotypes of *C. canephora* were assessed in three diverse environments over a 9-years period from 1996 to 2005. Genetic and environmental associations were assessed among 10 vegetative and five reproductive traits and yield. Genetic associations between yields over seven years and vegetative traits, except secondary branches per plant, were positive and significantly correlated with span ($r_G = 0.65^{**}$), girth ($r_G = 0.60^{**}$), diameter ($r_G = 0.55^{*}$) and number of primary branches ($r_G = 0.53^{*}$). The traits exhibited stronger genetic correlations with last 4 - 7 years yields ($r_G = 0.54^{*} - 0.68^{**}$) than with first 1 - 3 years yields ($r_G = 0.38 - 0.47^{*}$). Fruit-set observed in three fruiting seasons, when the trees were three, four and six years in the field, was consistently positive and significantly associated with yields over seven years ($r_G = 0.60^{**}; 0.63^{**}; 0.66^{**}$). However, genetic associations between yields over seven years and flowers per node observed in the three fruiting seasons was consistently negative and significant for two seasons ($r_G = -0.50^{*}; -39; -0.62^{**}$). First 1 - 3 years yield was a better predictor ($r^2_G = 0.79^{***}$) of yields over seven years, than first 1 - 2 years ($r^2_G = 0.42^{**}$) and first year ($r^2_G = 0.012$) yield. Selection for potential high yielding genotypes should, therefore, be based on an index involving span, girth, diameter and number of primary branches, first three years yield, fruit-set, and flowers per node. High positive environmental correlations were observed between bean yields and fruit-set, number of fruits per node, number of flowering and fruiting nodes, girth and number of primary branches. However, environmental conditions that reduced yields also increased flowers per node and promoted vegetative growth by increasing secondary branching, span, length, diameter and number of nodes per primary branch. Efficient selection for yield based on vegetative traits should, therefore, be undertaken under optimum growing conditions where there is a better balance between vegetative growth and yield.

Key words: Vegetative traits, reproductive traits, *Coffea canephora*, genetic correlations, environmental correlations, yields, indirect selection.

INTRODUCTION

Coffee is one of the main agricultural commodities traded worldwide. Its cultivation is mainly by smallholder farmers who hardly breakeven due mainly to low yields, high production cost and low world market prices. Increasing productivity, while reducing the cost of production is a main breeding objective of most producing countries. To meet that objective, research programmes to select high yielding varieties have been initiated in Ghana. Yield improvement of Robusta coffee involves the modification

of the genetic makeup and/or the modification of the growing environment of the crop. While methods for easy identification of genotypes with good morphological characteristics within the early years of cultivation and evaluation are available, identifying high yielding genotypes is often difficult to achieve and time consuming due to the long gestation period, biennial bearing and heterogeneous nature of Robusta coffee, in addition to large environmental component of variance for yield. A selection cycle is possible only after seven to eight years of cultivation. Hence, improving the yield of a population through recurrent selection takes a long time to achieve, while final testing of genotypes in multisite trials becomes

---

*Corresponding author. E-mail: ekwapong06@yahoo.com.

Table 1. Planting material and their sources of origin.

Clones from Cote d` Ivoire[†]	Clones from Togo[††]
A129	197
A115	149
A101	126
B170	181
B96	375
B36	107
B191	
E174	
E138	
E139	
E90	
E152	

[†]Selection based on 10-years yields; [††]Selection based on 3-years yields.

very expensive. Shortening the time required to obtain an accurate evaluation of yield potential would allow for faster release of new varieties at reduced cost, thus making the breeding programme more efficient. There have already been some reports of correlation studies among yield of coffee and juvenile and mature plant characteristics to address this problem (Ravohitarivo, 1980; Ameha, 1982; Bouharmont et al., 1986). However, most of these studies examined phenotypic correlations of total 5 - 10 years production with early yield and vegetative characteristics in *Coffea arabica*. The genetic correlation, which is the proportion of variance that two traits share due to genetic causes, is useful in studying the genetic relationships among traits under selection. Environmental correlations reveal associated changes in two traits caused by environmental and non-genetic factors (Falconer and Mackay, 1996). The few reports available on genetic correlations are mostly on Arabica coffee with genotypes planted at a single location (Walyaro and Van der Vossen, 1979; Walyaro, 1983; Leroy et al., 1994; Cilas et al., 1998, 2006). Generally, information on environmental associations among yield and agronomic traits are hardly used alongside yield data collected over locations and seasons to select high yielding genotypes. The magnitude and direction of these associations should be informative as a basis of indirect selection for yield and for environmentally modifying growing conditions. The objective of this study was, therefore, to estimate the magnitude and direction of genetic and environmental associations among yield and agronomic traits in Robusta coffee.

MATERIALS AND METHODS

Twelve clones obtained by individual selection, based on yield, from a population of three half-sib family groups introduced from Cote d`Ivoire, together with six clones introduced from Togo, were used

for the study (Table 1).

The study was conducted from 1996 to 2005 at three rain-fed sites (Tafo, Fumso and Bechem) within the forest zone of Ghana where coffee is grown, representing a range of soil types, fertility levels and climatic regimes (Table 2).

Single-node cuttings of the clones used were rooted in propagators and nursed in nursery bags for six months. Thirty-two plants of each genotype were randomly assigned to each of the three environments. At each location, the experimental design was a randomized complete block design with four replications. A single row of eight plants of each genotype were planted in each replication with plots measuring 2.44 x 19.51 m. Both inter-row and inter-plant spacings were 2.44 m giving a density of 1680 plants per hectare. Planting in the three environments was done in June 1996. No fertilizer was applied and crop management practices were similar for all locations. In order to assess genetic differences in number of stems produced, the plants were allowed to grow on one or two stems that had developed from the single-node cuttings. Stems were capped at 18 months from field planting by removal of the terminal bud and subsequently capped to 1.8 m and maintained at that height. The first capping resulted in each main stem developing into two branches at the point of capping.

Measurements of vegetative characteristics were taken three months after field planting on plant height, diameter of the main stem (girth) and number of primary branches and repeated each year after field planting until the plants were 48 months in the field, at the stage of maximum expansion. Vegetative measurements taken when the plants were 48 months in the field, was used for this study. Four random plants of each genotype in each plot were used. Traits assessed included girth(mm), taken at 10cm above the ground, crown diameter (span) in cm, taken as the length of the canopy measured at the widest portion of the tree canopy, number of stems, total number of primary branches counted per plant and per stem, and total number of secondary branches counted per plant. Length of primary branches (measured from the point of attachment to the main stem to the apex in cm), diameter of primary branches (10 cm from the main stem in mm) and number of nodes per primary branch were estimated as an average value of the six longest branches at the middle of the stem per plant. Internode length per branch was estimated as an average of length of primary branch divided by the number of nodes of primary branch for the six primary branches. Where there were more than one stem, stem diameter was calculated according to Stewart and Salazar (1992), and span was taken for only the biggest stem.

Table 2. Soil and climatic characteristics of experimental sites.

Characteristic	Site of experiment		
	Tafo	Fumso	Bechem
Soil type	Sandy loam	Coarse sandy to fine gravel	Humus fine sandy loam to fine clay
Altitude(above sea level)(m)	220	122	259
Latitude	6° 13'N	6° 6'N	7° 5'
Longitude	0° 22'W	1° 27'W	2° 2'
Mean total annual rainfall from 1996 - 2002 (mm)	1480	1320	1220
Mean monthly rainfall during flowering and fruit-setting period (Dec.- Feb.) from 1996 - 2002	44	29	17
Mean annual monthly raindays from 1996 - 2002	9.4	8.4	7.9
Mean monthly raindays during flowering and fruit-setting period (Dec.- Feb.) from 1996 - 2002	3.6	2.1	1.8
Mean annual average daily air temperature(°C) from 1996 - 2002	26.8	27.0	26.2
Mean monthly average daily air temperature(°C) (Dec.- to Feb.) from 1996 - 2002	27.1	27.2	26.2
Mean annual daily humidity at 1500 h (%) from 1996 - 2002	67.1	62.9	62.4
Mean daily humidity at 1500 h (%) (Dec.- to Feb.) from 1996 - 2002	56.7	56.0	48.3

Table 3. Form and generalized expectations of analysis of variance and covariance for two characters X and Y.

Source of variation	Degrees of freedom	Mean square	Expectation of mean square	Expectation of mean cross-produ
Replications/locations	(r-1)L			
Locations (L)	L-1	M_l	$\sigma^2_e + r\sigma^2_{gl} + rN(\sigma^2_E)$	$\sigma_{e(XY)} + r\sigma_{gl(XY)} + rN(\sigma_{E(XY)})$
Genotypes (G)	N-1	M_g	$\sigma^2_e + r\sigma^2_{gl} + rL(\sigma^2_g)$	$\sigma_{e(XY)} + r\sigma_{gl(XY)} + rL(\sigma_{g(XY)})$
G x L	(N-1) (L-1)	M_i	$\sigma^2_e + r\sigma^2_{gl}$	$\sigma_{e(XY)} + r\sigma_{gl(XY)}$
Error	(N-1) (r-1)L	M_e	σ^2_e	$\sigma_{e(XY)}$

r = number of replications = 4; L = number of locations = 3; N = number of genotypes = 18.

At flowering and fruiting time in December, 1998 to May, 1999, two plants from each plot, were randomly tagged. Three flowering primary branches at the middle of each plant were tagged for the determination of the number of flowering nodes per branch and number of flowers per node. Fruits that remained on the branches at six months from initial flowering were counted and used in estimating the number of fruiting nodes and fruits per node. Percent fruit-set (fruit-set) was estimated as the proportion of total flowers counted on the three flowering branches per tree that set fruit and remained on the branches at six months from flowering. To study genetic association among yields and reproductive traits as a factor of the age of the plants, the reproductive traits were further assessed in two more seasons: December, 1999 to May, 2000, and December, 2001 to May, 2002 using three trees per plot for each clone. The latter assessment was however at two locations, Tafo and Bechem.

Clean coffee yields were recorded on each tree for seven production years from October to January each year for the period 1998/1999 to 2004/2005. Transformation of cherry weight to clean coffee weight was done using the conversion factor 0.22 (Coste, 1992).

Analyses of variance and covariance were performed using MINITAB statistical software (MINITAB, 1997). The statistical model used for the combined analyses was:

$$Y_{ijk} = \mu + g_i + e_j + (ge)_{ij} + R_{jk} + \varepsilon_{ijk}$$

Where Y_{ijk} is the kth observation of any variable in the rth replications in environment j on genotype i; μ is the general mean; g_i and e_j represent the effects of the ith genotype and the jth environment; $(ge)_{ij}$ is the interaction effect between the genotypes and the environment; R_{jk} is the effect of the k-th replication within the j-th location, ε_{ijk} is the random error associated with the kth observation on genotype i in environment j. i = 18; j = 3; r = 4. The effects g_i's, e_j's, $(ge)_{ij}$'s and ε_{ijk}'s are assumed independently and randomly distributed with zero means and variances σ^2_g, σ^2_l, σ^2_{gl} and σ^2_e, respectively.

The form of the analysis of variance and covariance with expectations of mean squares and cross products is presented in Table 3.

Genotypic correlations (r_G) and environmental (r_E) correlations between traits were computed as:

$$r_G = cov_{g(XY)} / \sqrt{\sigma^2_{g(X)} \; \sigma^2_{g(Y)}} \; ; \quad r_E = cov_{E(XY)} / \sqrt{\sigma^2_{E(X)} \; \sigma^2_{E(Y)}}$$

Table 4. Genetic and environmental correlations between mean yields and vegetative traits (at 48 months) estimated in the combined analysis of variance and covariance for three locations in 18 Robusta coffee genotypes.

Trait	First three years yields	Last four to seven years yields	Yields over seven years
Girth	0.38	0.63**	0.60**
	0.26	0.67	0.49
Span	0.47*	0.68**	0.65**
	-0.98	-0.80	-0.91
Length of primary branches	0.25	0.47*	0.42
	-0.85	-0.53	-0.69
Diameter of primary branches	0.46*	0.54*	0.55*
	-0.99*	-0.89	-0.96
Number of nodes /primary branch	0.23	0.16	0.18
	-0.99*	-0.86	-0.95
Inter-node length / primary branch	0.01	0.26	0.20
	-0.56	-0.14	-0.34
Number of primary branches /plant	0.43	0.55*	0.53*
	0.56	0.87	0.75
Number of secondary branches /plant	-0.34	-0.25	-0.29
	-0.93	-0.99*	-0.99*
Number of stems / plant	0.24	0.41	0.38
	-0.97	-0.75	-0.87
Number of primary branches /stem	0.10	0.03	-0.07
	0.65	0.92	0.62

* = p ≤ 0.05, ** = p ≤ 0.01; genetic correlations (upper values); environmental correlations (lower values).

Where $cov_{g(xy)}$ was the estimated genotypic covariance component for traits X and Y, $\sigma^2_{g(X)}$ and $\sigma^2_{g(Y)}$ the genotypic variance component for traits X and Y respectively; $cov_{E(xy)}$ was the estimated macro-environmental (location) covariance component for traits X and Y , $\sigma^2_{E(X)}$ and $\sigma^2_{E(Y)}$ the macro-environmental variance component for traits X and Y respectively (Falconer and Mackay, 1996). Significance test for the correlations was by standard procedure (Steel et al., 1997).

RESULTS

Genetic correlations

Genetic correlation coefficients between yield and vegetative traits from the combined data across the three locations are shown in Table 4. Bean yields over seven years showed significant positive genetic correlation (r_G =0.53*- 0.65**) with girth, span, diameter and number of primary branches. These traits exhibited stronger correlation with last 4 - 7 years yields (r_G =0.54*- 0.68**) than with first 1 - 3 years yields (r_G =0.38 - 0.47*). An inverse relationship between yields and number of secondary branches was observed. Associations among the

vegetative traits showed significant (r_G =0.67**- 0.77***) genetic correlations among span, girth and diameter of primary branches (Table 5). Number of secondary branches per plant was inversely associated with traits that showed significant genetic correlations with yield namely: girth, span, diameter and number of primary branches. However, only the association with span was significant (r_G = -0.48*).

Genetic correlations between yields and the reproductive traits show that in 1998/1999 alone, when the trees were only two and half to three and half years old, genetic correlations between first two years yields and fruit-set and fruits per node had already coefficients of 0.66 and 0.68 respectively (not shown). Genetic correlation coefficients between bean yield and the reproductive traits in 1998/1999, 1999/2000 and 2001/ 2002 fruiting seasons from the combined data across the three locations are shown in Table 6. All associations in 1998/1999 were positive, except with flowers per node. Fruit-set was significantly (r_G = 0.59**- 0.65**) associated with first 1-3 year yields, last 4-7 year yields and yields over seven years. Number of fruits per node and fruiting nodes, however, significantly (r_G = 0.48*- 0.49*) correlate positively with early yields but significant correlations with late yields

Table 5. Genetic and environmental correlations among vegetative traits (at 48 months) estimated in the combined analysis of variance and covariance for three locations in 18 genotypes of Robusta coffee.

Trait	1	2	3	4	5	6	7	8	9
1. Girth	-								
2. Span	0.67**	-							
	-0.08								
3. Length of primary branches	0.32	0.78***	-						
	0.29	0.93							
4. Diameter of primary branches	0.70***	0.77***	0.38	-					
	-0.24	0.98	0.86						
5. Number of nodes /primary branch	-0.05	0.35	0.39	0.24	-				
	-0.17	0.99*	0.89	0.92					
6. Internode length / primary branch	0.32	0.39	0.56*	0.13	-0.55*	-			
	0.81	0.70	0.91	0.78	0.62				
7. Number of primary branches per plant	0.39	0.13	0.11	0.02	-0.10	0.16	-		
	0.95	-0.40	-0.04	-0.54	-0.49	0.37			
8. Number of secondary branches / plant	-0.32	-0.48*	-0.39	-0.30	-0.43	0.04	-0.02	-	
	-0.59	0.85	0.60	0.93	0.90	0.22	-0.82		
9. Number of stems/ plant	0.36	0.00	-0.11	0.08	-0.41	0.26	0.67**	0.18	-
	0.00	1.00***	0.96	1.00***	1.00***	0.86	-0.36	0.81	
10. Number of primary branches / stem	-0.05	0.05	0.17	-0.15	0.28	-0.11	0.22	0.12	-0.55*
	0.90	-0.50	-0.15	-0.63	-0.58	0.27	0.99*	-0.88	-0.43

* = $p \leq 0.05$, ** = $p \leq 0.01$, *** = $p \leq 0.001$; genetic correlations (upper values); environmental correlations (lower values).

and yields over seven years were not observed. Significant inverse associations were observed between number of flowers per node and late yield as well as yields over seven years (r_G = -0.54* and -0.50* respectively). Similar trend were observed between the traits recorded in 1999/2000 and 2001/2002 fruiting seasons (Table 6). Genetic correlations among the reproductive traits in two seasons (Table 7) showed significant genetic association between fruit-set and fruits per node (r_G = 0.70*** - 0.77**) and between fruiting and flowering nodes (r_G = 0.76*** - 0.85***).

Genetic correlation coefficients between overall seven year yields and the first year, first 1 - 2, 1 - 3, and last 4 - 7 years yields were positive and highly significant (r_G = 0.89*** - 0.99***) with first 1 - 3 years and last 4 - 7 years yields (Table 8).

Environmental correlations

Environmental correlations between coffee bean yields and the vegetative traits observed (Table 4) showed that, environmental conditions contributing towards low coffee berry yields favoured span, length and inter-node length of primary branches, and number of stems per plant and significantly (r_E = 0.99*) increased diameter and number of nodes and the formation of more secondary branches. Environmental correlations among the vegetative traits (Table 5) were mostly positive. However, inverse associations were observed between number of primary branches per plant/ per stem and the traits span, length and diameter of primary branches, number of nodes per

primary branch and number of secondary branches; and between girth and span, diameter of primary branches, number of nodes per primary branch and number of secondary branches. Significant positive environmental associations (r_E = 1.00**) were observed between number of stems per plant and the traits span, diameter of primary branches and number of nodes per primary branch.

Environmental conditions which favored fruit-set, number of fruits per node, fruiting and flowering nodes in 1998/1999 and 1999/2000 fruiting seasons had positive effects on all the yield traits (Table 6). However, environmental conditions that promoted the formation of more flowers per node in both fruiting seasons resulted in low yields. Environmental correlations among the traits (Table 7) showed that, fruit-set, fruits per node, fruiting and flowering nodes were environmentally positively associated but negatively associated with flowers per node.

Environmental associations among the yield traits (Table 8) were all positive. However, only environmental conditions that determined first two years yields significantly (r_E = 0.1***) determined overall seven year yields. Good environmental growing conditions must therefore be maintained at the early growth stages of the plants to promote early bearing and subsequent higher yields at the matured plant stage.

DISCUSSION

Knowledge of correlations among characters is useful in

Table 6. Genetic and environmental correlations between mean yields and reproductive traits (in three production seasons) estimated in the combined analysis of variance and covariance for three locations in 18 Robusta coffee genotypes.

Trait	Year	First three years yields	Last four to seven years yields	Yields over seven years
Fruit-set	1998/1999	0.65**	0.59**	0.63**
		0.97	0.97	0.99*
Number of fruits/node		0.48*	0.21	0.30
		0.92	0.61	0.80
Number of fruiting nodes		0.49*	0.17	0.23
		0.99*	0.91	0.99*
Number of flowers/node		-0.29	-0.54*	-0.50*
		-0.70	-0.96	-0.85
Number of flowering nodes		0.19	-0.04	-0.02
		0.99*	0.83	0.95
Fruit-set	1999/2000	0.75***	0.46*	0.60**
		0.99*	0.96	0.99*
Number of fruits/node		0.56*	0.15	0.29
		0.99*	0.92	0.97
Number of fruiting nodes		0.51*	0.17	0.29
		0.99*	0.89	0.95
Number of flowers/node		-0.20	-0.42	-0.39
		-0.96	-0.99*	-0.99*
Number of flowering nodes		0.02	-0.05	-0.05
		0.60	0.26	0.42
Fruit-set	2001/2002	0.30	0.73***	0.66**
Number of fruits/node		0.10	0.37	0.29
Number of flowers /node		-0.27	-0.63**	-0.62**

* = p ≤ 0.05, ** = p ≤ 0.01, *** = p ≤ 0.001; genetic correlations (upper values); environmental correlations (lower values). Values in 2001/2002 represent genetic correlations at Tafo and Bechem only.

Table 7. Genetic and environmental correlations among reproductive traits (in two production seasons) estimated in the combined analysis of variance and covariance for three locations in 18 Robusta coffee genotypes.

Trait	1	2	3	4	5
1. Fruit-set		0.70***	0.35	-0.44	0.12
		0.80	0.99*	-0.86	0.95
2. Number of fruits/ node	0.75***		0.10	0.33	-0.38
	0.99*		0.89	-0.37	0.95
3. Number of fruiting nodes	0.58**	0.54*		-0.36	0.85***
	0.98	0.99*		-0.76	0.99*
4. Number of flowers/node	-0.25	0.41	-0.08		-0.36
	-0.98	-0.95	-0.92		-0.65
5.Number of flowering nodes	0.13	0.21	0.76***	-0.07	
	0.53	0.45	0.67	-0.33	

* = p ≤ 0.05, ** = p ≤ 0.01, *** = p ≤ 0.001; 1998/1999 correlations (upper triangle); 1999/2000 correlations (lower triangle); genetic correlations (upper values); environmental correlations (lower values).

determining the success of indirect selection of one trait for the other in an improvement programme for any crop (Falconer and Mackay, 1996). Indirect selection of one trait for the other depends on many factors. Selection for yield in a perennial crop like coffee should be advantageous with traits that are determined at the juvenile or

Table 8. Genetic and environmental correlations among yield traits estimated in the combined analysis of variance and covariance for three locations in 18 Robusta coffee genotypes.

Trait	1	2	3	4	5
1. Yields over seven years	-				
2. Last four to seven years yield	0.99[***]	-			
	0.98				
3. First year yield	0.11	0.03	-		
	0.92	0.97			
4. First two years yield	0.65[**]	0.75[***]	0.43	-	
	1.00[***]	0.98	0.94		
5. First three years yield	0.89[***]	0.81[***]	0.31	0.79[***]	-
	0.97	0.90	0.83	0.98	

** = $p \leq 0.01$, *** = $p \leq 0.001$; genetic correlations (upper values); environmental correlations (lower values).

early reproductive stages of the plant. The significant positive genetic associations between bean yields and girth, span, diameter and number of primary branches, as well as the significant genetic associations observed among span, girth and diameter of primary branches suggest the possibility of increasing yields at both the juvenile and adult plant stages through the selection of vigorous young plants with thick and strong main stems and primary branches as well as wider span, characteristic of genotypes with erect primary branches. The significant associations observed in this study agrees with previous findings that, Robusta coffee yields (Bouharmont et al., 1986; Leroy et al., 1994; Cilas et al., 2006), as well as Arabica coffee yields (Walyaro and Van der Vossen, 1979; Walyaro, 1983; Cilas et al., 1998) are positively correlated with young plant vigour. Coffee trees were four years old and in their second fruiting season when the traits were observed. It is likely that the positive effects of good vegetative growth of stems on yields are due to the availability of reserves in such stems which provide yield enhancing assimilates. Stem reserves have been shown to play an important role as source of assimilate for fruit and grain yields in other plant studies (Blum et al., 1997; Ehdaie and Waines, 2006).

The positive genetic correlations and negative environmental associations between yields and span, length, diameter, nodes and inter-node length of primaries and number of stems per plant, show that environmental conditions that resulted in low yields contributed towards increased vegetative growth. Indirect selection of vegetative traits for yield in any breeding programme should, therefore, be undertaken under optimum conditions where there is a better balance between vegetative growth and yield.

The production of more secondary branches could be expected to result in more flowering and fruiting nodes and more fruits. However, the inverse genetic association between secondary branching and yields as well as the vegetative traits associated with yields (span, girth, diameter and number of primaries) and the significant

inverse environmental association between secondary branching and yields clearly indicate that secondary branching in coffee does not translate into high yields. Selecting plants with low secondary branching should therefore be advantageous for yields in coffee. Reducing secondary growth by pruning could also improve coffee yields significantly.

Having established that fruit-set and fruits per node of coffee trees were associated with yields, genetic associations further studied in 1999/2000 and 2001/2002 production seasons (when the second and fourth yields were recorded) established that fruit-set was in fact the most important reproductive trait that determined coffee yields at both the early and late reproductive stages of the coffee plant. Studies on genetic correlations of Robusta and Arabica coffee yields and agronomic traits have been undertaken before (Walyaro and Van der Vossen, 1979; Leroy et al., 1994; Cilas et al., 1998, 2006). In these previous reports, however, association between reproductive characters and yields were not matched with data on variation in age of plants. In associations examined in the present study, the influence of the age of the plants on how fruits per node, fruit-set and flowers per node affected yield was thus isolated.

The inverse genetic associations between number of flowers per node and fruit-set and yields imply that, comparatively more flowers were produced at the nodes by low yielding plants than by high yielding ones. Walyaro and Van der Vossen (1979), working on Arabica coffee observed similar negative genetic association of number of flowers per node with fruit-set and yields. Environmental conditions that promoted the production of more flowers per node also resulted in poor fruit-set and low yields as observed with secondary branching. In coffee flower buds borne in the axils of leaves on primary branches have the potential of either developing into new shoots or inflorescence. According to Wrigley (1988), more of the buds would produce shoots depending on the strength of the stimuli to vegetative growth. It is therefore likely that, any genetic or environmental condition that leads to the initiation of more auxiliary buds will also result in the formation of more flowers and/or secondary branches. For coffee, drought was observed to be an important stimulus for flower bud initiation as well as flowering (DaMatta and Ramalho, 2006). Evidence from comparative studies has shown that, drought-sensitive plants maintain lower internal water status than drought-tolerant ones under water-deficit conditions (Meinzer et al., 1990; Pinheiro et al., 2005). Drought-sensitive plants are, therefore, more likely to have higher stimulus for flower bud initiation, hence, maintaining higher flowering and secondary branching than drought-tolerant plants. It also implies that, flower bud initiation and therefore flowering and secondary branching would be higher in drier environments than under optimum conditions, hence, the genetic and environmental associations observed in this study between number of flowers per node, number of secondary branches per tree, fruit-set and yields.

The significant positive environmental associations between yields and fruit-set and fruiting nodes confirm these observations and highlight the importance of the environment in determining yields. The three environments used for the study range from high to low rainfall areas, varying seasonally in rainfall amount and distribution, most especially, during the flowering and fruit setting periods. Precipitation during fruit-set and fruit development was found to affect yields in coffee (Barros et al., 1999). Fruit-set and fruiting nodes on primary branches and hence yields were higher in Tafo, a high rainfall environment, than at Bechem, a low rainfall environment (data not shown). It is therefore important that coffee is grown in areas where there is enough rainfall during the fruit-setting and fruit expansion periods. In areas where rainfall is below optimum, coffee plants could benefit from irrigation, especially, during the fruit setting period. Cannell (1973) observed that the number of flowering and fruiting nodes on which flowers and fruits are borne increased by the application of fertilizers, especially nitrogen, as well as by mulching and irrigation. Adequate nitrogen supply had also been suggested for improved tolerance to drought (Ramalho et al., 1999, 2000). Fruit-set, fruits per node and hence yields should therefore improve with the application of fertilizers. The optimum environment for Robusta coffee cultivation must therefore be where rainfall is well distributed without a prolonged dryness, especially, after flowering to facilitate fruit setting; coupled with regular pruning to reduce secondary branches, and observance of good cultural practices for soil fertility improvement.

Conclusion

The high genetic correlations among fruit-set, first three years yields, and yields over seven years and between yields and the vegetative traits indicate that, selection for

high yielding genotypes is feasible when coffee plants are five years in the field based on fruit-set, first three years yields, span, girth, diameter and number of primary branches.

The main vegetative trait that affected yield environmentally, namely, secondary branching, was determined at the juvenile or early fruiting stages of the plant. Four reproductive traits, namely number of flowering nodes, flowers per node, fruiting nodes and fruit-set that affected yield environmentally were also determined at the early fruiting stages. With the environmental associations observed, it is possible to increase yield through simple agronomic practices. Such practices must reduce secondary branching and increase the diameter of the main stems, the number of primary branches, the number of flowering and fruiting nodes, fruits per node and fruit-set.

In this study, the genotypic and environmental variances and covariances of the traits were estimated from a population of 18 preselected genotypes and could be specific to the studied population. Nonetheless, these results should form an important basis for a pre-selection index for high yielding plants and for environmentally modifying growing conditions for increased coffee productivity.

ACKNOWLEDGEMENTS

This publication, Paper No. CRIG/09/2010/029/001 is published with the kind permission of the Executive Director, CRIG.

REFERENCES

Ameha M (1982). Heterosis in crosses of indigenous coffee (*Coffea arabica* L.) selected for yield and resistance to coffee berry disease. I. At first bearing stage. Ethiopian J. Agric. Sci. 4: 33-43.

Barros RS, Maestri M, Rena AB (1999). Physiology of growth and production of the coffee tree – a review. J. Coffee Res. 27: 1-54.

Bouharmont P, Lotodé R, Awemo J, Castaing X (1986). La selection generative du caféier Robusta au Cameroun. Analyse des resultants d´un essai d´hybrides diallèle partiel implanté en 1973. Café Cacao Thé 30(2): 93-112.

Blum A, Golan G, Mayer J, Sinmena B (1997). The effect of dwarfing genes on sorghum grain filling from remobilized stem reserves under stress. Field Crops Res. 52: 43-54.

Cannell MGR (1973). Effects of irrigation, mulch, and N fertilizers on yield components of Arabica coffee in Kenya. Exper. Agric. 9: 225-232.

Cilas C, Bouharmont P, Boccara M, Eskes AB, Baradat P (1998). Prediction of genetic value for coffee production in *Coffea arabica* from a half-diallele with lines and hybrids. Euphytica 104: 49- 59.

Cilas C, Bar-hen A, Montagnon C, Godon C (2006). Definition of architectural ideotypes for good yield capacity in *Coffea canephora*. Annals of Botany 97(3): 405-411.

Coste R (1992). Coffee, the plant and the product. 2nd ed. The Macmillan Press Ltd. London and Basingstone p. 206.

DaMatta FM, Ramalho JDC (2006). Impacts of drought and temperature stress on coffee physiology and production: a review. Brazilian J. Plant Physiol. 18(1): 55- 81.

Ehdaie B, Waines JG (1996). The genetic variation for contribution of preanthesis assimilates to grain yield in spring wheat. J. Genet. Breed. 50(1): 47-55.

Falconer DS, Mackay TFC (1996). Introduction to quantitative genetics. 4th ed. Longman Inc., London p. 464.

Leroy T, Montagnon C, Cilas C, Charrier A, Eskes AB (1994). Reciprocal recurrent selection applied to *Coffea canephora* Pierre. II. Estimation of genetic parameters. Euphytica 74: 121-128.

Meinzer FC, Grantz DA, Goldstein G, Saliendra NZ (1990). Leaf water relations and maintenance of gas exchange in coffee cultivars grown in drying soil. Plant Physiol. 94: 1781-1787.

MINITAB (1997). Minitab Statistical Software, Release 12, Minitab Inc, USA.

Pinheiro HA, DaMatta FM, Chaves ARM, Loureiro ME, Ducatti C (2005). Drought tolerance is associated with rooting depth and stomatal control of water use in clones of *Coffea canephora*. Annals of Botany 96(1): 101-108.

Ramalho JC, Campos PS, Quartin VL, Silva MJ, Nunes MA (1999). High irradiance impairments on photosynthesic electron transport, ribulose-1, 5-bisphosphate carboxylase/oxygenase and N assimilation as a function of N availability in *Coffea arabica* L. plants. J. Plant Physiol. 154: 319-326.

Ramalho JC, Pons TL, Groeneveld HW, Azinheira HG, Nunes MA (2000). Photosynthetic acclimation to high light conditions in mature leaves of *Coffea arabica* L.: Role of xanthophylls, quenching mechanisms and nitrogen nutrition. Australian J. Plant Physiol. 27: 43- 51.

Ravohitarivo CP (1980). Etude de la variabilité des descendances et des problémes liés à l'amelioration des Caféiers cultivés diploids. Thèse de doctorat de 3 cycle, université de Madagascar p. 105.

Steel RGD, Torrie JH, Dickey DA (1997). Principles and procedures of statistics: a biometrical approach (3rd edition). The McGraw-Hill Co., Inc., New York, USA p. 666.

Stewart JL, Salazar R (1992). A review of measurement options for multipurpose trees. Agroforestry Systems 19: 173-183.

Walyaro DJ (1983). Considerations in breeding for improved yield and quality in Arabica coffee (*Coffea arabica* L.). Doctorial thesis, University of Agriculture, Wageningen, The Netherlands p. 121.

Walyaro DJ, Van der Vossen HAM (1979). Early determination of yield potential in Arabica coffee by applying index selection. Euphytica 28(2): 465-472.

Wrigley G (1988). Coffee. Tropical Agriculture Series. Longman Scientific and Technical, Singapore p. 639.

14

Morphological distinctiveness and metroglyph analysis of fifty accessions of West African okra (*Abelmoschus caillei*) (*A.* Chev.) Stevels

Sunday E. Aladele

National Centre for Genetic Resources and Biotechnology (NACGRAB), P. M. B. 5382, Moor Plantation, Ibadan, Nigeria. E-mail: sundayaladele@yahoo.com.

The morphological uniqueness and metroglyph analysis of 50 accessions of West African okra (Abelmoschus caillei (A Chev.) Stevels) were assessed under three agro-ecological environments at Abeokuta, Ibadan and Mokwa in Nigeria. They were grown in a Randomized Complete Block Design with three replications; data were collected on 5 randomly selected plants from each plot. Data on twenty-one agronomic characters were collected; eleven traits were subjected to metroglyph analysis to investigate the extent of distinctiveness among the 50 accessions. The eleven agronomic characters representing the most commonly used in distinguishing between okra genotypes were used to construct a metroglyph chart. All accessions were grouped into seven distinct groups. The index scores ranged from 19.0 to 42.3 with groups I and VII showing the two extremes among the 50 accessions evaluated. The fruit surface of Akure-2-2 and Akure-1-1 were smooth while Akure-2-9 and Akure-2-4 showed slightly prickly fruits. This suggests that accessions from Akure town in Ondo state of Nigeria were domesticated to large extent and possibly are more related. The metroglyph showed that accessions with very poor yield had short plant height, an indication that for West African genotype okra to produce substantial yield, it must possess a strong stem, hence early planting is encouraged.

Key words: *Abelmoschus caillei*, accessions, genotype, index score, metroglyph.

INTRODUCTION

The value of a germplasm collection depends not only on the number of accessions it contains, but also upon the diversity present in those accessions (Ren et al., 1995). Knowledge of genetic diversity and relationships among okra (*Abelmoschus caillei*) germplasm may play significant role in breeding programmes for biotic and abiotic stress of okra. Within species variation among 30 West African genotypes were found to be considerably large based on phenotypic assessment (Ariyo, 1993). Nigeria, being the second largest producer of okra may have considerable level of genetic diversity as in many other important crop species (Gulsen et al., 2007). Understanding the genetic structure and germplasm diversity of okra being kept in the gene banks all over the world will bring valuable information for okra breeding programmes. *A. caillei* is commonly grown in the high rainfall zone of West Africa and mainly in subsistence systems (Schipper, 1998). High degree of morphological variation has been reported in the previous studies on West African okra (Omonhinmin and Osawaru, 2005). Many local cultivars occur in Africa and Asia, they differ from each other in growth habit (branching or non-branching, large or dwarf, late or early, hairy or glaborous, light green, dark green or red) and fruit characteristics (upright or pendulous, slender or wide, with 5 to 10 ridges). Pod shapes range from round to ridged and short to long. The plant and pods may have small spines on them that create allergies in some people (Splittstoessor, 1990).

Characterization and quantification of genetic diversity has long been a major goal in evolutionary biology. Information on the genetic diversity within and among closely related crop varieties is essential for a rational use of plant genetic resources. Bisht et al., (1995) reported that pigmentation and pubescence of stem, leaf, pods and seeds were important components of variability in okra germplasm. From the results he obtained from Principal

Component Analysis, it was clear that; days to flowering, plant height and various pod characteristics were important components of variability among genotypes. Plant breeders cannot develop a new variety without the use of genetic material with some levels of variation. The role of genetic resources in the improvement and development of cultivated plants has been well recognized (Tigerstedt, 1994). This is why collection, evaluation and storage of germplasm become the most important steps in a plant breeding. Characterization of the collected germplasm is indispensable in a plant breeding programme. It helps breeders in selecting suitable parents for crossing experiments to develop new varieties (Hartwig, 1972; Frankel, 1976).

The objectives of this study are to assess the distinctiveness of 50 accessions of West African Okra (A. caillei) using morphological data and to establish the level of uniformity among the similar accessions on the basis of metroglyph analysis.

MATERIALS AND METHODS

All the collections made were planted on a single row-plot per pod to determine the photosensitivity and phenotypic similarities and differences. Planting was done in June 2004 at National Centre for Genetic Resources and Biotechnology (NACGRAB) research field, Ibadan, Nigeria. Visual field assessment was done before, during and after flowering to choose the test entries for the diversity studies. Entries that flowers before September were rejected while those with diverse traits were selected among the photosensitive and late maturity types. Characters considered include stem colour, leaf colour, stem and fruit pubescence, branching habits and fruit shape. Some entries that showed segregation were rejected on the assumption that they might have been crossed with Abelmoschus esculentus. During flowering, waterproof paper covering was employed to cover the flower at the point of bloom to avoid cross-pollination among accessions. Rouging was carried out on all suspected off types on each row. The trials were located at three environments to assess the performance of the 50 genotypes at different ecological zones. These locations were:

1. Ibadan – Oyo state - NACGRAB Research field
2. Abeokuta – Ogun state- University of Agriculture Research field.
3. Mokwa – Niger state –National Cereals Research Institute satellite research station.

The experiment was designed as Randomized Complete Block Design (RCBD). Three locations, 3 replications of 2 rows per plot at 4.2 m long and with between rows interval of 1.0 m and within row interval of 0.6 m was followed. The experiment was conducted for two years during 2005 and 2006 rainy season. All the necessary cultural practices such as thinning, weeding, fertilizer applications and spraying of insecticides were applied as at when due. Thinning was done to reduce the trial to one plant per stand at the three locations. Compound fertilizer of NPK was applied in two doses at the rate of 60 kg N/ha. Morphological data were collected on 5 randomly selected plants based on International Board for Plant Genetic Resources (IBPGR) recommended descriptor procedures for okra (Palve et al., 1986).

Morphological data collected

1. Stem pubescence (1- glabrous, 2-slightly prickly, 3-conspicous)

2. Stem colour (1-green, 2-green with red, 3-yellowish-green, 4-purple)
3. Fruit pubescence (1-downy, 2-slightly prickly, 3-pricky)
4. Fruit colour (1-green, 2-dark green, 3-yellowish green)
5. Leave length (cm)
6. Leave width (cm)
7. Days to 50% flowering
8. Number of fruits per plant
9. Number of seed per fruit
10. 100 seed weight (g)
11. Fruit yield per plant
12. Stem diseases (1- Resistant, 2- Moderately Resistant, 3-Suscepible)
13. Fruit diseases (1- Resistant, 2- Moderately Resistant, 3-Suscepible)
14. Leaf diseases (1- Resistant, 2- Moderately Resistant, 3-Suscepible)
15. Peduncles length (cm)
16. Plant height (From the base of the plant to the tip of the last leaf)
17. Fruit positioning (3-Erect, 5 Horizontal, 7-pendulous.)
18. Internodes length (cm)
19. Stem diameter (cm)
20. Fruit length (cm)
21. Fruit width (cm)

Morphological variation among 50 accessions were studied using metroglyph and index score was used to determine the distinctiveness of each accession using the mean values of nine agronomic characters (Ariyo, 1988). Class intervals were used to divide all the 50 accessions into 5 groups. The symbol "@" was used as the baseline to represent score-1 for all the nine characters considered. Shading progressively was used to differentiate score interval for days to 50% flowering, while the direction of rays was made unique for individual character for the rest traits. Pod yield per plant and height at maturity, which were the two most variable characters, were employed in locating the glyph for individual accession. The X coordinate, being pod yield per plant and Y coordinate, the plant height. Other agronomic characters were represented on the glyph by shading or by rays at different positions depending on the glyph. The rays may be long or short depending on the value of the index score.

RESULTS

The morphological evaluation of the 50 accessions of West African okra (A. caillei) is presented in Table 1. It shows that twenty-eight accessions had their fruits in horizontal position while only 6 accessions had their fruits positioned in erect form. There were 15 accessions with fruit positioning that fall between erect and horizontal; these are called semi-erect and only Akure 2-9 had a pendulum fruit position. The stem colour ranged from light purple to dark red and from green with some traces of purple to deep green. There were more purple stems within the 50 accessions than green stem colours. The fruit colours of the majority of the accessions were green but with some yellow and red colour patches. The greenish colour ranged from light green to very dark green. In some few cases the green colour was almost suppressed by yellow colour as in CEN 007, ABC-1, NCRI-5, CEN05, OJAOBA-1, Ojaoba-2 and Ado Ekiti-2. Most of the 50 accessions had glaborous stems. Those with slightly pric-

Table 1. Morphological characteristics of 50 accessions of West Africa okra.

Accession name	Stem colour	Stem pubescence	Fruit colour	Fruit pubescence	Fruit position
CEN 010	LP	Glabrous	DG	Downy	Erect
NGAE – 96-002	Purple	Glabrous	DG	SP	Horizontal
NGAE – 96-012 – 1	LP	Glabrous	Green	Downy	Horizontal
NGAE – 96-012 – 2	Purple	Conspicous	L G	Pricky	SE
NGAE – 96-012 – 3	Green	Glabrous	G w R	Downy	SE
CEN 016	Purple	SP	DG	Pricky	SE
CEN 012	Purple	Conspicous	Green	Pricky	Horizontal
CEN 007	Green	Glabrous	Y w G	Downy	SE
NGAE – 96-04	Green	Glabrous	Green	SP	SE
CEN 015	LP	Glabrous	D G	Downy	SE
OAA 96/175-5328	Green	SP	D G	Pricky	Horizontal
AGA97/066-5780	Purple	Glabrous	D G	Downy	Horizontal
ADO-EKITI-1	Green	Glabrous	D G	Pricky	Horizontal
CEN 001	Purple	Glabrous	G w Y	Pricky	Horizontal
CEN 009	LP	Glabrous	Green	Downy	Horizontal
NGAE – 96-0062 -1	LP	SP	G w R	Pricky	Erect
NGAE – 96-0062 – 2	Purple	Glabrous	DG	SP	SE
NGAE – 96-0066	Purple	Glabrous	Green	SP	Horizontal
NGAE – 96-0061	LP	Glabrous	G w R	SP	SE
NGAE – 96-0060	Purple	Glabrous	Green	Downy	Horizontal
NGAE – 96-0067	G w P	Glabrous	G w R	SP	Horizontal
NGAE – 96-0064	P w G	SP	G w R	Pricky	Horizontal
CEN 006A	Purple	Glabrous	Green	SP	Horizontal
NGAE – 96-0063	DR	SP	Green	Pricky	Erect
NGAE – 96-011	G w P	Glabrous	DG	SP	Horizontal
CEN 005	G w P	Glabrous	Y G	SP	Horizontal
NGAE – 96-0068	G w P	Glabrous	Green	Pricky	Horizontal
NGAE – 96-0065	P w G	Glabrous	Green	Downy	Horizontal
ABC -1	Green	Glabrous	Y G	Downy	SE
NCRI – 02	Purple	Glabrous	G w R	Downy	SE
NCRI -05	Purple	Glabrous	Y G	SP	SE
NGAE – 96-0069	Purple	Glabrous	G w R	SP	Horizontal
OJAOBA – 1	Green	Glabrous	Y G	SP	Horizontal
OJAOBA-2	Green	Glabrous	Y G	SP	Horizontal
OJAOBA-3	Green	Glabrous	Green	SP	Erect
OJAOBA-4	P w G	Glabrous	Green	Downy	Erect
ADO-EKITI-2	Green	Glabrous	Y G	Pricky	Horizontal
ADO-EKITI -3	Purple	Glabrous	Green	Downy	SE
ADO-EKITI -5	Purple	Glabrous	DG w R	Downy	Horizontal
IFE -1	Green	Glabrous	G w R	SP	SE
IFE -2	Purple	Glabrous	G w R	Downy	Horizontal
AKURE -2-2	Purple	Glabrous	G w R	Downy	Horizontal
AKURE -2 -9	Purple	Glabrous	G w R	SP	Pendulum
AKURE -1 -1	Green	Glabrous	LG	Downy	Erect
AKURE -2-4	Purple	Glabrous	G w R	SP	SE
OWODE-1	Green	Glabrous	LG	Pricky	Horizontal
OWODE-2	Purple	Glabrous	G w R	Pricky	Horizontal

Table 1. Contd.

OWODE-3	Purple	Glabrous	Green	Downy	Horizontal
OWODE-4	P w G	SP	Green	Pricky	SE
OWODE-5	Purple	SP	DG w R	Pricky	Horizontal

LP - Light Purple, GwP- Green with Purple, PwG - Purple with Green DG- Dark Green, LG- Light Green, GwR-Green with Red, YG- Yellowish Green, GwY-Green with Yellow, DGwR- Dark Green with Red, SP- Slightly Pricky, SE- Semi-Erect.

Table 2. Class intervals and index Score and Sign for 9 characters.

Characters	Range	Score 1	Sign	Score 2	Sign	Score 3	Sign	Score 4	Sign	Score 5	Sign
Days to 50% flowering	132 - 150	< 135	@	135 - 138	@	139 - 141	@	142 - 145	@	> 145	@
100 – Seed weight	2.37 - 6.01	< 4.0	@	4.0 - 4.4	@	4.5 - 4.8	@	4.9 - 5.0	@	> 5.0	@
Fruit length (cm)	6.3 - 11.1	< 8.0	@	8.0 - 8.9	@	9.0 - 9.5	@	9.6 - 10.0	@	> 10.0	@
Fruit width (cm)	2.7 - 5.23	< 3.50	@	3.51-3.70	@-	3.71 - 3.90	@--	3.91-4.10	@---	> 4.10	@—
Fruit / plant	3 - 26.3	< 10	@	10.1-13.0	@	13.1 - 15	@	15.1-17.0	@	> 17.0	@
Leaf length	10.3 -28.3	< 14.0	@	14.1 - 16	@	16.1 - 18.0	@	18.1 - 20	@	> 20.0	@
Seed/ pod	40.0 - 100	< 60	@	60 - 70	-@	71 - 80	--@	81 - 90	---@	> 90	—@
Stem diameter	1.5 - 3.2	< 1.80	@	1.80-2.00	@	2.01 - 2.20	@	2.21-2.40	@	> 2.40	@
Number of internode	3.1 – 6.98	< 3.5	@	3.5 – 4.4	@	4.5 - 5.4	@	5.5 - 6.5	@	> 6.5	@

@ = Days to 50% flowering; @ = 100-seed weight (g); @ = Fruit length (cm); @- = Fruit width (cm); @ = Fruits / plant; @ = Leaf length (cm); @ = Seeds / pod, @ = Stem diameter (cm); @ = Number of internodes.

kly stems include: CEN 016, OAA 96/175-5328, NGAE-96-0062-1, NGAE-96-0064, NGAE-96-0063, Owode-4 and Owode-5. Eighteen accessions have downy fruits while 14 accessions have prickly fruits. The rest 17 accessions are slightly prickly with varying degree of roughness compared with the prickly type. All the 4 accessions from Akure had glabrous stem with purple stem except Akure-1-1 which has green stem. All accessions were evaluated except three were tolerant to major diseases affecting leaves, stem and fruits. The three accessions that were slightly susceptible to stem borer and cotton stainers attack based on field assessment include AGA97/066-5780, CEN009 and OJAOBA-3.

The grouping of the variability of the eleven characters into 5 classes on the basis of their class intervals as well as their index scores and signs is presented in Table 2. The eleven characters that were used for the metroglyph include the main agronomic characters such as fruit length, fruit width, leaf length, days to 50% flowering, plant height, number of internodes, hundred seed weight,

stem diameter and pod yield per plant. These were the most commonly used traits in distinguishing between okra genotypes (Ariyo, 1988). The mean values of the eleven traits considered for the metroglyph as well as the index score for each accession are presented in Tables 3 and 4. The metroglyph of the 50 accessions showed variation pattern in the eleven agronomic traits (Figure 1). The metroglyph chart grouped all the accessions into seven distinct groups with specific differences from other group. The index scores ranged from 19.0 to 42.3 with groups I and VII showing two extremes among the 50 accessions evaluated.

DISCUSSION

The scatter diagram can be distinguished into seven broad groups which reflect the uniqueness of each group based on the means of the different morphological traits measured.

Group I: short plants and low pod yielding genotypes.

Table 3. Mean values and index score (in brackets) for 11 characters.

No	Flower (day)	Fruit/Plant (cm)	Fruit Length (cm)	Fruit width (cm)	Height (cm)	No. of Internode (cm)	Leaf Length (cm)	Peduncle length (cm)	Seed 100 Wt (g)	Seed Fruit	Stem Diameter (cm)	Yield Plant (g)
1	136(2)	12.2(2)	8.9(2)	4.20(5)	96(3)	4.70(3)	15.3(2)	16.0	4.06(2)	74(3)	2.03(3)	79.1(5)
2	138(2)	7.9(1)	8.6(2)	3.77(3)	77(2)	4.87(3)	15.4(2)	19.1	4.03(2)	84(4)	2.30(4)	42.8(2)
3	138(2)	10.8(2)	9.4(3)	3.87(3)	99(3)	3.30(1)	16.3(3)	23.0	4.37(2)	89(4)	2.63(5)	68.3(4)
4	137(2)	8.3(1)	9.2(3)	3.61(2)	94(3)	3.87(2)	16.1(3)	22.8	4.48(3)	73(3)	2.27(4)	42.6(2)
5	140(3)	9.4(1)	9.2(3)	3.77(3)	90(3)	4.97(3)	15.9(2)	21.1	4.52(3)	80(3)	2.40(4)	49.1(2)
6	142(3)	4.7(1)	9.6(3)	3.90(2)	69(1)	4.50(3)	14.9(2)	16.0	4.35(2)	54(1)	1.93(2)	24.7(1)
7	140(3)	15.9(4)	10.1(4)	3.67(2)	119(5)	4.93(3)	16.3(3)	23.2	4.93(4)	75(3)	2.57(5)	40.0(2)
8	140(3)	7.0(1)	9.5(3)	3.57(2)	89(3)	3.13(1)	16.6(3)	20.0	3.83(1)	87(4)	2.27(4)	42.2(2)
9	136(2)	15.2(4)	8.5(2)	4.03(4)	104(4)	4.37(2)	16.2(3)	18.0	4.28(2)	73(3)	2.47(5)	70.2(5)
10	142(4)	10.1(2)	9.6(4)	3.80(3)	118(5)	3.50(2)	15.9(2)	20.0	3.44(1)	67(2)	2.17(3)	45.4(2)
11	136(2)	17.0(5)	10.6(5)	3.47(1)	111(4)	5.27(4)	17.7(3)	24.6	4.16(2)	82(3)	2.80(5)	61.3(4)
12	139(3)	10.9(2)	8.5(2)	3.53(2)	82(2)	4.30(2)	15.2(2)	16.9	4.23(2)	71(3)	2.00(2)	44.5(2)
13	133(1)	13.6(3)	7.7(1)	3.98(4)	113(4)	6.47(4)	17.3(3)	20.3	4.53(3)	87(4)	2.20(3)	89.0(5)
14	136(2)	13.2(3)	9.2(3)	4.34(5)	113(4)	4.97(3)	16.5(3)	22.2	4.59(3)	87(4)	2.53(5)	107.2(5)
15	140(3)	13.8(3)	9.0(3)	3.97(4)	118(5)	4.57(3)	16.1(3)	21.2	4.42(2)	67(2)	2.20(3)	82.8(5)
16	135(2)	17.7(5)	8.4(2)	3.76(3)	130(5)	6.37(4)	17.1(3)	23.4	4.73(3)	91(5)	2.53(5)	111.2(5)
17	139(3)	23.1(5)	10.0(4)	3.96(4)	118(5)	6.93(5)	19.0(4)	21.9	4.62(3)	86(4)	2.60(5)	146.1(5)
18	136(2)	13.6(3)	10.2(5)	3.80(3)	112(4)	4.70(3)	19.4(4)	21.6	4.78(3)	83(4)	2.47(5)	100.7(5)
19	136(2)	14.4(3)	10.4(5)	4.05(4)	136(5)	6.57(5)	19.9(4)	21.0	4.82(3)	91(5)	2.40(4)	100.2(5)
20	136(2)	17.1(5)	9.2(3)	4.17(5)	117(5)	5.60(4)	24.3(5)	27.2	4.95(4)	94(5)	2.60(5)	120.3(5)
21	133(1)	12.7(5)	9.8(4)	3.78(3)	101(4)	5.00(3)	17.1(3)	19.3	4.94(4)	74(3)	2.00(2)	91.0(5)
22	138(2)	10.6(2)	10.2(5)	3.78(3)	111(4)	5.40(3)	15.8(2)	22.4	4.97(4)	93(5)	2.13(3)	61.6(4)
23	145(4)	5.9(1)	7.3(1)	3.61(2)	80(2)	4.57(3)	14.3(2)	18.2	4.76(3)	78(3)	1.80(2)	48.0(2)
24	138(2)	5.3(1)	8.0(2)	3.73(3)	76(2)	3.90(2)	12.8(1)	15.8	4.91(4)	71(3)	1.67(1)	34.7(1)
25	136(2)	15.2(4)	8.2(2)	3.72(3)	145(5)	5.63(4)	17.6(3)	21.8	4.34(2)	64(2)	2.53(5)	110.0(5)
26	136(2)	9.1(1)	8.3(2)	3.59(2)	89(3)	4.23(2)	13.2(1)	17.8	4.68(3)	62(2)	2.03(3)	32.9(1)
27	142(3)	12.1(2)	9.1(3)	3.13(1)	97(3)	5.20(3)	15.4(2)	20.4	4.42(2)	79(3)	2.07(3)	58.1(3)
28	134(1)	14.7(3)	10.3(5)	3.53(2)	107(4)	5.23(3)	16.6(3)	22.6	4.24(2)	74(3)	2.33(4)	93.9(5)
29	154(5)	6.5(1)	8.7(2)	3.42(1)	100(3)	4.87(3)	16.0(2)	21.3	4.38(2)	65(2)	1.97(2)	41.6(2)
30	148(5)	7.2(1)	8.3(2)	3.60(2)	87(3)	3.37(1)	15.0(2)	13.9	5.61(5)	66(2)	1.90(1)	26.9(1)
31	140(3)	8.9(1)	8.9(2)	3.98(4)	96(3)	4.63(3)	15.3(2)	18.7	5.13(5)	81(4)	2.07(3)	60.9(4)
32	137(2)	7.3(1)	9.7(4)	3.88(3)	102(4)	3.77(2)	15.8(2)	20.0	4.94(4)	83(4)	2.07(3)	58.1(3)
33	141(3)	6.3(1)	7.8(1)	3.48(1)	78(2)	3.57(2)	13.8(1)	17.5	4.15(2)	66(2)	2.13(3)	34.4(1)
34	145(4)	4.9(1)	8.6(2)	3.05(1)	77(1)	4.20(2)	12.5(1)	15.1	4.20(2)	72(3)	1.73(1)	35.7(1)

Table 3. contd.

35	140(3)	9.7(1)	7.5(1)	3.81(3)	91(3)	4.67(3)	15.8	4.08(2)	64(2)	2.03(3)	31.0(1)
36	134(1)	13.9(3)	8.8(2)	3.43(1)	98(3)	4.33(2)	17.0	3.93(1)	72(3)	2.03(3)	69.5(4)
37	146(5)	9.8(1)	6.8(1)	3.35(1)	85(2)	4.13(2)	18.7	4.43(2)	61(2)	1.80(2)	47.4(2)
38	143(4)	8.5(1)	7.8(1)	3.65(2)	80(2)	5.40(3)	18.2	4.09(2)	60(2)	1.80(2)	35.8(1)
39	144(4)	3.5(1)	7.9(1)	3.11(1)	64(1)	4.63(3)	12.6	3.98(1)	41(1)	1.57(1)	21.4(1)
40	142(4)	10.7(2)	8.1(2)	3.70(2)	72(1)	3.83(2)	15.7	4.42(2)	88(4)	1.67(1)	69.7(4)
41	137(2)	6.8(1)	5.5(1)	3.38(1)	92(3)	3.77(2)	16.0	4.94(4)	71(3)	1.93(2)	34.3(1)
42	139(3)	5.5(1)	8.2(2)	3.41(1)	80(2)	4.43(3)	18.5	4.79(4)	65(2)	1.77(1)	43.0(2)
43	140(3)	5.7(1)	9.5(3)	3.57(2)	99(3)	4.20(2)	19.7	4.29(2)	72(3)	1.90(2)	37.1(1)
44	136(2)	5.3(1)	8.0(2)	3.78(3)	81(2)	3.87(2)	16.3	4.94(4)	80(3)	1.67(1)	30.6(1)
45	141(3)	6.9(1)	7.8(1)	2.97(1)	112(4)	4.87(3)	21.7	2.87(1)	63(2)	2.06(3)	29.5(1)
46	142(4)	7.4(1)	7.8(1)	3.53(2)	98(3)	5.83(4)	19.9	4.95(4)	79(3)	1.90(2)	41.6(2)
47	137(2)	17.3(5)	10.1(5)	3.64(2)	123(5)	5.30(2)	23.5	4.56(3)	73(3)	2.43(4)	91.5(5)
48	141(3)	14.2(3)	8.9(2)	3.24(1)	81(2)	4.47(3)	18.3	4.40(2)	81(4)	2.00(3)	80.4(5)
49	138(2)	7.4(1)	9.9(4)	3.10(1)	89(3)	4.43(3)	17.5	4.71(3)	68(2)	1.73(1)	36.0(1)
50	136(2)	13.0(3)	8.5(2)	3.75(2)	106(4)	3.77(2)	20.6	4.61(3)	98(5)	2.23(3)	66.2(4)

Table 4. Frequency distribution of index score of the seven groups of *A. caillei.*

Group	No. in group	16	19	21	22	23	24	25	26	27	28	29	30	31	32	33	34	36	37	38	40	41	42	45	47	48	Total	Average score
I	02	1			1																						38	19.0
II	21		2	2	3	4	2	1	1	1	3	1	1	1													506	24.1
III	04							1	1	1		1	1														128	32.0
IV	03				1						1		1														90	30.0
V	09					1					1	3			3	1	2		1	1	1	1	1	1			284	31.5
VI	07															1	1	1		1	1	1	1	1	1		259	37.0
VII	04															1	3	1	1	1	1	1	1	1	1	1	173	43.3
Total	50	1	2	2	5	4	3	2	1	1	5	2	2	1	3	1	3	2	1	3	1	1	1	1	1	1		

This group, though matured relatively late, their yield was relatively low. Accession CEN 016 and ADO - EKITI-5 belonged to this group.

Group II: Low yield but with relatively medium plant height. Twenty-one genotypes fall within this group and they are generally more closely related than other groups.

Group III: Short to medium plant height with moderate pod yield. CEN 010, NGAE-96-0067, IFE-1 and OWODE-3 belong to this group. They are also different from each other in terms of yield and height. OWODE-3 was relatively taller and yielded more than IFE-1.

Group IV: Moderately tall with low yield. Three genotypes belong to this group which includes CEN012, CEN015 and AKURE-2-4. They are generally late maturing with robust stem and relatively long pod size.

Group V: Average plant height and moderate pod yield. This group was generally in between the other groups and they are unique for their

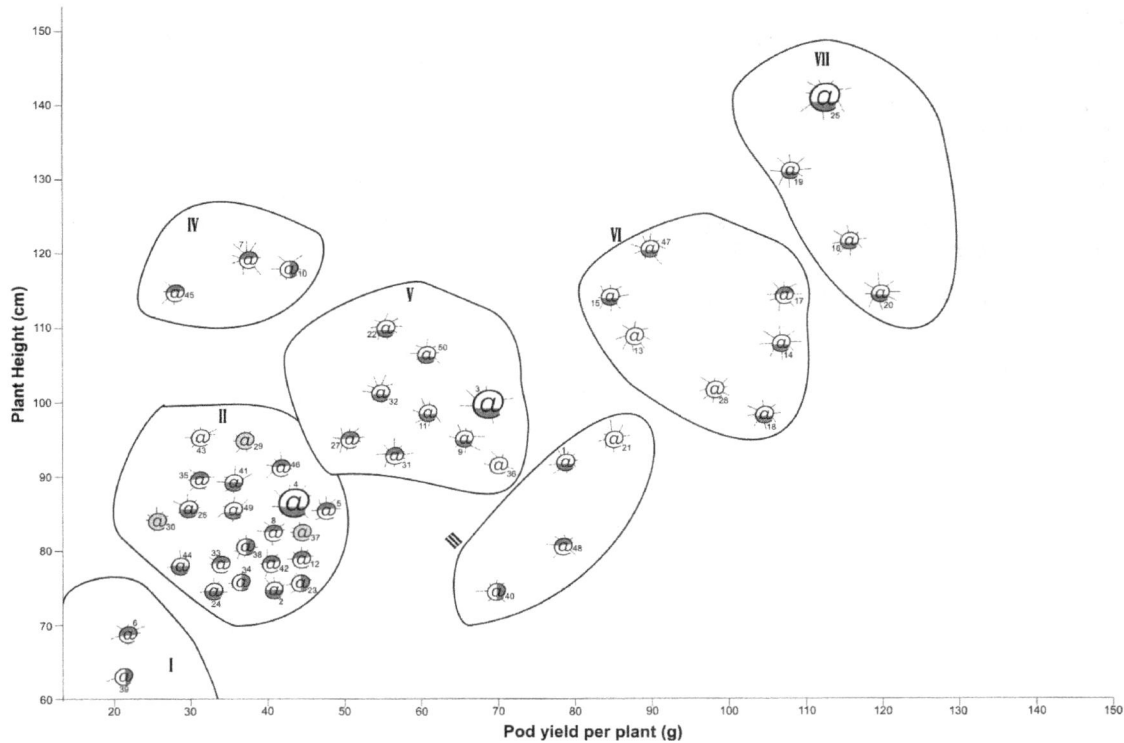

Figure 1. Scatter diagram of metroglyphs representing 50 accessions of *A. caillei* on eleven agronomic characters.

robust stem. They are more scattered and varied within the group in terms of height and yield unlike group II. Some were late while others were early. Eleven genotypes belong to this group as indicated by their code number on the glyph chart.

Group VI: High yielding genotypes with moderate plant height and medium maturity. This group consisted of seven genotypes and they possess medium to high value for all the agronomic characters considered for this study. Some of the group members were more related to other group members than within the group. This might be linked to their point of collection, which means they might be having some genetic relationships.

Group VII: This group is as unique as group I because it combines tall plant height with high pod yield. The four genotypes in this group were also different from each other in height but with relatively the same pod yield and varied maturity period. The genotypes in this group are NGAE-96-062-1 (16), NGAE-96-0061 (19), NGAE-96-0060 (60) and NGAE-96-011 (25).

The metroglyph showed that accessions with very poor yield had short plant height, an indication that for West African genotype okra to produce substantial yield, it must possessed strong stem, hence early planting is encouraged. The average index score of group I was 19.0, while that of group VII with four accessions in the group had index score of 42.3. This suggest how wide the differences in both yield and height among the groups. The index score of groups V which was on the glyph

chart was 32.5 and was almost the central point between the two extremes of groups I and VII. The index scores of groups III, IV and V with 32.0, 30 and 31.5 was a reflection on how close these groups might be. This suggests that breeders may contemplate using from any of these groups for crossing purposes and not all or two of them at the same time, though this still depends on the goal of the breeder.

It is interesting therefore to note the ability with which the metroplyph reduced the complex interrelationships among accessions to a simple pictorial scatter diagram which is easier to comprehend (Akoroda, 1983, Khan et al., 2007). The fruit surface of Akure-2-2 and Akure-1-1 were smooth while Akure-2-9 and Akure-2-4 showed slightly prickly fruits. This suggests that accessions from Akure town in Ondo state of Nigeria have been domesticated to large extent and possibly are more related. There is need for collec-tion and conservation of okra germplasm especially, West African genotypes to prevent total genetic erosion.

REFERENCES

Akoroda MO (1983). Principal Component Analysis and Metroglyph of Variation among Nigerian yellow yams Euphytica 32: 565 – 573.

Ariyo OJ (1993). Genetic diversity in West African Okra (*Abelmoschus caillei* (A. Chev.) Stevels) – Multivariate analysis of morphological and agronomic characteristics. Genet. Res. Crop Evol. 40: 25-32.

Ariyo OJ (1988). Metroglyph and index score analysis and variety distinctness in okra (*Abelmoschus esculentus*). Niger. J. Agron. 3:42-46.

Bisht IS, Mahajan RK, Rana RS (1995). Genetic diversity in South Asian okra *(Abelmoschus esculentus)* germplasm collection. Ann. Appl. Biol. 126: 239-550.

Frankel OH (1976). Natural Variation and its Conservation in Proceedings of an International Symposium on Genetic Control of Diversity in Plants at Lahore, Pakistan

Gulsen OS, Karagul S, Abak K (2007). Diversity and relationships among Turkish germplasm by SRAP and Phenotypic marker polymorphism. Biologia, Bratislava 62(1): 41-45.

Hartwig EE (1972). Utilization of soybean germplasm strains in a soybean improvement programme. Crop Sci. 12: 856-859.

Khan MR, Samad A, Begun S, Khaleda S, Alam AKMS, Rahman MZ(not cited. Please provide or delete) (2007). Metroglyph Analysis in Cotton (*Gossypium spp*). Bangladesh J. Sci. Ind. Res. 42(4): 449-454.

Omonhinmin CA, Osawaru ME (2005). Morphological characterization of two species of Abelmoschus: *Abelmoschus esculentus and Abelmoschus caillei*. Genet. Res. Newslett. 144: 51-55.

Palve SS, Rajput JC, Jamdagni BM (1986). Genetic Variability and Correlation studies in okra (*Abelmoschus esculentus*). Indian J. Agric. Res., 19(1): 20-22.

Ren J, McFerson J, Kresovich RLS, Lamboy WF (1995). Identities and Relationships among Chinese Vegetable Brassicas as Determined by Random Amplified Polymorphic DNA Markers. J. Am. Soc. Hort. Sci., 120(3): 548 - 555.

Splittstoessor WE (1990). Vegetable growing handbook. New York: Van Nostrand Reinhold. Pp. 248-250.

Stevels JCM (1988). Une nouvelle combinaison dans *Abelmoschus. (Malvaceae)* un gombo d'Afrique de l'Ouest et centrale. Bull. Mus. Nat. Hist, Nat., Paris u-ser. Lo., Section B., *Adansonia* 2: 137-144.

Tigerstedt PMA (1994). Adaptation, variation and selection in marginal areas. Euphytica 77: 171-174.

15

Multivariate analysis of agronomic and quality traits of hull-less spring barley (*Hordeum vulgare* L.)

Firdissa Eticha[1*], Heinrich Grausgruber[2] and Emmerich Berghoffer[3]

[1]National Wheat Research Project, Kulumsa Agricultural Research Center, P. O. Box 489, Asella, Ethiopia.
[2]Department of Applied Plant Sciences and Plant Biotechnology, BOKU–University of Natural Resources and Applied Life Sciences, Gregor-Mendel-Str. 33, 1180 Vienna, Austria.
[3]Department of Food Science and Biotechnology, BOKU–University of Natural Resources and Applied Life Sciences, Mutthgasse, Vienna, Austria.

A study was conducted to characterize a world collection of hull-less barley using multivariate traits. Significant variations were observed among genotypes in grain yield and grain physical characteristics. Genotypic as well as environmental effects were important for the variations occurred in chemical contents. Total phenolic content ranged from 166.0 to 295.0 mg/100 g ferulic acid equivalent. The total anthocyanin ranged from 3.0 to 284.5 ppm cyanidin glucoside equivalent. Yellow pigment content ranged from 3.9 to 8.7 ppm and protein content from 12.3 to 17.3%. Beta-glucan ranged from 3.5 - 7.4% for barley genotypes. Purple pigmented barley found to have high total anthocyanin content whereas the black pigmented barleys were superior in their total phenolics and yellow pigments. Waxy barleys contained higher levels of beta-glucan.

Key words: *Hordeum*, hull-less, barley, pigments, Ethiopia.

INTRODUCTION

Cultivated barleys can be classified according to caryopsis form in hulled (syn. covered) and hull-less (syn. naked) types. In the hulled form, the lemma and palea are fused to the pericarp, whereas in the hull-less forms, the chaff can be easily separated from the grain by threshing. Thus, for food uses hulled barley requires extensive processing, e.g. pearling, before use and hull-less barley is preferred over hulled barley for food production (Bhatty 1999). The hull-less trait is controlled by a single recessive gene (*nud, nudum*), located on the long arm of chromosome 7H (Lundqvist et al., 1997). The expression of *nud* precludes permanent adhesion of hulls to the kernel (Xue et al., 1997). No other member of *Triticeae* shows such a hull-caryopsis adhesion than hulled barley (Taketa et al., 2004, 2008). Cultivation of hull-less barley is as old as that of hulled barley but it is less common worldwide than that of hulled barley (Atanassov et al., 2001; Pandey et al., 2006). Compared to hulled barley, hull-less barley has higher contents of

protein and limiting amino acids, lysine and threonine (Baidoo and Liu, 1998; Bhatty, 1999). Moreover, it has lower levels of fibre components but contains considerably higher levels of beta-glucan (Xue et al., 1997; Baidoo and Liu, 1998). However, hull-less barley is also characterised by low grain yield, small grain size and superior growth requirements. This might explain why hull-less barley have been grown unsuccessfully in many regions of Ethiopia (Asfaw, 1989; Assefa and Labuschagne, 2004). Therefore, improvement in grain yield and related traits and grain quality for different end uses is essential in hull-less barley cultivar development. Barley kernel colour shows a great diversity, from white-yellowish to blue, purple and black. Seed colour depends on different pigments in different seed layers. Recently interest in some of these pigments, such as anthocyanins, increased due to their possible health promoting effects (Abdel-Aal and Hucl, 2003; Abdel-Aal et al., 2006; Hu et al., 2007). Moreover, grain colour can be used as a marker to detect genes conferring resistance to certain diseases if linkage and co-segregation with the respective genes exists (Bonman et al., 2005). Blue seed colour is due to anthocyanins in the

---

aleurone layer of the kernel (Wang et al., 1993). The genetics of blue colour was described by Finch and Simpson (1978). Aleurone colour is the result of five complementary factors, the non-blue aleurone xenia alleles (*blx*) 1 to 5. White aleurone results if the alleles are present in their re-cessive forms (Lundqvist et al., 1997). Colour intensity can be influenced by environmental factors and modifying genes. Information on the distribution of blue barleys in the world is rare. But they occur in different barley growing regions in Ethiopia (Negassa, 1985). Purple pigmentation of the lemma, palea and pericarp is controlled by two dominant complementary genes (*Pre1* and *Pre2*). Purple or red colouration of the hull and pericarp develops during the soft dough stage of grain fill and fades as the grain matures (Lundqvist et al., 1997). The pigmentation is due to anthocyanins (Woodward and Thieret, 1953). Black lemma and pericarp is controlled by the *Blp* locus at the long arm of chromosome 1H (Lundqvist et al., 1997). The black pigmentation is considered to be due to melanin-like pigments (Buckley, 1930; Woodward, 1941). The black colour develops be-fore maturation of the spike. Pigmented organs may in-clude all parts of the spike, awns, the upper portion of the stem, and upper leaves. The intensity of pigmentation is relatively stable over environments.

Multivariate analysis refers to all statistical methods that simultaneously analyse multiple measurements on each individual or object under investigation. More explicitly, any simultaneous analysis of more than two variables can be considered as multivariate analysis (Hair et al., 1998). Multivariate data analysis facilitates a graphic display of the underlying latent factors and interface between individual samples and variables (Nielsen and Munck, 2003). Principal component analysis (PCA) has been widely used in plant sciences for reduction of variables and grouping of genotypes. Kamara et al. (2003) used PCA to identify traits of maize (*Zea mays* L.) that accounted for most of the variance in the data. Granati et al. (2003) used PCA to investigate the re-lationship among *Lathyrus* accessions. Žáková and Benková (2006) identified traits that were the main sources of variation of genetic diversity among 106 Slovakian barley accessions. Cartea et al. (2002) and Salihu et al. (2006) used PCA and cluster analysis to group kale populations and winter wheat genotypes, respectively. In the current study, a set of data comprising agronomic and quality traits of a world collection of 81 hull-less barley genotypes were subjected to multivariate data analysis, namely, PCA, cluster analysis and CDA.

The main objectives of the study were to (1) characterize and classify diverse hull-less barley genotypes based on their overall similarity in agronomic and qualitative data and (2) identify the genotypes that best combine both agronomic and quality characters for the future use in hull-less barley breeding.

MATERIALS AND METHODS

Plant materials

A broad range of hull-less spring barley genotypes comprising breeding lines, landraces and cultivated varieties were investigated. Descriptions of the investigated germplasm in regard to name, gene bank accession code, donor institution and country of origin is given in Table 1.

Experimental site and trial management

The field trials were planted in row-column designs with two replicates at Raasdorf, Austria (16°35'E, 48°14'N) in spring 2006 and 2007. The entries were grown under organic farming conditions without application of external inputs. Sowings were done on April 3 in 2006 and on March 16 in 2007.

Agronomic traits

Data were collected for heading date (DH, days after April 30), grain yield (GYLD, g m^{-2}), thousand kernel weight (TKW, g), hectolitre/test weight (HLW, kg hL^{-1}), and kernel plumpness (KP25, %). HLW was measured by a ¼ l chondrometer (Institut für Laborbedarf, Wien, Austria). TKW was determined by using a Contador seed counter (Pfeuffer GmbH, Kitzingen, Germany). KP25 was determined by sieving 100 g grain with a Sortimat laboratory machine (Pfeuffer GmbH, Kitzingen, Germany); the percentage of grains with a width >2.5 mm was recorded.

Chemical analysis

Extraction of phenolics

Each grain sample was extracted twice and subsequently the dry matter content was determined. The solvent used for extraction was acidified methanol (85:15 MeOH: 1 m HCl). Grain samples of 2.5 ± 0.1 g were extracted with 20 ml solvent in 50 ml Erlenmeyer flasks. The mixtures were homogenized at ambient condition using a magnetic stirrer for 20 min and then stored in a refrigerator for 20 min at 4 °C. Subsequently the mixture was transferred into plastic tubes and centrifuged at 4000 rpm for 5 min. The centrifuged samples were placed into the refrigerator for another 20 min at 4 °C before the supernatants were filtered into 25 ml volumetric flasks fitted with funnels and folded filters Ø 125 μm. The supernatants were filled to equal volume of 25 ml with the solvent and stored under room temperature in dark places.

Anthocyanins

The total anthocyanin content (TAC) was determined following Abdel-Aal and Hucl (1999). The acidified MeOH extracts were filled in cuvettes of 1 cm thickness and measured at 525 nm in a type U-1100 spectrophotometer (Hitachi, Tokyo, Japan). The reading was first adjusted to zero with an empty microcuvette and afterwards by a cuvette with acidified MeOH solely. According to the calibration curve the results were calculated into mg cyanidin-3-glucoside equivalents per kg dry matter and/or parts per million (ppm).

Total phenolics

The total phenolic content (TPC) was determined spectrophotometrically using the Folin-Ciocalteu reagent according

Table 1. Description of the investigated hull-less spring barley genotypes.

No.	Genotype	Origin[1]	Donor	Head Rows	Grain Colour[2]
1	00/900/19/3/1	DE	SZ Ackermann, Irlbach, DE	2	
2	00/900/19/3/7	DE	SZ Ackermann, Irlbach, DE	2	
3	00/900/19/3/12	DE	SZ Ackermann, Irlbach, DE	2	
4	00/900/19/3/13	DE	SZ Ackermann, Irlbach, DE	2	
5	00/900/19/6/4	DE	SZ Ackermann, Irlbach, DE	2	*Blp*
6	00/900/19/6/8	DE	SZ Ackermann, Irlbach, DE	2	*Blp*
7	00/900/19/6/11	DE	SZ Ackermann, Irlbach, DE	2	*Blp*
8	A 032	GB	BGC, Okayama, JP	6	*Pre*
9	A 330	GT	BGC, Okayama, JP	2	*Blx*
10	BVAL 358117	ET	AGES, Linz, AT	2, 6	*Blp*
11	BVAL 358163	ET	AGES, Linz, AT	2, 6	*Blp*
12	C 051	CN	BGC, Okayama, JP	6	*Blp*
13	C 359	CN	BGC, Okayama, JP	6	*Pre*
14	C 651	CN	BGC, Okayama, JP	6	*Blx*
15	C 661	CN	BGC, Okayama, JP	6	*Blp*
16	CDC Candle	CA	VUKROM, Kromeriz, CZ	2	
17	Digersano	IT	ISC, Fiorenzuola d'Arda, IT	2	
18	Dometzkoer Paradies (U 347)	CZ	BGC, Okayama, JP	2	
19	E 048	ET	BGC, Okayama, JP	2	
20	E 056	ET	BGC, Okayama, JP	2	*Blx*
21	E 339	ET	BGC, Okayama, JP	2	*Blx*
22	E 359	ET	BGC, Okayama, JP	6	*Blx*
23	E 360	ET	BGC, Okayama, JP	2	*Blp*
24	E 515	ET	BGC, Okayama, JP	2	*Blx*
25	E 550	ET	BGC, Okayama, JP	6	*Blx*
26	E 552	ET	BGC, Okayama, JP	6	*Pre*
27	E 604	ET	BGC, Okayama, JP	2	*Blp*
28	E 632	ET	BGC, Okayama, JP	6	*Blx*
29	E 639	ET	BGC, Okayama, JP	2	*Blx*
30	E 649	ET	BGC, Okayama, JP	6	*Blx*
31	GE 037	AT	Arche Noah, Schiltern, AT	6	
32	GE 040 sel BA	AT	BOKU-DAPP, Vienna, AT	6	*Blx*
33	GE 090	AT	Arche Noah, Schiltern, AT	2	
34	HB 803	CA	VUKROM, Kromeriz, CZ	2	
35	HOR 345	DE	IPK, Gatersleben, DE	2	
36	HOR 346	DE	IPK, Gatersleben, DE	2	
37	HOR 816	DE	IPK, Gatersleben, DE	2	
38	HOR 2172	DE	IPK, Gatersleben, DE	2	
39	HOR 2199	DE	IPK, Gatersleben, DE	6	*Pre*
40	HOR 2593	DE	IPK, Gatersleben, DE	6	*Pre*
41	HOR 3647	DE	IPK, Gatersleben, DE	2	
42	HOR 3710	DE	IPK, Gatersleben, DE	6	*Pre*
43	HOR 3727	DE	IPK, Gatersleben, DE	6	*Pre*
44	HOR 3756	DE	IPK, Gatersleben, DE	2	
45	HOR 3803	DE	IPK, Gatersleben, DE	2	
46	HOR 4024	IN	IPK, Gatersleben, DE	2	*Pre*
47	HOR 4076	NP	IPK, Gatersleben, DE	6	*Pre*
48	HOR 4769	UN	IPK, Gatersleben, DE	6	*Pre*
49	HOR 4940	US	IPK, Gatersleben, DE	6	*Pre*
50	HOR 10955	NP	IPK, Gatersleben, DE	6	

Table 1. Continued.

51	HOR 11402	CN	IPK, Gatersleben, DE	6	*Pre*
52	HORA (BVAL 350010)	DE	AGES, Linz, AT	2	
53	I 002	IN	BGC, Okayama, JP	2	
54	I 026	IR	BGC, Okayama, JP	2	
55	I 311	IN	BGC, Okayama, JP	6	*Blx*
56	I 329	IN	BGC, Okayama, JP	6	*Blx*
57	ICARDA Black Naked	SY	SL, Reichersberg, AT	6	*Blp*
58	J 203	JP	BGC, Okayama, JP	2	
59	KM 1910B	CZ	VUKROM, Kromeriz, CZ	2	
60	KM 2074	CZ	VUKROM, Kromeriz, CZ	2	
61	KM 2384	CZ	VUKROM, Kromeriz, CZ	2	
62	Lawina	DE	GF Darzau, DE	2	
63	Merlin	CA	VUKROM, Kromeriz, CZ	2	
64	Mihori Hadaka 3 (J 373)	JP	BGC, Okayama, JP	6	
65	Murasaki Hadaka (J 307)	JP	BGC, Okayama, JP	2	*Blp*
66	N 023	NP	BGC, Okayama, JP	6	*Blp*
67	N 040	NP	BGC, Okayama, JP	6	*Blx*
68	N 308	NP	BGC, Okayama, JP	6	*Blx*
69	N 623	NP	BGC, Okayama, JP	6	*Blx*
70	N 624	NP	BGC, Okayama, JP	6	
71	Nackta (HOR 6936)	DE	IPK, Gatersleben, DE	2	
72	Namoi	AU	DPS, Aidelaide Univ., AU	2	
73	Priora	IT	ISC, Fiorenzuola d'Arda, IT	2	
74	Rimpaus Nackte (HOR 1629)	DE	IPK, Gatersleben, DE	2	
75	Rondo	IT	ISC, Fiorenzuola d'Arda, IT	6	
76	SNG 04	AT	BOKU-DAPP, Vienna, AT	2	*Blp*
77	T 045	TR	BGC, Okayama, JP	2	
78	T 247	TR	BGC, Okayama, JP	2	
79	Taiga (BVAL 358017)	DE	AGES, Linz, AT	2	
80	Torrens	AU	DPS, Aidelaide Univ., AU	2	
81	U 047	CZ	BGC, Okayama, JP	2	
82	U 363	SU	BGC, Okayama, JP	2	
83	U 647	CZ	BGC, Okayama, JP	2	
84	U 687	SU	BGC, Okayama, JP	2	
85	Wanubet	US	VUKROM, Kromeriz, CZ	2	
86	Washonubet	US	VUKROM, Kromeriz, CZ	2	

[1]ISO 3166 country codes: AU, Australia; AT, Austria; CA, Canada; CN, China; CZ, Czech Republic; DE, Germany; ET, Ethiopia; GB, United Kingdom; GT, Guatemala; IN, India; IR, Iran; IT, Italy; JP, Japan; MN, Mongolia; NP, Nepal; NI, Nicaragua; PK, Pakistan; SE, Sweden; UN, unknown/Germany; US, United States of America; SU, Soviet Union; SY, Syria.
[2]Grain colour is white/yellow unless indicated: *Blp*, black pericarp; *Blx*, blue aleurone; *Pre*, purple/red pericarp.

to Singleton et al. (1999). The reaction mixture contained 0.1 ml acidified MeOH extract, 0.5 ml Folin-Ciocalteu reagent (1: 10 Folin-Ciocalteu: H_2O) and 0.8 ml 7.5% Na_2CO_3. The latter was added 2 min after the extract and the Folin-Ciocalteu reagent were mixed. The blank sample was prepared simultaneously with 0.1 ml H_2O instead of extract. The mixture was heated in a water bath at 50°C for 5 min and cooled to ambient temperature before measuring the absorbance at 760 nm in a type U-1100 spectrophotometer (Hitachi, Tokyo, Japan). Two readings were recorded for each extract and the results were expressed as mg ferulic acid equivalents per 100 g dry matter (dm) according to the calibration curve.

Yellow pigments

Yellow pigment concentration (YP) was determined following ICC Standard Method 152. In brief, 2 ± 0.1 g of wholemeal flour was dispersed in 20 ml of distilled water-saturated n-butanol (1:6 v/v H_2O: butanol) in Erlenmeyer flasks. The suspension was well mixed and subsequently the flasks were stored overnight under room temperature and in dark for 18 to 20 h. Afterwards the suspension was filtered into brown jars using folded filter papers with a sieve size of Ø 110 µm. The extracts were measured at 440 nm wavelength in a type U-1500 spectrophotometer (Hitachi, Tokyo, Japan) against the standard solvent. Results were expressed according to the calibration curve as mg beta-carotene equivalents per 100 g dry matter (dm) and/or as parts per million (ppm).

Beta-Glucan

Beta-Glucan content (GLUC) was determined enzymatically using Megazyme kits (ICC Standard Method 166; Megazyme Int. Ireland Ltd., Wicklow, Ireland) and Near Infrared Spectroscopy (NIRS) using a Matrix-I FT-NIR Spectrometer (Bruker Optik GmbH, Ettlingen, Germany). For the enzymatic determination of beta-glucan, 100 ± 0.1 mg milled sample was suspended and hydrated in a sodium phosphate buffer solution of pH 6.5 and incubated with purified lichenase enzyme (specific, endol-(1→3), (1→4)-beta-D-glucan 4-glucanohydrolase, EC 3.2.1.73).

The glucose produced was assayed using glucose/peroxidise (GOPOD) reagent. The beta-glucan contents were determined in two to four replicated measurements. In 2007 determination of beta-glucan was carried out by NIRS on wholemeal grain flour in two replicates per sample. NIRS spectra were transferred into predicted beta-glucan contents using the calibration developed by Schmidt (2007). Both results of the two methods were combined and the results were reported as % on dry weight basis.

Protein content

Protein content (PROT) was determined by the Dumas (combustion) method (ICC Standard Method 167) using a CN-2000 analyzer (Leco Instrumente GmbH, Mönchengladbach, Germany). The nitrogen detected after combustion of each sample was transferred into protein by multiplying with a conversion factor 5.7. That is, protein content (%) = N × 5.7. Duplicate analysis was done for each grain sample and means were used for further statistical analysis.

Data analysis

Principal component analysis

PCA was performed using procedure PRINCOMP (SAS 1999a). Combined analysis of variance (ANOVA) has shown that there was significant genotype by environment interaction (G×E) effects. Therefore, annual means for the single traits were considered for PCA. PCA relies upon the Eigen vector decomposition of the covariance or correlation matrix (Granati et al., 2003). In the present study the correlation matrix was used for PCA. Any trait that does not have a significant correlation with PC scores is considered unimportant in classifying genotypes (Kamara et al., 2003). Correlation analysis was performed using procedure CORR (SAS, 1999b) to determine relationships between PC scores and the original data as well as between agronomic and quality traits.

Cluster analysis

Genotype scores (Eigen vectors) of the seven PCs which had Eigen values greater than unity were subjected to hierarchical cluster analysis using procedure CLUSTER and Ward's minimum variance method as a clustering algorithm (SAS, 1999c). Ward's minimum method is a hierarchical clustering procedure in which similarity used to join clusters is calculated as the sum of squares between the two clusters summed over all variables (Hair et al., 1998). It minimizes the within cluster sums of squares across all partitions. ANOVA and Tukey-Kramer mean comparison test were performed between clusters (groupings) by the procedure GLM (SAS, 1999d).

Canonical discriminant analysis

CDA was carried out using procedure CANDISC (SAS, 1999c) using grain colour for defining groups of genotypes. The first two canonical variables were plotted by procedure GPLOT (SAS, 1990; Friendly, 1991) to show the distribution of genotypes on two dimensional planes.

RESULTS

Significant variations were observed among genotypes in grain yield and grain physical characteristics (data not shown). Between the two experimental years, DH was highly correlated (r=0.69, P<0.0001). A significant correlation between the two contrasting years was also observed for TKW (r=0.70, P<0.0001), KP25 (r=0.57, P<0.0001) and HLW (r=0.56, P<0.0001), demonstrating that these traits are mainly qualitatively inherited and, therefore, relatively stable across environments. Plant height varied from 30 - 133 cm in 2006 indicating that there is enough variation for selecting short stalked genotypes with improved lodging resistance. Due to the drought stress in 2007, higher values of GYLD were generally obtained in 2006. The significant G×E interaction is underlined by a non significant correlation of the yield data of both test years (r=0.03, P=0.75).

Summary of data for the chemical content analysis was presented in Table 2. Genotypic as well as environmental effects were important for the variations occurred in chemical contents. High variation in TPC was observed among genotypes ranging from 155 to 291 mg 100 g^{-1} in 2006 and from 143 to 350 mg 100 g^{-1} in 2007. Genotypic values were higher in 2007 and most probably associated with drought stress. Extremely high variation was observed ¡among genotypes for TAC. The range was slightly higher in 2006 (2 - 342 ppm) than in 2007 (3 - 304 ppm). Considerable variation was also observed for YP. The highest value obtained in 2006 was 8.2 ppm; in 2007 it was 10 ppm. PROT varied from 10.7 to 16.7% in 2006 and from 11.1 to 18.6% in 2007. GLUC was stable across the years, ranging from 3.3 to 7.4% in 2006 and from 3.6 to 7.4% in 2007.

Principal component analysis

Seven principal components (PC) had Eigen values >1 and accounted for 72.7% of the total variance in the data (Table 3). The proportions of the total variance attributable to the first three PC were 18.6, 14.4 and 11.0%. The importance of traits to the different PC can be seen from the corresponding Eigen vectors which are presented in Table 4. The results showed that hectolitre weight (HLW) and total anthocyanin content (TAC) had the highest loadings in PC1 indicating their significant importance for this component. On the other hand, other traits are less important to PC1. Thousand kernel weight (TKW), grain grading (KP25) and beta-glucan content

Table 2. Chemical composition of hull-less spring barley.

Genotype	2006					2007				
	TPC	TAC	YP	PROT	GLUC	TPC	TAC	YP	PROT	GLUC
00/900/19/3/1	–	–	–	14.1	3.4	–	–	–	15.4	4.2
00/900/19/3/7	205	5	5.3	12.3	3.7	268	10	4.5	14.0	5.9
00/900/19/3/12	–	–	–	14.6	3.3	–	–	–	14.2	4.2
00/900/19/3/13	–	–	–	14.2	3.5	–	–	–	14.8	4.3
00/900/19/6/4	236	5	7.1	12.5	3.3	222	6	7.8	14.5	3.6
00/900/19/6/8	239	8	5.3	12.9	4.1	238	6	7.6	15.3	3.6
00/900/19/6/11	260	6	6.2	13.5	3.5	220	6	7.8	14.1	3.6
A 032	205	75	4.7	13.1	5.9	299	86	5.8	15.8	4.5
A 330	209	11	4.4	15.8	5.5	148	10	4.4	17.3	5.5
BVAL 358117	225	14	7.9	14.4	4.9	279	13	9.5	17.2	4.7
BVAL 358163	228	23	6.6	14.8	5.6	249	24	8.4	15.8	5.2
C 051	254	74	5.6	12.7	5.0	207	42	6.8	15.6	5.1
C 359	219	46	4.6	13.6	5.6	264	50	5.6	14.0	4.3
C 651	186	22	5.8	13.6	5.0	272	54	6.2	15.8	4.8
C 661	190	37	5.9	12.4	5.8	247	24	7.1	15.0	5.6
CDC CANDLE	210	5	6.6	13.1	6.3	174	5	6.1	13.7	6.8
DIGERSANO	256	12	3.5	12.9	3.8	275	8	5.2	15.2	4.8
DOMETZKOER PARADIES	254	11	5.5	13.9	5.9	266	5	5.9	14.3	6.8
E 048	171	5	5.4	13.9	5.0	179	5	5.4	16.2	5.2
E 056	260	21	5.5	14.2	5.7	248	17	5.7	17.4	6.2
E 339	210	6	5.6	14.5	6.0	265	6	6.0	17.5	5.3
E 359	205	27	5.1	13.0	4.8	149	22	4.6	12.0	4.8
E 360	181	12	5.6	13.8	4.6	271	16	6.8	16.8	4.6
E 515	199	25	3.8	13.9	5.5	279	28	6.2	15.4	6.0
E 550	207	20	5.1	15.3	5.4	250	20	5.5	16.8	6.3
E 552	200	26	6.2	14.3	4.5	178	57	5.8	15.0	5.5
E 604	214	39	4.4	15.8	5.6	350	27	8.2	16.9	4.5
E 632	186	28	5.8	13.5	4.9	162	17	6.2	16.0	5.4
E 639	191	20	5.7	12.9	5.4	178	8	5.8	16.5	4.8
E 649	267	31	6.7	14.9	4.6	251	18	6.0	15.9	4.5
GE 037	207	19	4.8	14.6	4.6	233	13	5.7	14.6	4.9
GE 040 sel BA	224	27	4.3	14.2	4.2	319	25	4.8	15.8	5.2
GE 090	225	7	5.2	14.9	5.0	180	4	5.4	13.9	5.1
HB 803	259	32	5.6	14.6	7.2	294	13	6.0	15.6	5.9
HOR 345	232	6	6.5	14.3	6.0	158	6	5.0	15.5	5.9
HOR 346	219	5	5.2	14.5	5.7	157	5	4.4	15.6	5.6
HOR 816	219	12	7.1	15.1	5.5	239	6	5.9	14.3	5.7
HOR 2172	179	6	6.3	14.5	5.9	254	7	6.8	15.4	6.4
HOR 2199	285	135	5.1	14.1	5.5	305	115	6.4	16.7	5.4
HOR 2593	266	301	6.4	13.4	3.6	259	123	6.7	17.4	5.4
HOR 3647	191	11	7.2	15.0	5.7	260	6	5.6	15.1	5.1
HOR 3710	291	117	4.6	13.5	5.3	276	113	7.1	15.3	4.7
HOR 3727	252	342	4.9	12.6	4.9	228	172	6.9	14.2	4.8
HOR 3756	199	12	4.6	15.0	5.5	257	6	5.3	15.4	5.2
HOR 3803	200	7	6.4	14.3	3.6	167	6	5.9	15.0	5.4
HOR 4024	237	265	3.6	11.1	5.5	321	304	4.8	13.5	5.1
HOR 4076	192	24	5.4	13.8	4.9	143	15	5.6	15.7	4.9
HOR 4769	289	130	5.4	13.2	5.0	258	73	5.9	15.3	4.7
HOR 4940	222	278	6.0	12.9	4.5	205	132	6.4	17.2	5.5
HOR 10955	187	8	4.9	14.9	5.4	228	4	4.9	13.4	5.5

Table 2. Continued.

HOR 11402	260	249	5.1	13.5	4.5	214	121	6.2	17.7	7.0
HORA	274	4	7.2	12.4	4.3	256	5	5.5	15.5	4.1
I 002	255	25	6.6	13.8	4.5	148	6	4.5	15.4	6.0
I 026	204	12	6.7	13.7	5.8	245	7	4.3	13.9	5.2
I 311	212	28	4.3	13.1	5.3	261	24	5.5	15.8	5.3
I 329	207	28	5.1	15.0	4.9	172	14	5.4	14.8	5.3
ICARDA BLACK NAKED	268	20	5.5	14.4	4.1	285	12	10.0	15.0	4.7
J 203	203	4	3.4	13.5	5.2	260	4	4.8	15.6	5.2
KM 1910B	155	6	6.6	11.3	4.2	261	5	5.4	14.9	4.6
KM 2074	205	6	7.4	10.7	3.8	246	8	7.3	15.8	4.4
KM 2384	203	6	5.7	11.2	4.3	227	6	5.1	16.2	3.9
LAWINA	255	5	4.6	13.4	4.6	274	4	5.5	16.5	4.7
MERLIN	223	5	7.9	14.3	5.4	221	3	6.6	15.0	5.6
MIHORI HADAKA 3	178	5	4.9	14.3	4.9	210	4	4.7	15.6	5.0
MURASAKI HADAKA	228	26	6.9	14.7	5.0	341	26	9.1	17.7	5.2
N 023	256	30	5.7	13.5	6.0	173	22	6.2	13.6	5.1
N 040	190	21	3.4	13.9	5.5	236	16	4.3	13.8	4.8
N 308	164	20	4.5	13.3	5.0	224	20	5.0	12.4	5.0
N 623	183	22	4.4	13.0	4.9	149	13	5.2	16.6	4.6
N 624	195	23	5.0	12.7	4.7	286	22	6.9	15.3	5.2
NACKTA	280	13	5.0	13.3	3.7	271	5	5.2	15.9	4.5
NAMOI	224	4	6.5	13.4	4.4	189	3	5.8	14.6	5.4
PRIORA	200	4	4.5	14.8	5.1	199	3	4.6	15.5	5.0
RIMPAUS NACKTE	194	4	5.7	14.0	5.1	263	7	6.8	15.7	4.0
RONDO	216	13	4.0	15.2	4.8	215	9	5.3	15.6	4.5
SNG 04	289	25	5.4	15.0	5.7	245	31	7.8	11.1	4.5
T 045	206	9	4.0	13.4	4.4	240	6	5.3	13.9	4.6
T 247	188	9	5.5	13.4	4.8	157	6	5.4	14.3	4.2
TAIGA	240	11	5.4	13.9	4.1	277	8	5.6	14.3	5.1
TORRENS	252	11	5.7	13.1	4.3	274	6	6.1	14.1	4.6
U 047	215	9	5.5	16.0	6.4	253	8	6.0	18.6	4.9
U 363	207	12	6.8	14.0	5.4	233	7	5.9	15.6	6.1
U 647	291	6	4.5	16.7	5.0	248	5	5.6	16.1	5.1
U 687	192	7	6.5	14.4	5.5	157	4	5.4	16.0	5.7
WANUBET	218	2	6.5	13.6	7.4	257	4	6.4	14.4	7.4
WASHONUBET	208	8	8.2	13.8	6.8	270	8	5.7	14.4	6.8
CHECK (LAWINA+TAIGA)	248	8	5.0	13.8	4.3	276	6	5.6	15.0	5.0
Mean	221	37	5.5	13.8	4.9	236	27	6.0	15.3	5.1
Minimum value	155	2	3.4	10.7	3.3	143	3	4.3	11.1	3.6
Maximum value	291	342	8.2	16.7	7.4	350	304	10.0	18.6	7.4
CV	15	188	19.2	7.9	17.4	20	173	19.2	8.3	14.8

(GLUC) are the main traits of PC2. For PC3 traits such as TKW, KP25 and TAC were the most important, whereas PROT and GLUC were mainly contributing to PC4 and yellow pigment content (YP) affected PC5. Multiple traits contributed to the sixth PC in varying proportions in different seasons. The seventh PC was derived mainly from the variance due to differences in GLUC. In general, it is possible to see some traits that inconsistently contributed to the total variance in different seasons.

Correlation analysis was performed between the original data and the PC scores to determine the contribution of each trait to the total variance (Table 5). PC1 showed a significant correlation with HLW and TAC. The second principal component PC2 was positively associated with TKW, KP25 and GLUC and negatively with TPC and TAC. PC3 had a significant correlation with traits such as TKW, KP25 and TAC. In the fifth PC the most important traits which showed significant correlation

Table 3. Principal component analysis of 81 hull-less spring barley genotypes: Eigen values and percent variation accounted by the first seven principal components.

Principal component	Eigen value	Variance (%)	Cumulative variance (%)
PC1	3.71	18.6	18.6
PC2	2.89	14.4	33.0
PC3	2.20	11.0	44.0
PC4	1.78	8.9	52.9
PC5	1.60	8.0	60.9
PC6	1.38	6.9	67.8
PC7	1.03	5.2	73.0

Table 4. Eigen vectors (loadings) of the first seven principal components.

Trait × Year[1]	Eigen vector						
	PC1	PC2	PC3	PC4	PC5	PC6	PC7
DH 06	-0.37	0.02	0.23	-0.06	0.01	-0.01	0.04
DH 07	-0.38	-0.10	0.26	-0.10	-0.03	0.04	0.05
GYLD 06	0.17	-0.07	0.13	-0.28	0.11	0.33	0.23
GYLD 07	0.11	0.13	-0.37	-0.05	0.45	0.13	-0.01
HLW 06	0.31	-0.03	0.13	-0.10	-0.33	0.22	0.26
HLW 07	0.43	0.07	-0.15	0.01	0.04	0.05	-0.05
TKW 06	0.05	0.38	0.34	0.04	0.02	0.16	-0.24
TKW 07	-0.12	0.39	0.37	-0.07	0.12	0.07	-0.06
KP25 06	0.31	0.17	0.26	-0.08	0.22	-0.01	-0.05
KP25 07	0.12	0.35	0.29	-0.11	0.20	-0.18	0.11
PROT 06	-0.10	0.17	-0.11	0.39	-0.19	0.37	-0.42
PROT 07	-0.12	-0.13	0.27	0.26	-0.28	0.11	0.19
TPC 06	0.01	-0.25	0.15	0.17	0.32	0.09	-0.52
TPC 07	0.08	-0.23	0.14	0.10	0.18	0.54	0.27
TAC 06	0.24	-0.29	0.29	0.24	0.04	-0.32	-0.09
TAC 07	0.30	-0.28	0.26	0.19	0.04	-0.23	-0.01
YP 06	-0.24	0.05	-0.05	0.17	0.42	-0.24	0.32
YP 07	-0.14	-0.27	0.05	0.17	0.39	0.23	0.10
GLUC 06	0.08	0.28	-0.04	0.44	0.02	0.13	0.25
GLUC 07	0.04	0.21	-0.06	0.52	-0.05	-0.15	0.26

[1]Abbreviations for traits followed by year: GYLD: grain yield; HLW: hectolitre/test weight; KP25: percentage kernel plumpness >2.5 mm; TKW: thousand kernel weight; GLUC, beta-glucan content; PROT, protein content; TAC, total anthocyanin content; TPC, total phenolics content; YP, yellow pigment content.

with the scores were KP25, TPC and YP. For the last two PC many traits have shown different correlation coefficients depending on the season.

Cluster analysis

Based on cluster analysis, the 81 hull-less spring barley genotypes were separated into seven major groups which each have two or more subgroups. Figure 1 illustrates the seven clusters formed by hierarchical clustering. Table 6 summarizes the number of genotypes

in each cluster and their proportion in terms of their countries of origin. ANOVA revealed significant differences between clusters for all traits. Means of clusters and Tukey-Kramer mean comparisons are presented in Table 7.

The first cluster (Clus1) is composed by the largest number of genotypes (n=25). Yellow grained genotypes comprised 80% and blue aleurone types 16% of the group. One black grained genotype was also included in the group. The yellow grained genotypes originated mainly from Western Europe, e.g. Germany and Italy, whereas the pigmented forms originate almost all from

Table 5. Pearson correlation coefficients of significant (P<0.05) correlations between trait × year means and loadings of the first seven principal components.

Trait × Year[1]	PC1	PC2	PC3	PC4	PC5	PC6	PC7
DH 06	-0.72		0.34				
DH 07	-0.74		0.39				
GYLD 06	0.32			-0.38		0.39	0.23
GYLD 07		0.23	-0.55		0.56		
HLW 06	0.59				-0.41	0.26	0.26
HLW 07	0.83		-0.22				
TKW 06		0.64	0.51				-0.25
TKW 07	-0.23	0.65	0.54				
KP25 06		0.29	0.38		0.27		
KP25 07		0.59	0.43		0.25		
PROT 06		0.29		0.52	-0.24	0.44	-0.43
PROT 07	-0.24		0.40	0.35	-0.35		
TPC 06		-0.42	0.23	0.23	0.40		-0.53
TPC 07		-0.40			0.23	0.64	
TAC 06	0.47	-0.49	0.42	0.32		-0.38	
TAC 07	0.58	-0.48	0.39	0.25		-0.27	
YP 06	-0.47			0.22	0.53	-0.28	0.32
YP 07	-0.26	-0.46		0.23	0.50	0.27	
GLUC 06		0.47		0.59			0.26
GLUC 07		0.36		0.69			0.26

[1]Abbreviations for traits followed by year: GYLD: grain yield; HLW: hectolitre/test weight; KP25: percentage kernel plumpness >2.5 mm; TKW: thousand kernel weight; GLUC, beta-glucan content; PROT, protein content; TAC, total anthocyanin content; TPC, total phenolics content; YP, yellow pigment content

Ethiopia. The group is mainly characterized by high values for HLW, TKW and PROT (Table 7).

The second cluster (Clus2) is a small group of only three varieties and is tightly linked to Clus1. This group consists of 'waxy' starch cultivars *Wanubet* and *Washonubet* from USA and *CDC Candle* from Canada. This cluster is especially distinct to the other ones concerning GLUC and YP. For these traits the three varieties exhibit the highest contents over all environments. It is worth mentioning that the other two 'waxy' cultivars, that is, *Merlin* and *HB 803*, were grouped in the neighbouring Clus1.

The third cluster (Clus3) comprises the German variety *Hora* together with Czech and German breeding lines. The latter genotypes show black grain pigmentation. This group of genotypes are late in heading (DH) and exhibited the lowest mean values for GLUC. Together Clus1, 2 and 3 form a major cluster of mainly yellow grained barley of European and/or American origin.

Contrary, the next three clusters (Clus4, 5 and 6) form a major group of accessions originating from diverse countries, amongst others from centres of barley domestication and/or diversification. Clus4 is heterogeneous in regard to grain pigmentation and origin, however, includes the German and Italian cultivars *Lawina* and *Nackta*, and *Digersano*, respectively. This is interesting since all other standard European varieties and/or breeding lines are concentrated in the first major group formed by Clus1, 2 and 3. Genotypes of Clus4 are characterized by high values of HLW, medium to high GYLD and TPC, and low content of GLUC. For all other traits the genotypes show average performance. Clus4 forms another bigger group together with Clus5. In the latter cluster mainly pigmented Asian and Ethiopian barley accessions are present. Like Clus4 this group is characterized by intermediate values for agronomic and quality traits. Concerning KP25 Clus5 shows significantly lower values of plump kernels. From a breeding perspective, one of the most interesting groups is Clus6. This cluster contains yellow, blue and black pigmented genotypes from Ethiopia, Nepal, Japan and Australia. The accessions are characterized by early heading, medium to high GYLD, high HLW and KP25. Unlike in other clusters the agronomic traits have contributed to a greater extent for the classification of this cluster.

The last and very distinct cluster, Clus7, consists of only purple pigmented barely accessions. Concerning their origin it must be supposed that all accessions originate from the Himalaya region. Germany and the USA represent only countries which carried out collection missions to the Himalaya and stored these genotypes in their national gene banks. Clus7 genotypes are

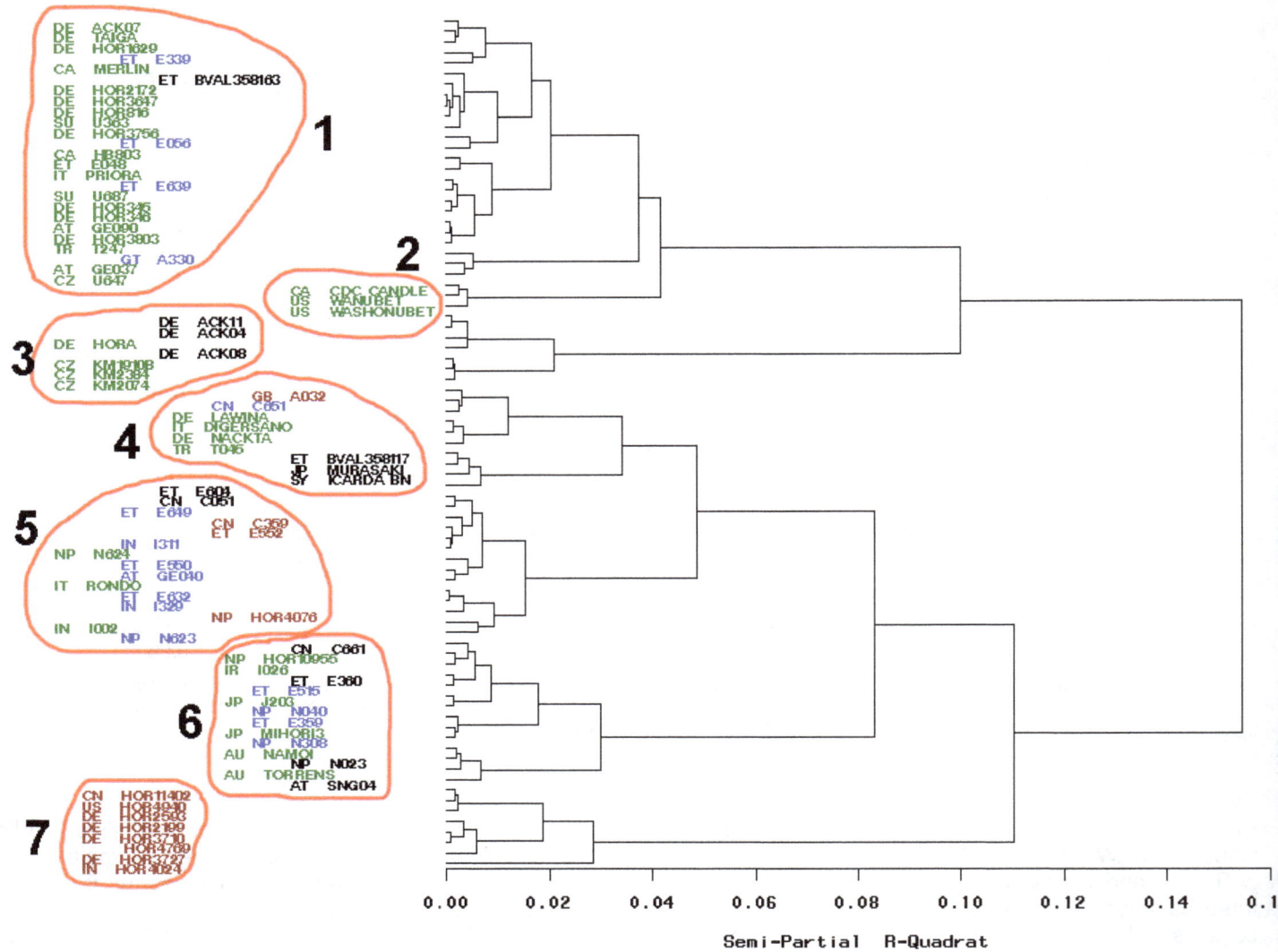

Figure 1. Dendrogram of the cluster analysis based on the genotypic scores of the first seven principal components. Colours of genotype names indicate their grain colour (green, white/yellow; blue, blue aleurone; red, purple pericarp; black, black pericarp). ISO3166 country codes are indicated before the genotype name (for abbreviation of country codes and genotype names see Table 9.2). Breeding lines of Saatzucht Ackermann are abbreviated as follows: ACK07, 00/900/19/3/7; ACK11, 00/900/19/6/11; ACK04, 00/900/19/6/4; ACK08, 00/900/19/6/8.

Table 6. The number and proportion of genotypes in each cluster with respect to their countries of origin.

Origin[1]	Clus1 N	Clus1 %	Clus2 N	Clus2 %	Clus3 N	Clus3 %	Clus4 N	Clus4 %	Clus5 N	Clus5 %	Clus6 N	Clus6 %	Clus7 N	Clus7 %
AT	2	50							1	25	1	25		
AU											2	100		
CA	2	67	1	33										
CN							1	20	2	40	1	20	1	20
CZ	1	25			3	75								
DE	10	50			4	20	2	10					4	20
ET	5	36					2	14	4	29	3	21		
GB							1	100						
GT	1	100												
IN									3	75			1	25

Table 6. Contd.

IR											1	100		
IT	1	50			1	50								
JP					1	33					2	67		
NP									4	50	4	50		
SU	2	100												
TR	1	50			1	50								
UN					1	50							1	50
US			2	67									1	33
Total	25	31	3	4	7	9	10	12	14	17	14	17	8	10

[1] ISO3166 country codes: AT, Austria; AU, Australia; CA, Canada; CN, China; CZ, Czech Republic; DE, Germany; ET, Ethiopia; GB, United Kingdom; GT, Guatemala; IN, India; IR, Iran; IT, Italy; JP, Japan; NP, Nepal; SU, Soviet Union; UN, unknown; US, United States of America.

Table 7. Means of agronomic and quality traits of the seven clusters. Means with the same letter are not significantly different from each other at 0.05.

Cluster	DH06[1]	DH07	GYLD06	GYLD07	HLW06	HLW07	TKW06	TKW07	KP25 06	KP25 07
	42 abc	25 ab	221 b	201 b	74 a	77 ab	40 a	39 a	28 abc	32 a
	46 ab	26 ab	224 ab	323 a	67 b	73 bc	32 b	32 cd	19 c	14 ab
	47 a	32 a	276 ab	182 b	71 ab	71 c	32 b	36 ab	18 c	23 ab
	40 bc	26 ab	357 a	209 b	77 a	77 ab	33 b	32 bc	24 bc	13 ab
	36 cd	20 bc	205 b	192 b	75 a	77 ab	30 b	27 d	16 c	10 b
	34 d	17 c	315 ab	258 ab	75 a	80 a	36 ab	33 bc	48 a	32 a
	36 cd	22 bc	264 ab	185 b	76 a	80 a	34 ab	32 cd	41 ab	25 ab

Cluster	YP06	YP07	PROT06	PROT07	TPC06	TPC07	TAC06	TAC07	GLUC06	GLUC07
	5.8 ab	5.7 a	14.4 a	15.5 ab	214 b	220 b	11 b	8 b	5.3 b	5.4 b
	7.1 a	6.1 a	13.5 ab	14.2 b	212 b	234 ab	5 b	6 b	6.8 a	7.0 a
	6.5 ab	6.6 a	12.1 c	15.2 ab	225 ab	239 ab	6 b	6 b	3.9 c	4.0 c
	5.2 b	7.0 a	13.9 ab	16.0 a	232 ab	289 a	23 b	24 b	4.7 bc	4.7 bc
	5.2 b	5.7 a	13.9 ab	15.6 ab	215 b	215 b	30 b	24 b	4.9 bc	5.1 b
	5.1 b	5.6 a	13.7 ab	14.1 b	209 b	231 ab	17 b	15 b	5.2 b	5.1 b
	5.1 b	6.3 a	13.0 bc	15.9 a	263 a	258 ab	227 a	144 a	4.9 bc	5.3 b

[1] For abbreviations followed by years see descriptions under Tables 4 and 5.

characterized by the highest content of TAC and TPC. High values were also observed for HLW and KP25, whereas the yield level is only moderate to low. Correlations between agronomic and quality traits were found to be in most cases non-significant or not consistent across the environments. The only relationship showing significant negative correlations in both years was between PROT and GYLD (r = -0.23*; r = -0.47**).

Figure 2 displays the relationships between selected traits and beta-glucan content within the seven clusters. It is obvious that with the exception of Clus3 in each cluster genotypes with GLUC >5% and high levels of TAC, TPC, YP or PROT are available. These genotypes could be used in a first step in a breeding programme for 'healthy' hull-less barley varieties by pyramiding the respective traits. Concerning GYLD major improvements are necessary to reach yields comparable to spring malting barley varieties adapted to the growing conditions of eastern Austria which range from 50 to 60 dt ha^{-1}.

Canonical discriminant analysis

CDA using grain colour as classifying variable resulted in three significant discriminant functions (Table 8). Since the first two functions explained together 88% of the multivariate variation only these two functions were plotted in Figure 3. High scores for the first canonical function were associated with TAC as the most important parameter for differentiating the genotypes. In addition to TAC, TPC and TKW contributed to the separation of genotypes with the first function (Can1). Concerning the second canonical function (Can2) YP, GLUC and HLW were the most important traits for differentiation between

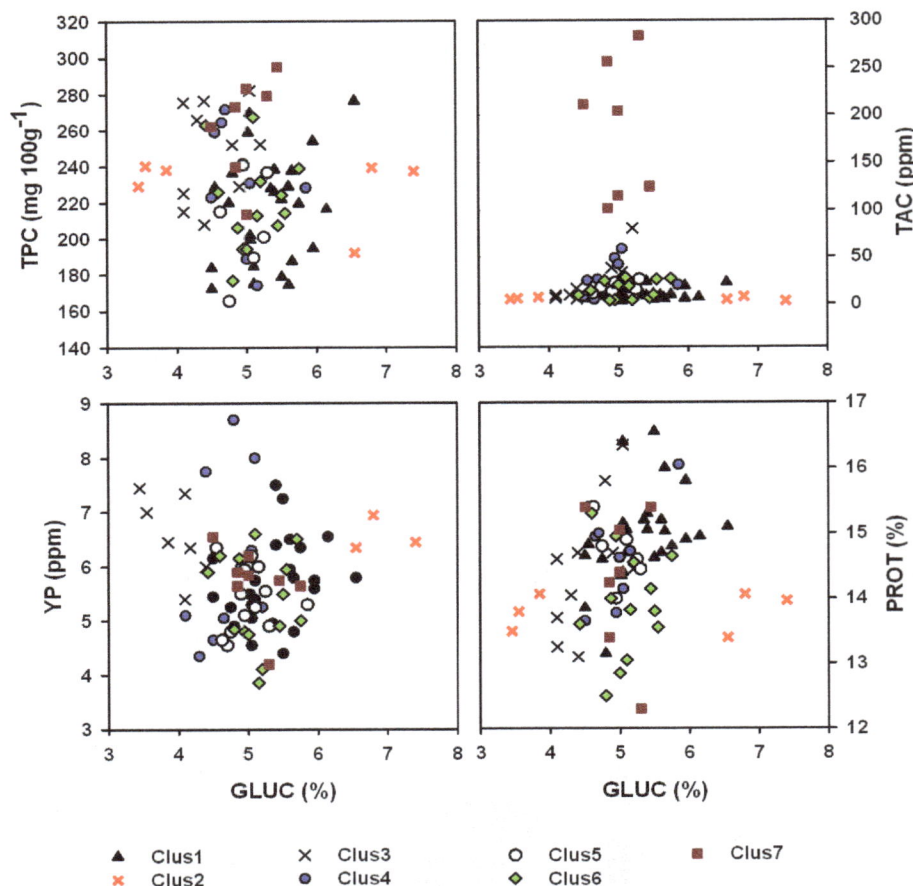

Figure 2. Scatter plot of quality traits (PROT, protein content; TPC, total phenolic content; TAC, total anthocyanin content; YP, yellow pigment content) within clusters in relation to beta-glucan content (GLUC).

Table 8. Canonical discriminant analysis of hullless spring barley.

CAN	Canonical correlation	Eigen value	Proportion	Likelihood ratio (Wilks' lamda)	Pr>F
1	0.899	4.19	0.58	0.033	<.0001
2	0.826	2.15	0.30	0.172	<.0001
3	0.676	0.85	0.12	0.542	0.0014

groups. From Figure 3, it is obvious that along Can1 only three groups of hull-less barley can be differentiated clearly: yellow (*blx*), blue (*Blx*) and black (*Blp*), and purple (*Pre*) seeded genotypes. Blue aleurone and black pericarp types are not significantly separated by Can1, however, a clear separation of these two groups is possible by Can2.

DISCUSSION

Effective application of multivariate analysis on

agronomic and quality characters can result in meaningful grouping of genotypes. On the basis of genetic diversity in regard to important agronomic and quality traits, the investigated hull-less spring barley germplasm was grouped into seven groups by PCA and following cluster analysis. This procedure revealed that European and American varieties and breeding lines differentiated more or less significantly from African and Asian material. Exceptions were the Italian varieties *Digersano* and *Rondo*, the German varieties *Nackta* and *Lawina* and the Australian varieties *Torrens* and *Namoi*. In the latter cases the early heading and maturity date of

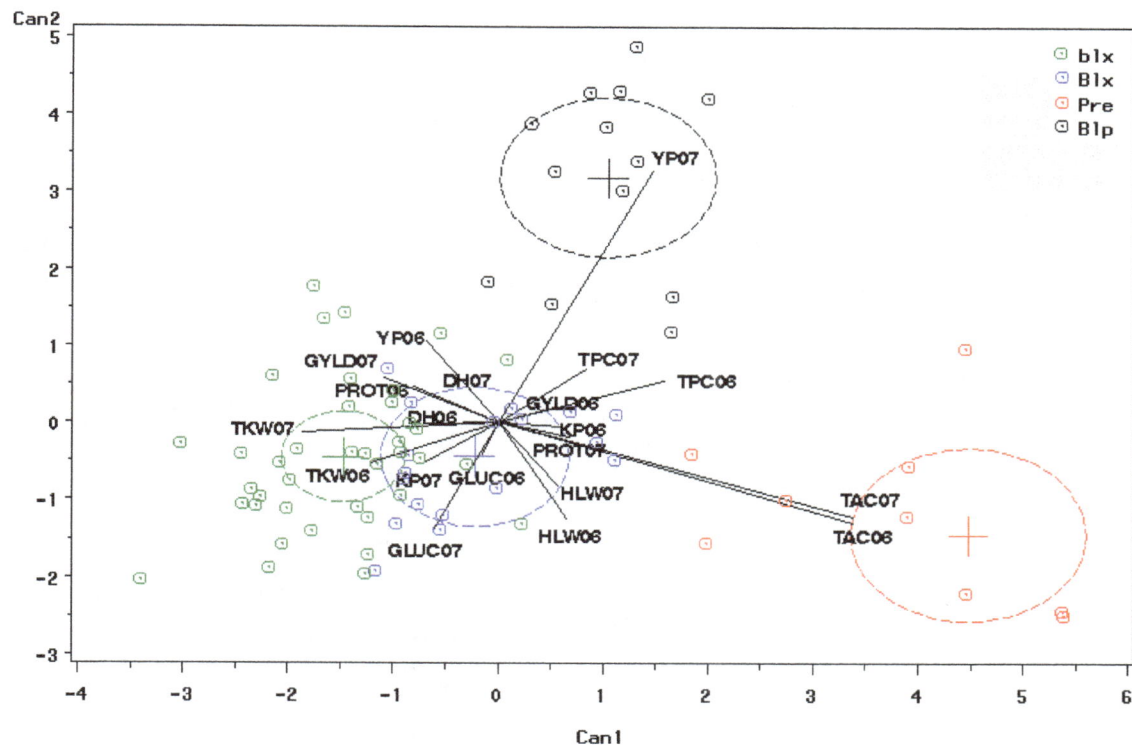

Figure 3. Distribution of hullless spring barley genotypes according to canonical discriminant analysis. The length of each variable (for abbreviations see footnote of first chapter page) vector is proportional to its contribution to separating the grain colour classes (*blx*: non blue aleurone; *Blx*: blue aleurone; *Blp*: black pericarp; *Pre*: purple/red pericarp), and the direction of the vector indicates its relative contribution to the Can1 and Can2 linear combinations. Single genotypes are marked as dots. Mean values of kernel colour groups are indicated by the plus symbol, and their 95% confidence interval by dashed circles.

the Australian barleys are more similar to the Ethiopian and Asian genotypes. The phenotype of *Rondo*, 6-row *uzu*-dwarf barley, resembles also more some of the Japanese genotypes. The narrow genetic background of waxy barley is demonstrated by the fact that three of these genotypes formed a small distinct cluster, while the other two genotypes were located in the neighbouring cluster. An efficient separation using different multivariate analysis techniques was demonstrated for the purple pericarp accessions. Both PCA combined with cluster analysis and CDA clearly separated the purple pigmented genotypes. Recently PCA has been used by various authors for the reduction of multivariate data into a few artificial varieties which can be further used for classifying material. This approach is especially valuable for the screening of a large number of genetic resources by a large number of descriptor variables (Cartea et al., 2002; Granati et al., 2003; Kamara et al., 2003; Salihu et al., 2006).

Canonical discriminant analysis revealed that TAC was the best variable for distinguishing purple pericarp, blue aleurone and non blue aleurone types of hull-less barley. Zeven (1991) reported that in wheat purple and blue

colour types are caused by anthocyanins located in the pericarp or aleurone, respectively. Contrary to our results Abdel-Aal et al. (2006) indicated that blue aleurone wheat had significantly higher TAC than purple types. However, Kim et al. (2007) found non-significant differences between TAC levels of blue (337.6 ± 152.4 µg g^{-1}) and purple (312.7 ± 244.3 µg g^{-1}), but significantly lower TAC values for black (84.5 ± 37.2 µg g^{-1}) genotypes of Korean hull-less barley. Abdel-Aal et al. (2006) also found that blue barley *Tankard* had a TAC value of 35 µg g^{-1} which is similar to our results for some of the blue types. Significantly higher amounts of TAC were reported for black rice and purple corn (Abdel-Aal et al. 2006). Hence, various TAC values were reported depending on genotypes and crop species suggesting the requirement for further studies by including more genotypes and species.

Black pericarp barley was efficiently separated by the other grain colour classes by YP. Hitherto, no references are available that black barley generally contains higher amounts of yellow pigments. Since both methods for the determination of TAC and YP are simple spectrophotometrical methods, it is necessary as a following step to quantify and identify the exact anthocyanins and carotenoids of blue aleurone and black pericarp barley by e.g. HPLC. The correlation analysis between agronomic and quality traits was found to be non-significant between almost all the traits. The absence of correlation and the negative correlation among agronomic and quality traits could be a big challenge and/or impediment in the future to develop hull-less barley cultivars that are endowed with quality traits like phenolics and soluble fibre (beta-glucan). Our results showed that there were no correlation of GLUC with GYLD, HLW, TKW and KP25. However, Hang et al. (2007) reported that GLUC was found to be positively correlated with KP25, but showed a weak negative correlation with GYLD in 27 barley genotypes. Similar to our results, Peterson et al. (1995) reported that the correlations between GLUC and GYLD, HLW, and TKW were generally nonexistent or inconsistent across years or nurseries in oat.

The knowledge of the phenolic metabolic pathways has now opened the possibility of directly developing new varieties with specifically modified phenolic profiles. In particular, the use of molecular techniques to 'fine tune' the control of phenolic metabolism to up-regulate desirable metabolic routes or to down-regulate undesirable ones is now a very real possibility (Parr and Bolwell 2000). Based on the present results it was recommended to make crosses among genotypes in Clus1, Clus5, Clus6 and Clus7 in breeding programmes. Classifying genotypes according to their agronomic and quality traits with sophisticated multivariate techniques can reduce the cost of time and money in crop improvement. However, stability analysis of different traits on the already established groups of the current study requires further investigations based on sufficient data that cover different years and experimental locations.

REFERENCES

Abdel-Aal ESM, Hucl P (1999). A rapid method for quantifying total anthocyanins in blue aleurone and purple pericarp wheat. Cereal Chem. 76: 350-354.

Abdel-Aal ESM, Hucl P (2003). Composition and stability of anthocyanins in blue-grained wheat. J. Agric. Food. Chem. 51: 2174-2180.

Abdel-Aal ESM, Young JC, Rabalski I (2006). Anthocyanin composition in black, blue, pink, purple, and red cereal grains. J. Agric. Food Chem. 54: 4696-4704.

Asfaw Z (1989). Relationship between spike morphology, hordeins and altitude within Ethiopian barley, Hordeum vulgare L. (Poaceae). Hereditas 110: 203-209.

Assefa A, Labuschagne MT (2004). Phenotypic variation in barley (Hordeum vulgare L.) landraces from North Shewa in Ethiopia. Biodivers Conserv. 13: 1441-1451.

Atanassov P, Borories C, Zaharieva M, Monneveux P (2001). Hordein polymorphism and variation of agromorphological traits in a collection of naked barley. Genet. Res. Crop Evol. 48: 53-360.

Baidoo SK, Liu YG (1998). Hull-less barley for swine: ileal and faecal digestibility of proximate nutrients, amino acids and non-starch polysaccharides. J. Sci. Food Agric. 76: 397-403.

Bhatty RS (1999). The potential of hull-less barley. Cereal Chem. 76: 589-599.

Bonman JM, Bockelman HE, Jackson LF, Steffenson BJ (2005). Disease and insect resistance in cultivated barley accessions from USDA National Small Grains Collection. Crop Sci. 45: 1271-1280.

Buckley GFH (1930). Inheritance in barley with special reference to the color of the caryopsis and lemma. Sci. Agric. 10: 460-492.

Cartea ME, Picoagea A, Soengas P, Ordás A (2002). Morphological characterization of kale populations from northwestern Spain. Euphytica 129: 25-32.

Finch RA, Simpson E (1978). New colours and complementary colour genes in barley. Z Pflanzenzüchtg 81: 40-53.

Friendly M (1991). SAS® system for statistical graphics. SAS Institute, Cary, NC p. 697.

Granati E, Bisignano V, Chiaretti D, Crino P, Polignano BG (2003). Characterization of Italian and exotic Lathyrus germplasm for quality traits. Genet. Res. Crop Evol. 50: 273-280.

Hair FF, Anderson RE, Tatham RL, Black WC (1998). Multivariate data analysis, 5th Ed. Prentice-Hall Inc, Upple Saddle River, NJ, USA pp. 239-325.

Hang A, Obert D, Gironella AIN, Burton CS (2007). Barley amylose and beta-glucan: Their relationships to protein, agronomic traits, and environmental factors. Crop Sci. 47: 1754-1760.

Hu C, Cai Y, Li W, Corke H, Kitts DD (2007). Anthocyanin characterization and bioavailability assessment of a dark blue grained wheat (Triticum aestivum L. cv. Hedong Wumai) extract. Food Chem. 104: 955-961.

Kamara AY, Kling JG, Menkir A, Ibikunle O (2003). Agronomic performance of maize (Zea mays L.) breeding lines derived from low nitrogen maize population. J. Agric. Sci. 141: 221-230.

Kim MJ, Hyun JN, Kim JA, Park JC, Kim MY, Kim JG, Lee SJ, Chun SC, Chung IM (2007). Relationship between phenolic compounds, anthocyanins content and antioxidant activity in coloured barley germplasm. J. Agric. Food Chem. 55: 4802-4809.

Lundqvist U, Franckowiak J, Konishi T (1997). New and revised descriptions of barley genes. Barley Genet. Newsl. 26: 22-516.

Negassa M (1985). Pattern of phenotypic diversity in an Ethiopian barley collection and the Arisi-Bale highlands as a center of origin of barley. Hereditas 102: 139-150.

Nielsen JP, Munck L (2003). Evaluation of malting barley quality using exploratory data analysis. I. Extraction of information from micro-malting data of spring and winter barley. J. Cereal Sci. 38: 173-180.

Pandey M, Wagner C, Friedt W, Ordon F (2006). Genetic relatedness and population differentiation of Himalayan hull-less barley (Hordeum vulgare L.) land-races inferred with SSRs. Theor. Appl. Genet. 113: 715-729.

Parr AJ, Bolwell GP (2000). Phenols in plant and in man. The potential for possible nutritional enhancement of the diet by modifying the phenols content or profile. J. Sci. Food Agric. 80: 985-1012.

Peterson DM, Wesenberg DM, Burrup DE (1995). beta-Glucan content and its relationship to agronomic characteristics in elite oat germplasm. Crop Sci. 35: 965-970.

Salihu S, Grausgruber H, Ruckenbauer P (2006). Agronomic and quality performance of international winter wheat genotypes grown in Kosovo. Cereal Res. Commun. 34: 957-964.

SAS (1990). SAS/GRAPH® software, SAS Institute, Cary, NC, USA. 6: 2.

SAS (1999a). SAS/STAT® user's guide, SAS Institute, Cary, NC, USA. 8: 3.

SAS (1999b). SAS® procedures guide, SAS Institute, Cary, NC, USA. 8: 1.

SAS (1999c). SAS/STAT® user's guide, Vers. SAS Institute, Cary, NC, USA 8: 1.

SAS (1999d). SAS/STAT® user's guide, SAS Institute, Cary, NC, USA. 8: 2.

Singleton VL, Orthofer R, Lamuela-raventos RM (1999). Analysis of total phenolics and other oxidation substrates and antioxidants by means of Folin-Ciocalteu reagent. Methods Enzymol. 299: 152-178.

Taketa S, Kikuchi S, Awayama T, Yamamoto S, Ichii M, Kawasaki S (2004). Monophyletic origin of naked barley inferred from molecular analysis of a marker closely linked to the naked caryopsis gene (nud). Theor. Appl. Genet. 108: 1236-1242.

Taketa S, Amano S, Tsujino Y, Sato T, Saisho D, Kakeda K, Nomura M,

Suzuki T, Matsumoto T, Sato K, Kanamori H, Kawasaki S, Takeda K (2008). Barley grain with adhering hulls is controlled by an ERF family transcription factor gene regulating a lipid biosynthesis pathway. Proc. Nat. Acad. Sci. 105: 4062-4067.

Wang X, Olsen O, Knudsen S (1993). Expression of the dihydroflavonol reductase gene in an anthocyanin-free barley mutant. Hereditas 119: 67-75.

Woodward RW (1941). Inheritance of melanin-like pigment in the glumes and caryopses of barley. J. Agric. Res. 63: 21-28.

Woodward RW, Thieret JW (1953). A genetic study of complementary genes for lemma, palea, and pericarp in barley (Hordeum vulgare L.). Agron. J. 45: 183-185.

Xue Q, Wang L, Newman RK, Newman CW, Graham H (1997). Influence of the hull-less, waxy starch and short-awn genes on the composition of barleys. J. Cereal Sci. 26: 251-257.

Žáková M, Benková M (2006). Characterization of spring barley accessions based on multivariate analysis. Commun. Biom. Crop Sci. 1: 124-134.

Zeven AC (1991). Wheats with purple and blue grains: a review. Euphytica 56: 243-258.

16

Estimation of genetic divergence among some cotton varieties by RAPD analysis

L. Chaudhary[1]*, A. Sindhu, M. Kumar, R. Kumar[2] and M. Saini[3]

[1]Department of Biotechnology Engineering, Ambala College of Engineering and Applied Research, Mithapur, Ambala Cantt., Haryana, India- 133 101.
[2]Department of Botany and Plant Physiology, CCS Haryana Agricultural University, Hisar-125 004, India.
[3]Department of Plant, Soil and Agricultural System, Southern Illinois University, Carbondale-62901, USA.

Total genomic DNA from 15 cotton varieties were analysed to evaluate genetic diversity among them through random amplified polymorphic DNA (RAPD) analysis, with 30 random decamer primers using the polymerase chain reaction (PCR). A total of 370 bands were observed, with 12.3 bands per primer, of which 91.6% were polymorphic. OPM-16 produced the maximum number of fragments while the minimum number of fragments was produced with the primer OPM-18. Cluster analysis by the unweighted paired group method of arithmetic means (UPGMA) showed that 15 varieties can be placed in five groups with a similarity ranging from 0.48 - 0.86. Maximum similarity was observed between H-777, H-974 and H-1098 (0.86). Interestingly, these varieties have been developed at one breeding center. The analysis revealed that the intervarietal genetic relationship of several varieties is related to their center of origin. Most of the varieties have a narrow genetic base. These results were well in accordance with previous reported results. The RAPD analysis indicates that it may be a more efficient marker than morphological marker, isozyme and restriction fragment length polymorphism (RFLP) technology. The results obtained can be used in selecting divergent parents for breeding and mapping purposes.

Key words: Cluster, cotton varieties, diversity, RAPD, genetic similarities.

INTRODUCTION

Cotton is the most important textile fiber crop and is the second most important oil seed crop in the world. Cotton belongs to genus *Gossypium* and comprises of 50 different species, distributed in eight genomes. Of the 50 species, only four species are cultivated in India. *Gossypium arboreum and Gossypium harbaceum* belong to the old world diploid group, where as the new world tetraploid cultivated species are *Gossypium hirsutum and Gossypium barbedance*. India is among the top three cotton producing countries with an annual production of nearly 23.2 million bales. A large number of cotton varieties grown in India originated from intraspecific crosses of *G. hirsutum* and this practice resulted in a narrow genetic base for the new varieties. Previously morphological markers, with their complex and undeciphered genetic control, were used for individual identification. Morphological features are indicative of genotypes but are represented by only a few loci because there is not a large enough number of a character available. Moreover, they can also be affected by environmental factors and growth practices. Protein markers can be used to provide varietal profiles because the variations for these markers are ubiquitous and this variation can be understood in genetic terms. Although proteins are products of the primary transcripts of DNA, environmental factors can affect qualitative and quantitative levels of protein. Wendel et al. (1992) studied the genetic distances of a large number of accessions of upland cotton from different locations by isozyme analysis. However, isozyme analysis has certain limitations due to the availability of limited number of marker loci. Within the last few years Restriction Fragment Length Polymorphism (RFLP) technology have been applied to several cotton species to study the

*Corresponding author. E-mail: lakshmi_gpb@rediffmail.com.

Table 1. List of cotton varieties used in genetic analysis studies.

Sr. No.	Variety	Year	Pedigree	Center of origin
1.	HS-6	1991	(BN X K3199) BN	CCSHAU, HISAR
2.	H-777	1978	Reselection BN	CCSHAU, HISAR
3.	H-974	1993	(H14X 45 red AK) H777	CCSHAU, HISAR
4.	H-1098	1997	(LH354 X SBI 71)H777	CCSHAU, HISAR
5.	H-1117	2002	Selection	CCSHAU, HISAR
6.	F-414	1978	Bikaneri narma	PAU, RS, FARIDKOT
7	LH-900	1987	(LH223 -480) LH223-343	PAU, LUDHIANA
8.	F-505	1987	F414X A231	PAU, RS, FARIDKOT
9.	F-846	1994	F452X LH223-481	PAU, RS, FARIDKOT
10.	LH-1556	1996	(LH886 X LH900) LH952	PAU, LUDHIANA
11.	RST-9	1992	BN X PS 10-27-1	RAU, RS, SRIGANGANAGAR
12.	RS-810	2000	RS644 X Khandwa 3	RAU, RS, SRIGANGANAGAR
13.	RS-2013	2002	Selection	RAU, RS, SRIGANGANAGAR
14.	Bikaneri Narma	1978	Local selection	RAU, BIKANER
15.	Ganganagar Ageti	1982	RS89	RAU, RS, SRIGANGANAGAR

evolution, population genetics, phylogenetic relationships and genome mapping (Yu et al., 1997), but it creates low variation in cotton compared to other plant taxa (Brubaker and Wendel, 1994). The random amplified polymorphic (RAPD) technique of Williams et al. (1990) provides an unlimited number of markers which can be used for various purposes like cultivar analysis and species identification in most plants. DNA fingerprinting studies to assess genetic purity with RAPD have already been conducted in cotton (Soregaon, 2004). Multani and Lyon (1995) reported that RAPD markers could be used to distinguish closely related varieties. Keeping in view of these findings the present work was planned to study the genetic divergence of popular cultivated varieties of *G. hirsutum*.

MATERIALS AND METHODS

Plant material

The plant material used in the study consisted of 15 cotton varieties (Table 1). The plants were grown in pots in a greenhouse.

DNA isolation

Total genomic DNA was isolated with the modified CTAB method (Saghai-Maroof et al., 1984). Approximately 5 g leaf material was ground to a fine powder using liquid nitrogen and quickly transferred into 25 ml of prewarmed (60°C) isolation buffer in a capped polypropylene tube, incubated for 1 hour at 65°C in a water bath and mixed by gentle swirling after every 10 min. To these tubes, equal volume of Chloroform: Isoamyl alcohol (24:1) was added and the contents were shaken for 10 min by hand. The tubes were centrifuged for 10 min at 8000 rpm; the upper aqueous layer was extracted twice with fresh Chloroform: Isoamyl alcohol and the final aqueous layer were transferred to a centrifuge tube. To these tubes, 0.6 volume of pre-chilled isopropanol was added and shaken

gently for precipitation. By using a glass hook, DNA was spooled out in the form of whitish fibers and transferred to washing solution and dried. DNA was dissolved in an appropriate volume of 1X TE buffer.

For purification, RNase A was added to the tube (50 μg/ ml) and the mixture incubated for 1 h at 37°C. DNA was extracted with CI by centrifuging the tubes at 10,000 rpm for 5 min at room temperature. DNA was precipitated with 2 volume of pre-chilled absolute ethanol and was recovered by centrifuging the tubes at 5000 rpm for 10 min; the pellet was washed with 70% ethanol and dissolved in appropriate volume of IX TE buffer.

PCR and gel electrophoresis

The random decamer oligonucleotide primers for the PCR were obtained commercially from Operon Technologies, Alameda, Calif; 30 random primers (OPM-01 to 20 primers; OPA - 01 - 10) were used in this study. PCRs were carried out in 0.05 cm^3 reaction volumes each containing 50 ng of genomic template DNA, 0.2 μM of the particular primer, 100μM of each dNTP, 2 μl of Taq polymerase 10X buffer, 1 unit Taq polymerase (Perkin Elmer) and 2.5 mM MgCl$_2$. PCR amplification was performed on a PTC 100 Thermal Cycler (MJ Research, Inc. Watertown, MA, USA) under the following conditions: Initial denaturation at 95°C for 5 min, followed by 35 cycles of denaturing at 94°C for 1 min, annealing at 36°C for 1 min, extension at 72°C for 2 min and final extension at 72°C for 10 min. The amplification products were resolved on 1.0% agarose gel and visualized under UV light following staining with ethidium bromide.

Data analysis

The frequency of RAPD polymorphism was calculated based on presence (taken as 1) or absence (taken as 0) of common bands (Ghosh et al., 1997). The binary data were used to compute Pairwise similarity coefficient (Jaccard, 1908) on NTSYS-PC. A dendrogram based on similarity coefficient was generated by using the unweighted pair group of arithmetic means (UPGMA).

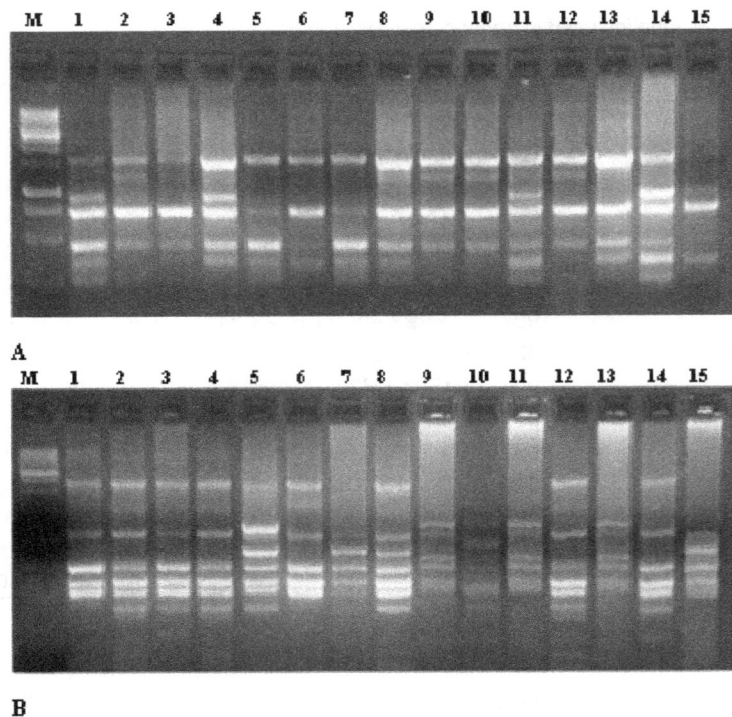

Figure 1. Amplified profile of 15 cotton varieties with primers (A) OPM-08 and (B) OPM-13. Lane M = 1kb size marker; Lane 1 = HS-6; Lane 2 = H-777; Lane 3 = H-974; Lane 4 = H-1098; Lane 5 = H-1117; Lane 6 = F-414; Lane 7 = LH-900; Lane 8 = F-505; Lane 9 = F-846; Lane 10 = LH-1556; Lane11 = RST-9; Lane 12 = RS-810; Lane 13 = RS-2013; Lane14 = Bikaneri Narma; Lane 15 = Ganganagar Ageti.

RESULTS AND DISCUSSION

DNA of 15 cotton varieties was amplified with 30 different random primers purchased from Operon Technologies. All 15 genotypes with the 30 primers revealed a unique banding pattern and so can be used for variety identification. This might be indicative of a wide genetic base of the cotton varieties studied. Different primers produced a different level of polymorphism among the different varieties (Figure 1A and B).

A total of 370 DNA fragments were amplified, with an average of 12.3 RAPD markers per primer. Out of 370 amplified fragments 31 (8.4%) were found to be monomorphic. The remaining 339 (91.6%) amplicons were polymorphic in one or other of the 15 varieties studied. The amplitude of polymorphisms was high even there was not a single primer (out of 30 studied) which could differentiate clearly all the varieties. The size of the amplified fragments also varied with different primers. The approximate size of the largest fragment produced was 3.0 kb and the smallest fragment produced was approximately 0.25 kb. Out of the 15 varieties studied, LH-1556 produced the maximum number of DNA amplified fragment (248), while LH-900 produced 153 bands, which is the minimum number. Other varieties produced common bands in the range of 189 - 236. The

variety H-974 and F-846 produced maximum number of common bands that is 223. A maximum of 14 fragments were amplified with primer OPM-16 and a minimum of 8 bands with primer OPM-18.

To estimate the genetic similarities of the cotton varieties a similarity matrix was obtained using Jaccard's similarity coefficient (1908) and is shown in Table 2. These similarity coefficients were used to generate a dendrogram (Figure 2) by UPGMA analysis in order to determine the grouping of different varieties. Maximum similarity was observed between H-777, H-974 and H-1098 (0.86). Interestingly, these varieties have been developed at one breeding center. On the basis of the RAPD data their genetic bases look very narrow. From the similarity matrix, the least similar variety is LH-900. Its similarity ranges from 0.48 - 0.57. The coefficient of similarity for most of the other varieties was found to be in the range 0.48 - 0.82. Similarly, Multani and Lyon (1995) studied a number of Australian cultivars and found 92.1 - 98.9% genetic similarity among nine cultivars of *G. hirsutum* L, while *G. barbedance* L. variety Pima S-7 showed about 57% similarity with *G. hirusutum* L. varieties. Tatineni et al. (1996) assessed genetic diversity among 19 cotton genotypes and compared the RAPD data with the taxonomic data. In their studies, 33.3% of the primers did not produce any polymorphism, while

Table 2. Jaccard similarity coefficient among 15 cotton varieties as revealed from RAPD marker analysis.

	HS-6	H-777	H-974	H-1098	H-1117	F-414	LH-900	F-505	F-846	LH-1556	RST-9	RS-810	RS-2013	Bikaneri Narma	Ganganagar Ageti
HS-6	1.00														
H-777	0.67	1.00													
H-974	0.65	0.86	1.00												
H-1098	0.65	0.82	0.83	1.00											
H-1117	0.69	0.60	0.55	0.60	1.00										
F-414	0.58	0.72	0.74	0.74	0.56	1.00									
LH-900	0.63	0.57	0.53	0.55	0.69	0.50	1.00								
F-505	0.60	0.75	0.75	0.76	0.59	0.74	0.57	1.00							
F-846	0.60	0.76	0.75	0.77	0.53	0.74	0.54	0.77	1.00						
LH-1556	0.61	0.75	0.78	0.81	0.58	0.72	0.51	0.76	0.76	1.00					
RST-9	0.62	0.76	0.75	0.75	0.55	0.74	0.48	0.74	0.77	0.82	1.00				
RS-810	0.58	0.72	0.71	0.74	0.61	0.72	0.54	0.73	0.75	0.82	0.81	1.00			
RS-2013	0.60	0.72	0.72	0.76	0.59	0.71	0.50	0.78	0.76	0.78	0.76	0.76	1.00		
Bikaneri Narma	0.60	0.70	0.70	0.72	0.58	0.70	0.52	0.70	0.72	0.70	0.72	0.68	0.76	1.00	
Ganganagar Ageti	0.63	0.76	0.73	0.74	0.57	0.70	0.55	0.75	0.74	0.74	0.73	0.74	0.77	0.75	1.00

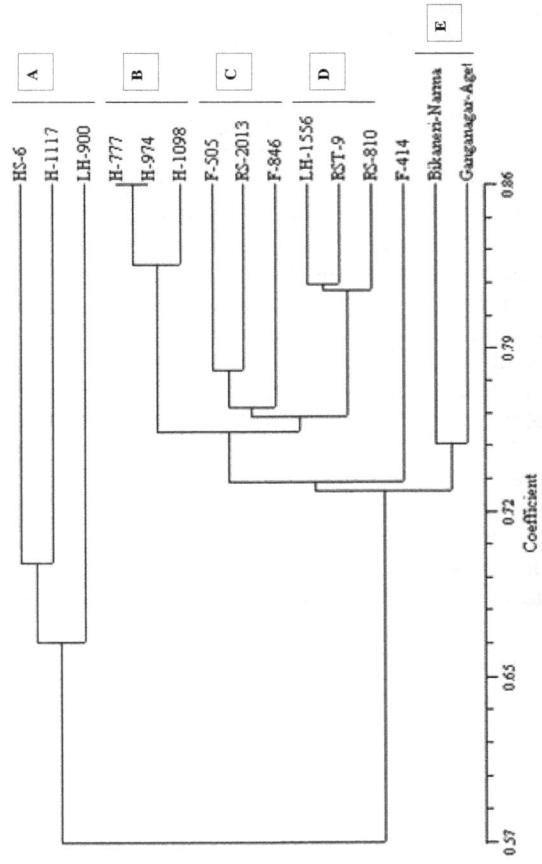

Figure 2. Dendrogram showing the genetic similarity among 15 cotton varieties as derived from RAPD data using the unweighted pair group method of arithmetic means (UPGMA).

Iqbal et al. (1997) reported the existence of 70 - 90% similarity coefficient in 22 *G. hirusutum* varieties but with very low level of genetic divergence. Brubaker and Wendel (1994) also demonstrated that the level of RFLP diversity was low in *G. hirsutum* L. cultivars as compared to other reported taxa. Cluster analysis using RAPD resulted in five main cluster groups. The dendrogram (Figure 2) assigned the varieties into groups which correspond well with their centers or sub centers of release and / or pedigree relationship. In cluster 'B' three varieties H-777, H-974 and H-1098 are more closely related as compared to any other variety. They have high estimates of genetic identity (0.86). In cluster 'A' out of three *G. hirsutum* varieties HS-6 and H-1117 clustered together indicating that they are more closely related as compared to LH-900. The cluster 'C' and 'D' comprises three varieties each namely F-505, RS-2013 and F-846 and LH-1556, RST-9 and RS-810, respectively. Their similarity ranges between 0.70 - 0.77 and 0.70 - 0.82, respectively. The variety F-414 showed a similarity index of 0.74 with the rest of the varieties of cluster B, C and D. The cluster 'E' comprises of two varieties Bikaneri-Narma and Ganganagar-Ageti. Earlier, subclustering had been reported in some elite cotton genotypes developed at one breeding station (Iqbal et al., 1997). The clustering of the varieties might be due to selection of the elite lines from a single population. Esbroeck et al. (1998) shown that genetic uniformity of U.S. cotton cultivars is greater today than it was 25 years ago. This was also reported in Basmati rice by Bligh et al. (1999). Moreover, breeders mostly shares the elite lines of other breeding stations in cotton improvement programs, making the breeding material identical which ultimately result in close kinship of the varieties. The genetic similarity obtained from the analysis will be useful in selecting divergent parents for breeding and mapping purposes.

REFERENCES

Bligh HFG, Blackhall NM, Edwards KJ, McClung AM (1999). Using amplified fragment length polymorphism and simple sequence length polymorphism to identify cultivars of brown and white milled ride. Crop Sci. 39: 1715-1721.

Brubaker CL, Wendel JF (1994). Re-evaluating the origin of domesticated cotton (*Gossypium hirsutum*; Malvaceae) using nuclear restriction fragment length polymorphisms (RFLPs). Am. J. Bot. 81: 1309-1326.

Esbroeck Van GA, Bowman DT, Calhoun DS, May OL (1998). Changes in the genetic diversity in cotton in the USA from 1970 to 1995. Crop Sci. 38: 33-37.

Ghosh S, Karanjawala ZE, Hauser ER (1997). Methods for precise sizing, automated inning of alleles and reduction or error rates in large scale genotyping using fluorescently labeled dinucleotide markers. Genome Res. 7: 165-178.

Iqbal MJ, Aziz N, Saeed NA, Zafar Y, Malik KA (1997). Genetic diversity evaluation of some cotton varieties by RAPD analysis. Theor. Appl. Genet. 94: 139-144.

Jaccard P (1908). Nauvelles recherches sur la distribution florale. Bull. Soc. Vaudoise Sci. Nat. 44: 223-270.

Multani DS, Lyon BR (1995). Genetic fingerprinting of Australian cotton cultivars with RAPD markers. Genome 38: 1005-1008.

Saghai-Maroof MA, Soliman KM, Jorgensen RA, Allard RW (1984). Ribosomal DNA spacer length polymorphism in barley: Mandelian inheritance, chromosomal location and population dynamics. Proc. Natl. Acad. Sci. (USA). 81: 8014-8018.

Soregaon CD (2004). Studies on genetic introgression in interspecific crosses of cotton (*Gossypium* spp) M.Sc. (Agri.) Thesis, UAS, Dharwad, India. p. 130.

Tatineni V, Cantrell RG, Davis DD (1996). Genetic diversity in elite cotton germplasm determined by morphological characteristics and RAPD. Crop Sci. 36: 186-192.

Wendel JF, Brubaker CL, Percival AE (1992). Genetic diversity in *Gossypium hirsutum* and the origin of upland cotton. Am. J. Bot. 79: 1291-1310.

Williams JGK, Kubelik AR, Levak KJ, Rafalski JA, Tingey SV (1990). DNA polymorphism amplification by arbitrary primers are useful as genetic markers. Nucleic Acid Res. 18: 6531-6535.

Yu ZH, Park YH, Lazo GR, Kohel RJ (1997). Molecular mapping of the cotton genome. Agron. Abstracts, ASA, Madison, Wiconsin. p. 147.

17

Genetic diversity and linkage disequilibrium of two homologous genes to maize *D8*: Sorghum *SbD8* and pearl millet *PgD8*

Yongle Li[1], Sankalp Bhosale[2], Bettina I. G. Haussmann[3], Benjamin Stich[4], Albrecht E. Melchinger[2] and Heiko K. Parzies[2*]

[1]Plant Breeding, Technische Universität München/Centre of Life and Food Sciences Weihenstephan, 85350 Freising, Germany.
[2]Institute for Plant Breeding, Seed Science, and Population Genetics, University of Hohenheim, 70593 Stuttgart, Germany.
[3]International Crops Research Institute for the Semi-Arid Tropics (ICRISAT), BP 12404, Niamey, Niger.
[4]Max Planck Institute for Plant Breeding Research, 50829 Cologne, Germany.

Yield and yield stability of sorghum [*Sorghum bicolor* (L.) Moench.] and pearl millet [*Pennisetum glaucum* (L.) R.Br.] are highly influenced by flowering time and photoperiodic sensitivity in the arid to semi-arid regions of West and Central Africa. Photoperiodic sensitivity is the key adaptation trait of local landraces because it assures flowering at the end of the rainy season, independent of variable dates of planting. Flowering time genes are mainly integrated into four pathways with close interaction among each other: Vernalization, autonomous, GA (gibberellic acid) and photoperiod. In the GA pathway, maize *D8*, wheat *RHT1* and rice *SLR* have been identified as homologous genes to the Arabidopsis *GAI*, which is a negative regulator of GA response. We have identified two homologous genes to *D8*: Sorghum *SbD8* and pearl millet *PgD8*. These genes were expressed in the root and leaves of sorghum and pearl millet as revealed by EST database search and reverse transcription PCR, respectively. The genetic diversity of *SbD8* was considerably lower than that of *PgD8*. The extent of linkage disequilibrium in *PgD8* is lower than that of maize *D8*. *SbD8* and *PgD8* polymorphisms might be appropriate for dissection of photoperiod sensitivity using association mapping approaches.

Key words: DELLA proteins, GA pathway, flowering time, photoperiod sensitivity, sorghum, pearl millet.

INTRODUCTION

Sorghum (*Sorghum bicolor* (L. Moench.) and pearl millet (*Pennisetum glaucum* (L.) R.Br.) are the major staple crops and sources of income for about 120 million people

in the arid to semi-arid regions of West and Central Africa (WCA). WCA is characterized by high climate variability which results in a high variability with respect to sowing date (Niangado, 2001). Local Guinea-race sorghum and pearl millet landraces have developed mechanisms of adaptation to these unpredictably changing growing conditions. Photoperiod sensitive flowering, the response of the plant to length of day, is one of the mechanisms present in a large portion of the local WCA cereal landraces. It can enhance adaptation to variable planting dates that are due to a scattered beginning of the rainy season, as is typical for WCA. It enhances simultaneous flowering of the cultivar in the target region, independent

---

*Corresponding author. E-mail: parzies@uni-hohenheim.de.

Abbreviations: GA, Gibberellic acid; **EST,** expressed sequence tag; **WCA,** West and Central Africa; **LD,** linkage disequilibrium; **InDels,** insertions or deletions; **SNP,** single nucleotide polymorphism.

of the individual planting date in different fields. This has particular advantages in terms of reducing bird damage and insect pressure, adjusting vegetative development. Therefore, sorghum and pearl millet cultivars with photoperiod sensitivity may have the potential to increase yield and yield stability (Haussmann et al., 2007). Most sorghum and pearl millet cultivars are considered as quantitative short day plants, but different cultivars differ in their responses to photoperiod. Some cultivars are day-neutral, whereas others show a high response to photoperiod. For sorghum and pearl millet, only little information is available on flowering time genes, which are candidate genes for photoperiodic sensitivity. In contrast, several models for the molecular mechanism of flowering time were proposed for the model plant *Arabidopsis thaliana* (Roux and Touzet, 2006; Bernier and Périlleux, 2005; Putterill et al., 2004; Simpson and Dean, 2002). In these models, flowering time genes are mainly integrated into four closely interacting pathways: vernalization, autonomous, gibberellic acid (GA) and photoperiod. In vernalization pathway, *VERNALIZATION 2* (*VRN*2) inhibits flowering via repressing *VERNALIZATION* 1 (*VRN*1). This inhibition is overcome by prolonged exposure of a long period of cold in winter (Amasino, 2005). In autonomous pathway, internal developmental signals are required instead of environmental factors. Both pathways regulate a strong repressor of the flowering gene *FLOWERING LOCUS C* (*FLC*) (Bernier and Périlleux, 2005; Putterill et al., 2004). Genes included in the GA pathway such as *GIBBERELLIC ACID INSENSITIVE* (*GAI*), *REPRESSOR OF GA1-3* (*RGA*), and *RGA-LIKE1-3* (*RGL*1-3), act as constitutive growth repressors whose growth-repressing activity is opposed by GA in modulating floral development of *A. thaliana* (Cheng et al., 2004; Tyler et al., 2004). The photoperiod pathway involves genes encoding the photoreceptors *PHYTOCHROME -A to -E* (PHYA to PHYE) and *CRYPTOCHROME 1-2* (CRY1-2). Furthermore, the circadian clock components are required for correct circadian time measurement, several clock-associated genes, such as *GIGANTEA* (*GI*), *F-BOX* 1 (FKF1) and *PSEUDO-RESPONSE REGULATORS7* (PRR7) are activators of *CONSTANS* (*CO*) (Bernier and Périlleux, 2005; Imaizumi and Kay, 2006).

A crucial feature of the flowering regulatory network is that all four pathways ultimately regulate a common set of key integrator genes, *SUPPRESSOR OF OVEREXPRESSION OF CO1* (*SOC*1) and *FLOWERING LOCUS T* (*FT*), which act on the floral meristem identity genes *APETALA 1* (*AP1*) and *LEAFY* (*LFY*) to initiate flowering (Boss et al., 2004; Henderson and Dean, 2004; Roux and Touzet, 2006). GA is a large family of tetracyclic diterpenoid plant growth factors, regulating seed germination, stem elongation, flowering and fruit development. In the facultative long-day plant Arabidopsis, GA is required for promoting flowering under short days (Wilson et al., 1992). Studies of GA signal transduction using genetic approaches have led to the

identification of positive and negative signaling components. *GAI* was identified via transposon insertional mutagenesis and was characterized as a transcription factor that negatively regulates GA responses in Arabidopsis (Peng et al., 1997). A 51-bp deletion in the highly conserved N-terminal DELLA domain of *GAI* was identified as a dominant gain-of-function mutant (*gai*) with a dwarf phenotype caused by a reduced GA response (Peng et al., 1997). Recently, it was proposed that GA overcomes the DELLA repressing function by binding to the GA receptors OsGID1 or AtGID1a,b,c (Nakajima et al., 2006; Ueguchi-Tanaka et al., 2007) which induces the degradation of DELLA-proteins (Fu et al., 2002; Harberd, 2003). The involvement of photoperiod in the GA pathway suggests that DELLA is a putative crucial factor integrating these two different flowering pathways (Thomas et al., 1999; Garcia-Martinez and Gil, 2001; Achard et al., 2007). In several plant species, genes homologous to *GAI* of Arabidopsis were identified: *D8* in maize (Harberd and Freeling, 1989; Winkler and Freeling, 1994), the "green revolution" genes *Rht* in wheat (Gale and Youssefian, 1985), *SLN1* in barley (Chandler et al., 2002), *SLR1* in rice (Ikeda et al., 2001), *VvGAI* in grape (Boss and Thomas, 2002), *Brrga1-d* in *Brassica rapa* L. (Muangprom et al., 2005) and *MdDELLAs* in apple (Foster et al., 2007). However, information on homologous genes to *GAI* in sorghum and pearl millet is missing. Geneticists and biochemists have identified many relevant genes of the flowering time pathways using artificially induced variations, such as mutants. However, the ability to modify photoperiodic sensitivity in plant breeding programs will depend on an increased level of understanding naturally occurring variation (Yano et al., 2001). Deciphering the genetic determinism of natural variation is of interest not only to evolutionary biologists studying the genetics of adaptation in wild species, but also to plant breeders because it could provide useful guidelines for quantitative trait loci (QTL) studies and identification of target genes for selection as well as marker assisted selection (Morgante and Salamini, 2003). A genetic variant of maize *D8* leads to an earlier flowering phenotype (Thornsberry et al., 2001). This allele is present at a high frequency in North America while being almost absent in tropical regions and thus believed to be involved in maize climatic adaptation through diversifying selection for flowering time (Camus-Kulandaivelu et al., 2006). Nevertheless, only for few species, information is available on the genetic diversity of genes homologous to *GAI*. The objectives of our study were to investigate in a diverse set of sorghum and pearl millet genotypes: (i) the presence, (ii) the expression and (iii) the molecular diversity of genes homologous to *D8*.

MATERIALS AND METHODS

Plant material

Twenty six inbred lines in selfing generation >S_6 of sorghum and 20 inbreds (S_4) of pearl millet were selected from ~200 lines of

Table 1. Accession name and photoperiodic reaction of sorghum and pearl millet.

Accession name of sorghum	Photoperiodic reaction	Accession name of pearl millet	Photoperiodic reaction
IER(8)-02-SB–FSDT-12B	Sensitive	PE00057-B-B-1*	Sensitive
IER(9)-02PR-3009KB	Sensitive	PE05433-B-B-1	Sensitive
IER(13)-99-CLO 634B	Sensitive	PE05460-B-B-1	Sensitive
IS 3534 B	Sensitive	PE00991-B-B-1*	Sensitive
Fara Fara-17	Sensitive	PE00194-B-B-1	Sensitive
CS05/06AF(Guinea Niger)	Sensitive	PE05371-B-B-1	Sensitive
90 SN 1	Sensitive	PE05327-B-B-1	Sensitive
CS 0 4, SK 5912	Sensitive	PE05943-B-B-1	Sensitive
Bahu Banza	Sensitive	PE05303-B-B-1	Sensitive
Nigeria(SVMD Sova2006)	Sensitive	PE11322-B-B-1	Sensitive
ESO3(Fambe B)	Insensitive	PE02790-B-B-1	Sensitive
S05 AF- IS6731bf(R 127)	Insensitive	PE08058-B-B-1	Sensitive
ISO/ 06 P10 a- CSM63 E	Insensitive	PE08039-B-B-1*	Insensitive
IER(7)-97SB-FSDT-150 B	Insensitive	PE05927-B-B-1	Insensitive
GRINKAN	Insensitive	PE05321-B-B-1	Insensitive
MALISOR-92-1	Insensitive	PE02943-B-B-1*	Insensitive
Lata 3 (Balla Berthe)	Insensitive	Souna3-B-B-1	Insensitive
IER(18)-02-SB-F5DT169	Insensitive	PE00307-B-B-1*	Insensitive
IER(10)-00-KO-F5DT19	Insensitive	PE03855-B-B-1	Insensitive
IER(11)-08-SB-F4FT298	Insensitive	PE05393-B-B-1	No data
ST 9007-5-3-1	Insensitive		
IER(12)08-SB-F4FT189	Insensitive		
ISO/06P11dGPN99271202	No data		
ISO/05P7C- L224/25	No data		
ISO/05 P 11 d- CSM 388	No data		
ISO/03-CGM19/9-1-1	No data		

*RNA was extracted for RT-PCR.

sorghum and pearl millet (Table 1). The sorghum and pearl millet inbreds with diverse photoperiod responses were obtained from the regional centers of the International Crops Research Institute for the Semi-Arid Tropics (ICRISAT) in Mali and Niger, respectively, and represent the genetic diversity present in WCA.

Photoperiod response assessment

We employed a simple non-destructive method to assess photoperiod response. We measured the vegetative phase of each inbred line by sowing on two different planting dates. The vegetative phase is the duration between sowing and initiation of flowering. The difference between vegetative phase duration at June and July plantings is used as an index to determine the photoperiod sensitivity of these inbred lines (White and Laing, 1989).

DNA and RNA extraction

From all sorghum and pear millet inbreds, DNA was isolated from fresh leaves of 3 - 4 week old plants, according to the protocol of Saghai-Maroof et al. (1984) with some modifications. From a random set of five pearl millet inbreds, RNA was extracted from fresh leaves of 5 - 6 week old plants using the RNeasy Plant Mini Kit of Qiagen Company (Hilden, Germany).

Genomic PCR and reverse transcription PCR

Primer pair D8-3 was obtained from the literature (Andersen et al., 2005). The other five primer pairs were designed based on the sequence of the D8 gene of Zea mays (GeneBank accession no. AF413114) to amplify conserved domains of D8 (Table 2) using Primer Premier Software (Premier Biosoft International, Palo Alto, CA, USA). All primer pairs were synthesized by Biomers (Ulm, Germany).

Touchdown PCR for genomic PCR amplification was carried out as follows: After an initial denaturation at 96 °C for 10 min, 20 cycles were conducted of 96 °C for 1 min, 60 – 50 °C (65 – 55 °C for D8-3) for 1min (decreasing with 0.5 °C per cycle) and 72 °C for 1 min, followed by 15 cycles of 96 °C for 1min , 50 °C (60 °C for D8-3) for 1 min, and 72 °C for 1 min, with a final extension step at 72 °C for 15 min. Reverse transcription (RT) PCR was carried out as described above plus a reverse transcription process where RNA and reagents were placed at 50 °C water bath for 30 min before touchdown PCR.

Processing DNA sequence

All genomic PCR and RT-PCR products were sequenced by QIAGEN (Hilden, Germany). Consensus sequence contigs were generated using SeqMan (DNAstar, Madison, WI, USA). Three fragments amplified by DELLA-3, VHVVD, and D8-3 were

Table 2. Names, amplified regions, melting temperature (Tm), sequences and results of primer pairs for amplified sequences.

Primers	Amplified regions and expected size (bp)	Tm (°C)	Forward sequences(5′→3′) Reverse sequences(5′→3′)	Amplification product in
DELLA-3	DELLA	51	GCTCCTCCAAGGACAAGATG	Pearl millet
	450-500	52	TAGTGCGACCGCCATCC	
VHVVD	NLS	57	TGGATGGCGGTCGCACTAG	Pearl millet
	300-350	56	TGGGCGAACTTCAGGTAGGG	
D8-3	SH2	50	CGATGACACGGATGACGA	Pearl millet
	350-400	54	AGGCATTGGAGCCCAGGT	
D8-S	5′UTR and DELLA	57	GCTATCCCAGAACCGAAACCG	Sorghum
	600-650	57	CGACGAGGAAGACGAAGACGA	
D8-1614	NLS	54	TCCACATCGTCCACCGTCAC	Sorghum
	550-600	50	GGGCGAACTTCAGGTAGG	

combined to an incomplete pearl millet *PgD8* gene. Two fragments amplified by *D8*-1614 and *D8*-S were combined to an incomplete sorghum *SbD8* gene. Consensus sequence for all inbreds were aligned using CLUSTAL alignment implemented in MegAlign (DNAstar, Madison, WI, USA) and Genebee multiple alignment program (Brodsky et al., 1992). Polymorphisms appearing in less than three inbreds were rechecked on chromatogram files to avoid PCR or scoring errors. The putative amino acid sequences were deduced from nucleotide sequence by EMBL-EBI Transeq tool. From six reading frames of putative amino acid sequences, the one with the most similarity to *D8* amino acid sequence and not including a stop codon was chosen.

Phylogenetic and molecular genetic diversity analyses

Cluster analysis was performed on deduced amino acid sequences of the DELLA and VHYNP regions of *PgD8*, *SbD8*, *D8* (*Z. mays*, GeneBank accession no. Q9ST48), *SLN1* (*Hordeum vulgare*, Q8W127), *SLR1* (*Oryza sativa*, Q7G7J6), *RHT1* (*Triticum aestivum*, Q9ST59) and *GAI* (*A. thaliana*, Q9LQT8) using Genebee (Brodsky et al., 1992). We chose the amino acid sequences of AP2 DNA binding domain and the two CBF subfamily signature motifs as reference and performed the same analysis for *SbCBF5* (*S. bicolor*, AY785898), *ZmCBF5* (*Z. mays*, DV523865), *HvCBF5* (*H. vulgare*, AY785855), *OsCBF5* (*O. sativa*, AY327040), *TaCBF5* (*T. aestivum*, EF028752) and *AtCBF5* (*A. thaliana*, CAA18178).

We calculated π which is the average number of nucleotide differences per site between two sequences (Nei, 1987). Tajima's test for selection was applied which is based on the differences between the number of segregating sites and the average number of nucleotide differences (Tajima, 1989). The squared allele frequency correlation r^2 was estimated for all pairs of polymorphic sites with allele frequencies ≥ 0.10. Chi-square test was used to determine significance of linkage disequilibrium (LD) between pairs of polymorphic sites. Sites containing alignment gaps and single nucleotide polymorphisms with more than two alleles were excluded from the molecular genetic analyses. All molecular genetic analyses were performed using DNASP software package (Rozas et al., 2003).

To investigate the expression status of SbD8, the nucleotide sequence of *SbD8* was "blasted" against the S. bicolor EST database of NCBI by BLASTN.

RESULTS

Amplification and alignment analysis of *SbD8* and *PgD8*

Sorghum DNA was successfully amplified by two primer pairs, *D8*-S and *D8*-1614 (Table 2). Twenty two of the 26 sorghum genotypes were amplified by *D8*-S; 24 of the 26 sorghum genotypes were amplified by *D8*-1614. Pearl millet DNA from 16 of the 20 genotypes were successfully amplified by three primer pairs VHVVD, DELLA-3 and *D8*-3. Alignment of the deduced amino acid sequences of *PgD8*, *SbD8* and *D8* (Figure 1) showed 51.5% homology. If non-amplified regions were neglected, homology was much higher (84.3%). Eight conserved regions were found. The first two domains (VHYNP and DELLA) are highly conserved within the DELLA protein subfamily (Figure 2), showing 80.9% homology.

Gene expression of *SbD8* and *PgD8*

The nucleotide sequences of *SbD8* and *PgD8* were used as queries to search the *Sorghum bicolor* and *Pennisetum glaucum* EST database in NCBI by BLASTN (http://www.ncbi.mln.nih.gov/). Six ESTs were found with significant homology to *SbD8* (Table 3).

The highest individual query coverage was only 61%, however, the total query coverage was 91.6% (data not shown). This high total query coverage together with a low E-value, high individual ESTs total score, and maximum identity suggested that *SbD8* was expressed in *S. bicolor*.

No EST was found with significant homology to *PgD8* in NCBI *P. glaucum* EST database.

Fifteen ESTs were found in Gramene

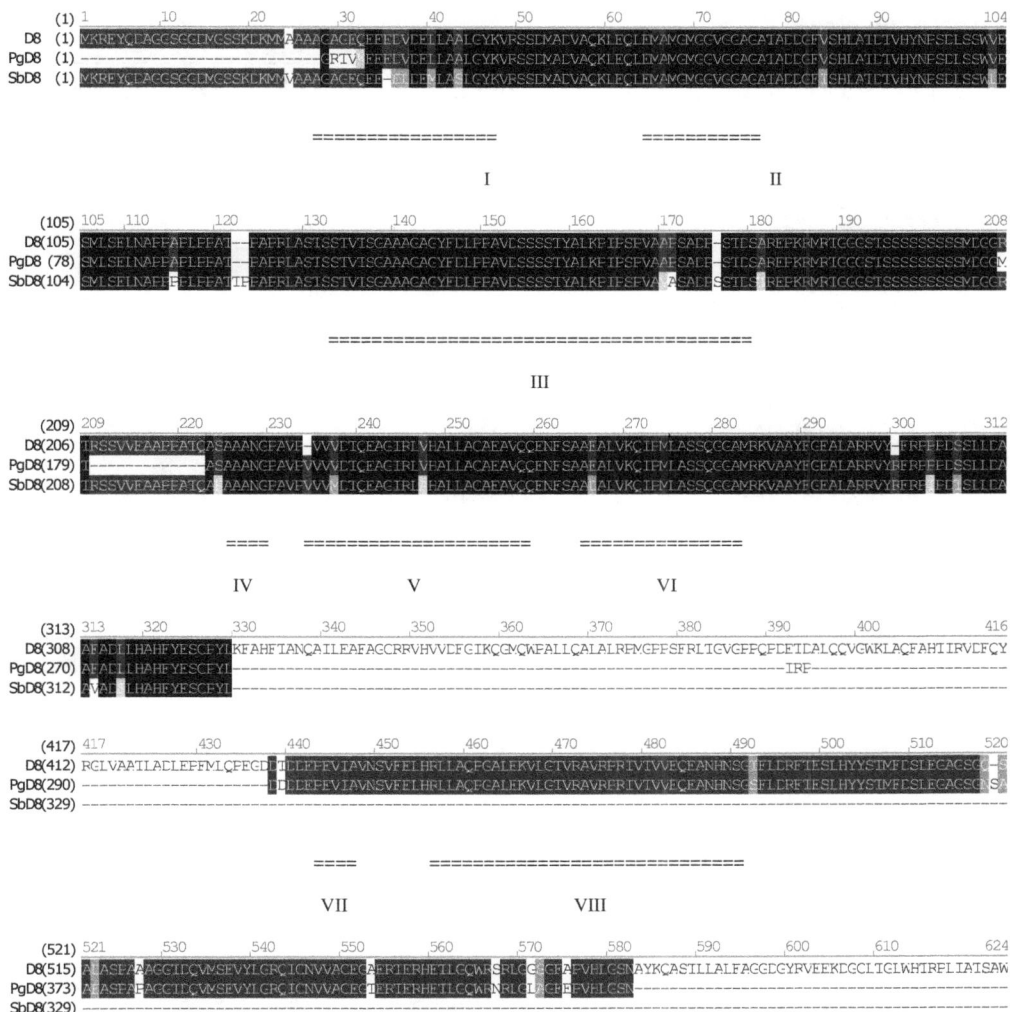

Figure 1. Alignment of the deduced amino acid sequences of *PgD8*, *SbD8* and *D8* (*Z. mays*, GenBank accession no. Q9ST48). Gaps in position 122, 123, 176, 234 and 300 indicated by single line (-) are introduced to maximize alignment. Other gaps indicated by single line (-) are the regions which have not been amplified. Identical amino acid residues are highlighted in black. Similar amino acid residues are highlighted in dark grey. Eight conserved sequence regions are indicated by double lines: Region I is the DELLA motif; Region II is VHYNP; Region III is PolyS/T; Region IV is valine-rich; Region V is LHR1; Region VI is a nuclear-localization signal (NLS); Region VII is the LXXLL motif; Region VIII is the putative Src homology 2 (SH2).

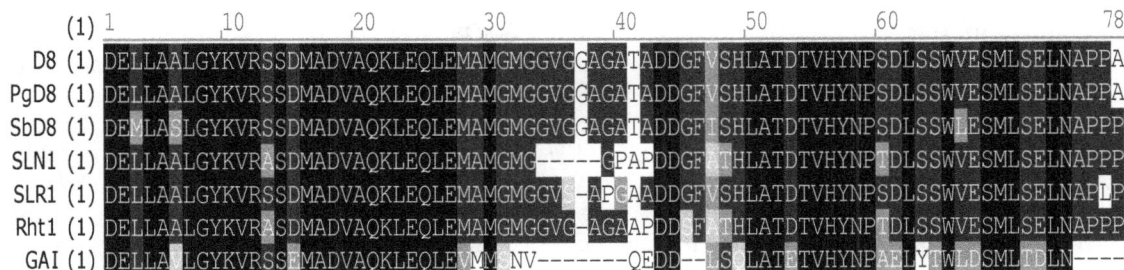

Figure 2. Deduced amino acid sequences alignment in the DELLA and VHYNP regions of *PgD8*, *SbD8* and another DELLA subfamilies: *D8* (*Z. mays*, GenBank accession no. Q9ST48), SLN1 (*H. vulgare* no. Q8W127), SLR1 (*O. sativa* no. Q7G7J6), RHT1 (*T. aestivum* no. Q9ST59) and GAI (*A. thaliana* no. Q9LQT8). Identical amino acid residues are highlighted in black. Gaps indicated by single line (-) are introduced to maximize alignment. Similar amino acid residues are highlighted in dark grey or grey.

Table 3. S.bicolor EST that hit *SbD8* by BLASTN.

Accession number	Description	Total score	Query coverage	E-value	Maximum identity
CN131369.1	Acid- and alkaline-treated roots *S. bicolor* cDNA, mRNA	1346	61%	0.0	99%
CD219739.1	Callus culture cell suspension *S. bicolor* cDNA, mRNA	1113	51%	0.0	99%
BE595338.1	Pathogen induced 1 (PI1) *S. bicolor* cDNA, mRNA	1219	55%	0.0	99%
BM318611.1	Pathogen induced 1 (PI1) *S. bicolor* cDNA, mRNA	1023	46%	1e-170	99%
BG411689.1	Embryo 1 (EM1) *S. bicolor* cDNA, mRNA	446	21%	6e-124	98%
CN137035.1	Oxdatively-stressed leaves and roots *S. bicolor* cDNA , mRNA	339	16%	1e-91	97%

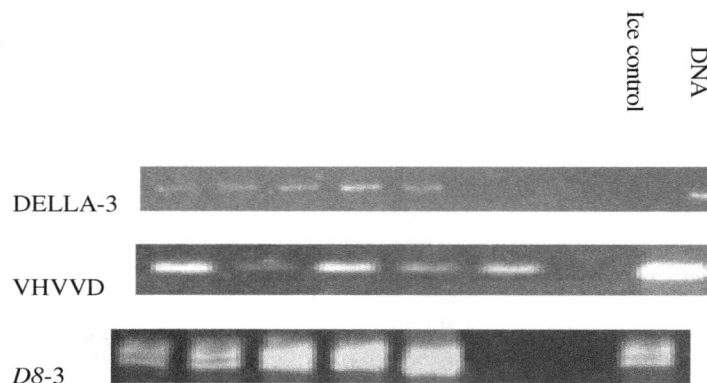

Figure 3. Gel Electrophoresis of reverse transcriptase PCR (RT-PCR) products. All 5 Pearl millet genotypes were amplified by three different primer pairs (DELLA-3, VHVVD, and *D8*-3). Ice control was a negative control for DNA contamination. DNA was a positive control.

(http://www.gramene.org) but high E-value (>0.5) and low coverage (<0.5%) prevented us from confirming that *PgD8* is expressed in pearl millet. Therefore, RT-PCR was performed to investigate the expression of *PgD8* (Figure 3). Pearl millet mRNA, extracted from 4-week old leaves of five inbreds, was amplified by RT-PCR with three primer pairs (VHVVD, DELLA-3, *D8*-3). The sequences of the RT-PCR product confirmed that *PgD8* is expressed.

Molecular genetic diversity and LD

Out of 1124 sites investigated, five were polymorphic in *SbD8* and out of 1228 sites, 32 were polymorphic in *PgD8* (Table 4). One 3-bp insertion or deletion (InDel) was observed near the SH2-like domain in the C-terminal of the open reading frame of *PgD8*. By contrast, no InDel were found in *SbD8*. All other polymorphism were SNPs. According to the value of π/bp, *PgD8* showed higher nucleotide diversity than *SbD8*. For sorghum as well as pearl millet, the nucleotide diversity for non-synonymous polymorphic sites was higher than for synonymous polymorphic sites. Tajima's D test was significant (P<0.05) for *PgD8*, but not for *SbD8*.

The number of pair-wise comparisons versus significant (P<0.05) pair-wise comparisons by Chi-square test in *SbD8* and *PgD8* were 15 versus 3 and 496 versus 381,

Table 4. *SbD8* and *PgD8* nucleotide diversity.

Region	No. of polymorphic sites	% sites with polymorphism	Nucleotide diversity π/bp
SbD8			
Synonymous	0	0	0
Non-synonymous	5	0.62	0.00204
Total	5	0.45	0.00163
PgD8			
Synonymous	6	0.67	0.00532
Non-synonymous	26	2.89	0.00792
Total	32	3.56	0.00704

respectively. LD between pairs of sites against nucleotide distance remained almost the same in *SbD8*, whereas in *PgD8* it declined exponentially

DISCUSSION

Photoperiod reaction and flowering time

The ability to recognize and respond to changes in day length is known as photoperiodism. For cultivars with photoperiodic sensitive reaction, that is, photoperiod-dependent flowering, flowering time is regulated by the daily duration of light. Photoperiod sensitivity is a key agronomical trait of local landraces of sorghum and pearl millet in WCA. This is due to the fact that photoperiod sensitivity assures flowering at the end of the rainy season, independent of the date of sowing, which is extremely important because of the high variability in the start of the rainy season (Vaksman et al., 1996; Niangado, 2001).

Amplification and alignment analysis of homologous genes to *D8*: *SbD8* and *PgD8* in sorghum and pearl millet

The successful genomic amplification of *SbD8* and *PgD8* suggested that the homologous genes to *D8* are also present in sorghum and pearl millet. This finding was further confirmed by alignment analysis of the deduced amino acid sequences of *PgD8*, *SbD8* and *D8*. Eight conserved regions were found (Figure 1). Region I: the highly conserved N-terminal DELLA motif. This region is absent in Arabidopsis mutant with dwarf phenotype caused by a reduced GA response (Peng et al., 1997 and 1999). Region II: a conserved domain that acts like the DELLA motif (Sun and Gubler, 2004; Gubler et al., 2002) and both of them contain putative phosphorylation sites for GA signal perception (Itoh et al., 2005). Region III: PolyS/T, a putative enhancer for suppressive activation of the DELLA proteins (Silverstone et al., 1998; Itoh et al., 2002). Region IV: Leucine heptad repeats, which are

found in transcription factors such as bZIP proteins and are important for protein–protein interaction (Bolle, 2004). Region V: a Valine-rich region characteristic for transcription factors. Region VI: a nuclear localization signal (NLS) Nakai and Kanehisa, 1992). Region VII: a LXXLL region (where X stands for any amino acid), that was identified in a number of steroid receptor co-activators (SRCs) and is responsible for SRC binding to steroid receptors in the nucleus (Heery et al., 1997; Torchia et al., 1997). Region VIII: a putative Src homology 2 (SH2) phosphotyrosine binding domain (Peng et al., 1999) which is present in a family of transcription factors called signal transducers and activators of transcription (STATs) in animals (Darnell, 1997). The function of SH2 is to mediate the binding of STATs to various receptor tyrosine kinases by which the STATs are then activated and translocated from the cytoplasm to the nucleus (Ikeda et al., 2001). In maize, a 6bp deletion flanking the SH2-like domain of *D8* was significantly associated with flowering (Thornsberry et al., 2001). This finding suggested that SH2 might play a role in controlling flowering time.

Molecular genetic diversity and LD

In our investigation, the nucleotide diversity π/bp of *SbD8* was 0.00163. In contrast, the nucleotide diversity π/bp of *PgD8* was higher (0.00704). This finding might be explained by (1) the lower inbreeding generation (S_4 vs. S_6) of pearl millet compared with sorghum and (2) the higher rate of polymorphism in allogamous species such as pearl millet compared to autogamous species such as sorghum (Rafalski, 2002). Thornsberry et al. (2001) reported a nucleotide diversity π/bp for *D8* in maize of 0.0018 which is similar to *SbD8* but lower than that of *PgD8*. The opposite result was expected as maize, is an allogamous species. This discrepancy might be explained by the fact that different gene fragments were examined in our study and that of Thornsberry et al. (2001).

The Tajima's D test values of *SbD8* and *PgD8* were 0.34062 and -1.95606, respectively. The value for *SbD8* was non-significantly (P=0.05) different from 0, indicating

(a) *SbD8*

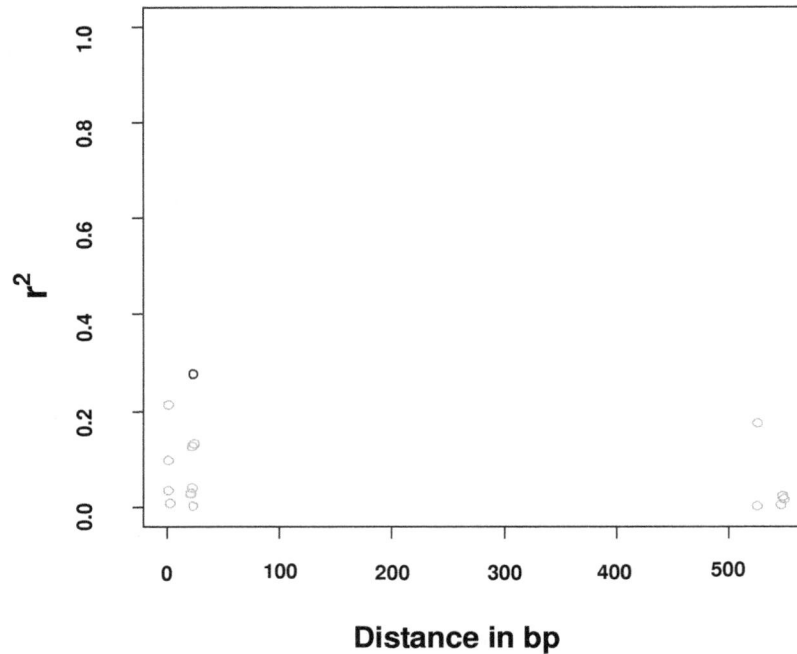

Distance in bp

(b) *PgD8*

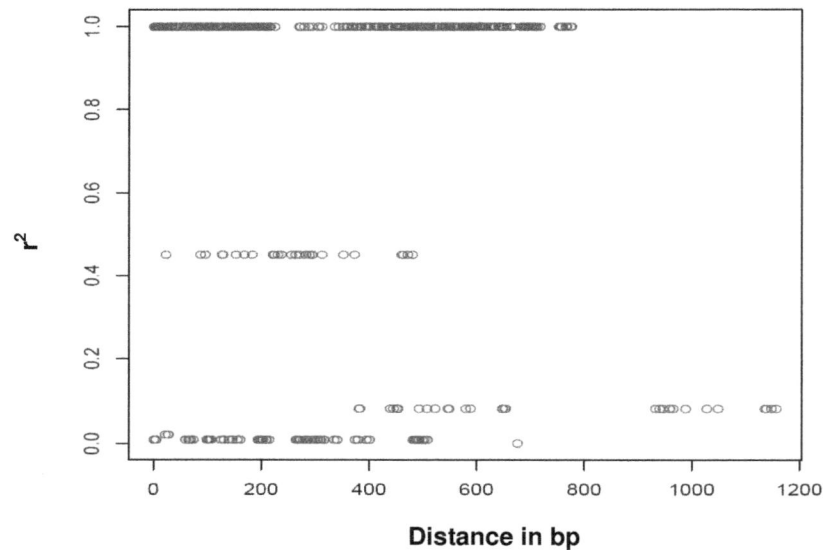

Distance in bp

Figure 4. Plots of squared correlation of allele frequencies (r^2) against nucleotide distance between polymorphic sites in *SbD8* and *PgD8*.

that polymorhisms are selectively neutral, whereas the value for *PgD8* was significant, suggesting that it has been a target of selection.

The r^2 values for *PgD8* declined to 0.1 or less within 1000 bp (Figure 4). In contrast, Remington et al. (2001) reported for *D8* that in a diverse set of 102 maize inbred lines LD decays within 2400 bp to r^2 values <0.1. This can

be explained by the bottleneck effect since selection in maize was more intensive and consequently only few allelic combinations were passed on to future generations. Since we have only investigated one gene in 20 inbred lines, a better understanding of LD and consequently mapping resolution in pearl millet would require a larger number of loci and genotypes. In sorghum, only

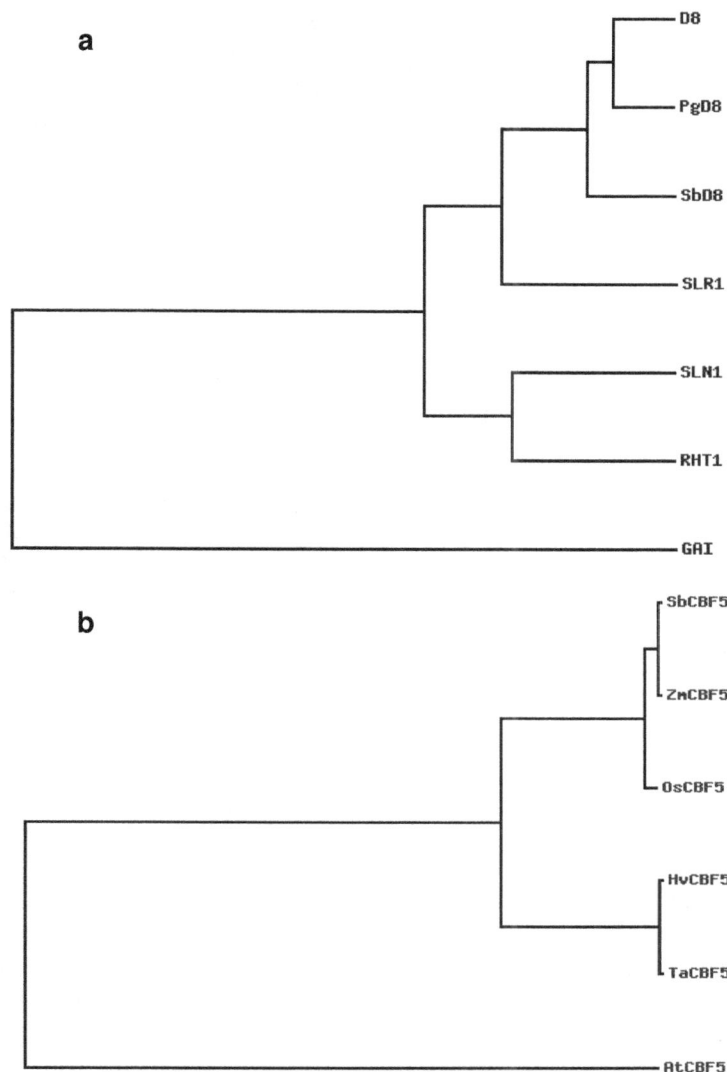

Figure 5. a) Cluster analysis of *PgD8*, *SbD8*, *D8* (*Z. mays*), *SLN1* (*H. vulgare*), *SLR1* (*O. sativa*), *RHT1* (*T. aestivum*) and *GAI* (*A. thaliana*) based on deduced amino acid sequences of DELLA and VHYNP region. Species that are connected by a single branch point (node) are sister taxa. The more nodes separating two species, the more distantly related they are. b) Cluster analysis of *SbCBF5* (*S. bicolor*), *ZmCBF5* (*Z. mays*), *HvCBF5* (*H. vulgare*), *OsCBF5* (*O. sativa*), *TaCBF5* (*T. aestivum*) and *AtCBF5* (*A. thaliana*) based on amino acid sequences of AP2 DNA binding domain and two CBF subfamily signature motifs.

few polymorphic sites were observed. Thus, no clear conclusions can be made regarding the decay of LD.

Cluster analysis of *SbD8* and *PgD8*

Full justification on deduced amino acid sequences of the DELLA and VHYNP regions, the cluster analysis of*PgD8*, *SbD8*, maize *D8*, barley *SLN1*, rice *SLR1*, wheat *RHT1* and Arabidopsis *GAI* revealed four clusters: Panicoideae (sorghum, maize and pearl millet), Oryzoideae (rice), Pooideae (wheat and barley) and Brassicaceae (Arabidopsis). In order to compare our result with the cluster analysis of another gene, we performed the same analysis on the amino acid sequence of CBF5, which is a transcriptional factor specifically bound to *cis-* elements of cold responsive gene (COR) under cold stress. The cluster analysis of the *D8* homologous and CBF5 were in good accordance with each other (Figure 5).

Furthermore, the result of our cluster analysis was in good accordance with that expected on the basis of the grass phylogeny (Grass Phylogeny Working Group, 2001). However, one difference is that in our study, maize is more closely related to pearl millet than sorghum, whereas the opposite has been reported on the basis of the grass phylogeny. This difference might be due to the fact that different parts of the genome were examined. We examined the nuclear genes Sbd8 and PgD8 whereas the Grass Phylogeny Working Group (2001) based their analyses on chloroplast genes rbcL and ndhF. This explanation was supported by findings of Gaut and Doebley (1997) who reported a divergence time between sorghum and maize of 16.5 million years by using nuclear genes mdh and waxy, whereas 9 million years was believed to be the divergence time of these two crops in the studies of the Grass Phylogeny Working Group using chloroplast genes.

IMPLICATIONS AND PERSPECTIVES OF THIS STUDY FOR SORGHUM AND PEARL MILLET BREEDING

Sorghum and pearl millet are crops with high drought tolerance and economic importance in arid to semi-arid regions of WCA. However, it is difficult and time consuming to select cultivars with photoperiod sensitivity in the field. Based on this study, LD decays within 1000 bp in PgD8 that is very useful information for identification of functional nucleotide polymorphisms using a LD-approach. The results of this study open up possibilities to develop advanced plant breeding techniques such as marker-assisted selection. Photoperiod sensitivity is of particular importance for local landraces of sorghum and pearl millet in WCA. Two parallel studies uncovered that Phytochrome Interacting Factors (PIFs) and DELLA are two crucial integration nodes in two previous distinguish flowering time pathways: photoperiod and GA pathway. (de Lucas et al., 2008; Feng et al., 2008). Maize D8 seems to affect the quantitative variation of flowering time and plant height according to mutagenesis and quantitative trait locus studies (Peng et al., 1999; Koester et al., 1993; Schön et al., 1994). Furthermore, Thornsberry et al. (2001) identified a set of intragenic polymorphisms associated with differences in flowering time using association approaches, nine (four SNPs, four InDels and one miniature transposable element) of which were validated by Andersen et al. (2005). When population structure was ignored, six of the nine D8 polymorphisms were significantly associated with flowering time. However, when population structure was taken into account, only a single polymorphisms (one InDel in the promoter region) was associated with flowering time. It is noteworthy that Thornsberry et al. (2001) used a worldwide collection of maize, whereas Andersen et al. (2005) used only European materials. Besides a 6-bp InDel near the SH2-like domain (position 3472) in the C-terminal of the open reading frame (ORF) had strong association with flowering time under long-day conditions, after correcting for population structure (Camus-Kulandaivelu et al., 2006). Interestingly, a 3-pb InDel near the SH2-like domain of PgD8 has been found in our study. Therefore, it may be possible to select sorghum and pearl millet lines with photoperiod sensitivity using SNPs or InDels as indirect (statistical) functional markers (IFMs). IFMs are derived from polymorphic sites within genes causally affecting phenotypic trait variation by association studies (Andersen and Luebberstedt, 2003).

In order to develop IFMs, association studies need to be conducted to confirm a significant correlation between SNPs and photoperiod sensitivity. In this regard, population structure must be taken into account since it can lead to spurious associations between SNP-marker and the phenotype (Yu et al., 2006; Zhao et al., 2007; Stich et al., 2008). Additionally, IFMs need to be verified by using a mapping population. Association studies by using the candidate gene approach have been proven to be a powerful tool in medical genetics. As more advanced statistical methods are now available to overcome problems caused by population structure, association studies can be applied to practical cereal breeding approaches in the near future.

In this study, two homologous genes to D8: Sorghum SbD8 and pearl millet PgD8 have been found and characterised. EST database search and reverse transcription PCR suggested that SbD8 and PgD8 were expressed in the root and leaves of sorghum and pearl millet, respectively. The genetic diversity of SbD8 was considerably lower than that of PgD8. The extent of linkage disequilibrium in PgD8 is lower than that of maize D8. The implications of discovering SbD8 and PgD8 for sorghum and pearl millet breeding have been discussed.

ACKNOWLEDGMENTS

We thank staff of the International Crops Research Institute for the Semi-Arid Tropics for providing plant material, phenotypic data and valuable advice. We further thank Dr. A. C. Thuillet and Dr. N. von Wirén for their helpful suggestions and Sabine Boger for her technical assistance. This research was supported by the Federal Ministry for Economic Cooperation and Development of Germany (BMZ).

REFERENCES

Achard P, Liao LC, Jiang C, Desnos T, Bartlett J, Fu X, Harberd HP (2007). DELLAs contribute to plant photomorphogenesis. Plant Physiol. 143: 1163-1172.

Amasino RM (2005). Vernalization and flowering time. Curr. Opin. Biotechnol. 16: 154-158.

Andersen JR, Schrag T, Melchinger AE, Zein I, Lueberstedt T (2005). Validation of Dwarf8 polymorphisms associated with flowering time in elite European inbred lines of maize (*Zea mays L.*). Theor. Appl. Genet. 111: 206-217.

Andersen JR, Luebberstedt T (2003). Functional markers in plants. Trends Plant Sci. 8: 554-560.

Bernier G, Périlleux C (2005). A physiological overview of the genetics of flowering time control. Plant Biotechnol. 3: 3-16.

Bolle C (2004). The role of GRAS proteins in plant signal transduction and development. Planta 218: 683-692.

Boss PK, Bastow RM, Mylne JS, Dean C (2004). Multiple pathways in the decision to flower: enabling, promoting and resetting. Plant Cell 16(Suppl.): S18-S31.

Boss PK, Thomas MR (2002). Association of dwarfism and floral induction with a grape 'green revolution' mutation. Nature 416: 847-850.

Brodsky LI, Ivanov VV, Kalaidzidis YL, Leontovich AM, Nikolaev VK, Feranchuk SI, Drachev VA (1995). GeneBee-NET:Internet-based server for analyzing biopolymers structure. Biochem. 60(8): 923-928.

Camus-Kulandaivelu L, Veyrieras JB, Madur D, Combes V, Fourmann M (2006). Maize adaptation to temperate climate: Relationship between population structure and polymorphism in the Dwarf8 gene. Genet. 172: 2449-2463.

Cheng H, Qin L, Lee S, Fu X, Richards DE, Cao D, Luo D, Harberd NP, Peng J (2004). Gibberellin regulates *Arabidopsis* floral development via suppression of DELLA protein function. Development 131: 1055-1064.

Chandler PG, Marion-Poll A, Ellis M, Gubler F (2002). Mutants at the Slender1 locus of barley cv Himalaya: molecular and physiological characterization. Plant Physiol. 129: 181-190.

Darnell JE (1997). STATs and gene regulation. Sci. 277: 1630-1635.

de Lucas M, Daviere JM, Rodriguez-Falcon M, Pontin M, Iglesias-Pedraz JM, Lorrain S, Fankhauser C, Blazquez MA, Titarenko E, Prat S (2008). A molecular framework for light and gibberellin control of cell elongation. Nature 451: 480-484.

Feng S, Martinez C, Gusmaroli G, Wang Y, Zhou J, Wang F, Chen L, Yu L, Iglesias-Pedraz JM, Kircher S (2008). Coordinated regulation of Arabidopsis thaliana development by light and gibberellins. Nature 451: 475-479.

Foster T, Kirk C, Jones WT, Allan AC, Espley R, Karunairetnam S, Rakonjac J (2007). Characterisation of the DELLA subfamily in apple (Malus x domestica Borkh). Tree Genet. Genomes 3(3): 187-197.

Fu X, Richards DE, Ait-Ali T, Hynes LW, Ougham H, Peng JR, Harberd NP (2002). Gibberellin-mediated proteasome-dependent degradation of the barley DELLA protein SLN1 repressor. Plant Cell 14: 3191-3200.

Gale MD, Youssefian S (1985). Dwarfing genes in wheat. Butterworths, London pp. 1-35.

Garcia-Martinez JL, Gil J (2001). Light regulation of gibberellins biosynthesis and mode of action. J. Plant Growth Regul. 20: 354-368.

Gaut BS, Doebley JF (1997). DNA sequence evidence for the segmental allotetraploid origin of maize. Proceedings of the National Academy of Sciences, USA 94: 6809-6814.

Grass Phylogeny Working Group (2001). Phylogeny and subfamilial classification of the grasses (Poaceae). Annals of the Missouri Botanical Garden 88: 373-457.

Gubler F, Chandler PM, White R, Llewellyn D, Jacobsen J (2002). GA signaling in barley aleurone cells: Control of SLN1 and GAMYB expression. Plant Physiol. 129: 191-200.

Harberd NP, Freeling M (1989). Genetics of dominant gibberellin insensitive dwarfism in maize. Genetics 121: 827-838.

Harberd NP (2003). Relieving DELLA restraint. Sci. 299: 1853-1854.

Haussmann BIG, Boureima SS, Kassari IA, Moumouni KH, Boubacar A. (2007). Two mechanisms of adaptation to climate variability in West African pearl millet landraces – a preliminary assessment. E-Journal of Semi-Arid Tropical (SAT) Research. Sorghum, millets and other cereals. http://ejournal.icrisat.org/ 3(1): 3.

Heery DM, Kalkhoven E, Hoare S, Parker MG (1997). A signature motif in transcriptional co-activators mediates binding to nuclear receptors. Nature 387: 733-736.

Henderson IR, Dean C (2004). Control of *Arabidopsis* flowering: the chill before the bloom. Develop. 131: 3829-3838.

Ikeda A, Ueguchi-Tanaka M, Sonoda Y, Kitano H, Koshioka M, tsuhara

Y, Matsuoka M, Yamaguchi J (2001). Slender rice, a constitutive gibberellin response mutant, is caused by a null mutation of the SLR1 gene, an ortholog of the height-regulating gene GAI/RGA/RHT/*k*. Plant Cell 13: 999-1010.

Imaizumi T, Kay SA (2006). Photoperiodic control of flowering: not only by coincidence. Trends Plant Sci. 11: 550-558.

Itoh H, Sasaki A, Ueguchi-Tanaka M, Ishiyama K, Kobayashi M, Hasegawa Y, Minami E, Ashikari M, Matsuoka M (2005). Dissection of the phosphorylation of rice DELLA protein, SLENDER RICE1. Plant Cell Physiol. 8: 1392-1399.

Itoh H, Ueguchi-Tanaka M, Sato Y, Ashikari M, Matsuoka M (2002). The gibberellin signaling pathway is regulated by the appearance and disappearance of SLENDER RICE1 in nuclei. Plant Cell 14: 57-70.

Koester R, Sisco P, Stuber C (1993). Indentification of quantitative trait loci controlling days to flowering and plant height in two near-isogenic lines of maize. Crop Sci. 33: 1209-1216.

Morgante M, Salamini F (2003). From plant genomics to breeding practice. Curr. Opin. Biotechnol. 14: 214-219.

Muangprom A, Stephen G, Sun TP, Thomas C (2005). A novel dwarfing mutation in a green revolution gene from *Brassica rapa*. Plant Physiol. 137: 931-938.

Nakai K, Kanehisa M (1992). A knowledge base for predicting protein localization Site in eukaryotic cell. Genomics 14: 897-911.

Nakajima M, Shimada A, Takashi Y, Kim YC (2006). Identification and characterization of *Arabidopsis* gibberellin receptor. Plant J. 46: 880-889.

Nei M (1987). Molecular Evolutionary Genetics, Columbia University Press, NY.

Niangado O (2001). The state of millet diversity and its use in West Africa. in: Cooper HD, Spillane C, Hodgin T (eds). Broadening the genetic base of crop production. IPGRI/FAO, Rome pp.147-157.

Peng JR, Richards DE, Hartley NM, Murphy GP, Devos KM, Flintham JE, Beales J, Harberd NP (1999). Green revolution' genes encode mutant gibberellin response modulators. Nature 400: 256-261.

Peng JR, Carol P, Richards DE, King KE, Cowling RJ, Murphy GP, Harberd NP (1997). The *Arabidopsis* GAI gene defines a signaling pathway that negatively regulates gibberellin responses. Genes Dev. 11: 3194-3205.

Putterill J, Laurie R, Macknight R (2004). It's time to flower: the genetic control of flowering time. BioEssays 26: 363-373.

Rafalski JA (2002). Application of single nucleotide polymorphisms in crop genetics. Curr. Opin. Plant Biol. 5: 94-100.

Remington DL, Thornsberry JM, Matsuoka Y, Wilson IM, Whitt SR, Doebley J, Kresovich S, Goodman MM, Buckler ES (2001). Structure of linkage disequilibrium and phenotypic associations in the maize genome. Proc. Natl. Acad. Sci. USA 98: 11479-11484.

Roux F, Touzet P (2006). How to be early flowering: an evolutionary perspective. Trends in Plant Sci. 11(8): 375-381.

Rozas J, Sánchez-DelBarrio JC, Messeguer X, Rozas R (2003). DnaSP, DNA polymorphism analyses by the coalescent and other methods. Bioinformatics 19: 2496-2497.

Saghai-Maroof MA, Soliman KM, Jorgensen RA, Allard RW (1984). Ribosomal DNA spacer-length polymorphisms in barley: mendelian inheritance, chromosomal location, and population dynamics. PNAS 81(24): 8014-8018.

Schön CC, Melchinger AE, Boppenmaier J, Brunklaus-Jung E, Herrmann RG, Seitzer JF (1994). RFLP mapping in maize – quantitative trait loci affecting test cross performance of elite European flint lines. Crop Sci. 34: 378-389.

Simpson GG, Dean C (2002). *Arabidopsis*, the Rosetta Stone of flowering time? Science 296: 285-289.

Silverstone AL, Ciampaglio CN, Sun TP (1998). The *Arabidopsis* RGA gene encodes a transcriptional regulator repressing the Gibberellin signal transduction pathway. Plant Cell 10: 155-170.

Stich B, Möhring J, Piepho HP, Heckenberger M, Buckler ES, Melchinger AE (2008). Comparison of mixed-model approaches for association mapping. Genetics 178: 1745-1754.

Sun TP, Gubler F (2004). Molecular mechanism of gibberellin signaling in plants. Annu. Rev. Plant Biol. 55: 197-223.

Tajima F (1989). Statistical method for testing the neutral mutation hypothesis by DNA polymorphism. Genetics 123: 585-595.

Thornsberry JM, Goodman MM, Doebley J, Kresovich S, Buckler ES

(2001). Dwarf8 polymorphisms associate with variation in flowering time. Nat. Genet. 28: 286-289.

Thomas SG, Phillips AL, Hedden P (1999). Molecular cloning and functional expression of gibberellin 2-oxidases, multifunctional enzymes involved in gibberellin deactivation. Proc. Natl. Acad. Sci. USA 96: 4638-4703.

Torchia J, Rose DW, Inostroza J, Kamei Y, Westin S, Glass CK, Rosenfeld MG (1997). The transcriptional coactivator p/CIP binds CBP and mediates nuclear-receptor function. Nature 387: 677-684.

Tyler L, Thomas SG, Hu J, Dill A, Alonso JM, Ecker JR, Sun TP (2004). DELLAproteins and gibberellin-regulated seed germination and floral development in Arabidopsis. Plant Physiol. 135: 1008-1019.

Ueguchi-Tanaka M, Nakajima M, Motoyuki A, Matsuoka M (2007). Gibberellin receptor and its role in gibberellin signaling in plants. Annual Review of Plant Bio. 58: 183-198.

Vaksman M, Traoé S, Niangado O (1996). Le photopériodisme des sorghos africains. Agriculture et Dévélopment 9: 13-18.

White JW, Laing DR (1989). Photoperiod response of flowering in diverse genotypes of common bean (Phaseolus vulgaris). Field Crops Res. 22: 113-128.

Wilson RN, Heckman, Somerville C (1992). Gibberellin is required for flowering in Arabidopsis thaliana under short days. Plant Physiol. 100: 403-408.

Winkler RG, Freeling M (1994). Physiological genetics of the dominant gibberellin nonresponsive maize dwarfs, Dwarf8 and Dwarf9. Planta 193: 341-348.

Yano M, Kojima S, Takahashi Y, Lin HX, Sasaki T (2001). Genetic control of flowering time in rice, a short-day plant. Plant Physiol. 127: 1425-1429.

Yu J, Pressoir G, Briggs WH., Bi VI, Yamasaki M, Doebley JF, McMullen MD, Gaut BS, Nielsen DM, Holland BJ, Kresovich S, Buckler ES (2006). A unified mixed-model method for association mapping that accounts for multiple levels of relatedness. Nat. Genet. 38: 203-208.

Zhao K, Aranzana MJ, Kim S, Lister C, Shindo C (2007). An Arabidopsis example of association mapping in structured samples. PLoS Genet. 3: 71-82.

18

Multivariate analysis of genetic divergence in twenty two genotypes of groundnut (*Arachis hypogaea* L.)

Sunday Clement Olubunmi Makinde[1]* and Omolayo Johnson Ariyo[2]

[1]Department of Botany, Faculty of Science, Lagos State University, Ojo Campus,
P.O Box 001, LASU Post Office Ojo, Lagos State, Nigeria.
[2]Department of Plant Breeding and Seed Technology, College of Plant Science
University of Agriculture Abeokuta, P. M. B. 2240 Abeokuta, Ogun State, Nigeria.

Twenty-two groundnut genotypes collected from different germplasm centers were cultivated in botanical nursery of the Lagos State University Ojo-Campus during the raining season of 2009. The data collected on 33 characters were subjected to multivariate analysis to study the variability within the genotypes and to determine the efficiency of the methods in classifying genotypes. The first three axes each of factor analysis and principal component analyses (PCA) captured 42 and 55% respectively of the total variance and jointly identified final plant height, leaflet length, stem pigmentation, nodes on the main stem and number of leaves per plant at flowering as characters contributing most to variation. The first three axes of the canonical and discriminant analyses accounted for 85 and 90% of the total variation respectively and identified in addition to the above characters, pod beak, hairiness of mature leaflet, pod constriction, lateral branch pattern and peg colour as important. Genotype clustering using single linkage clustering technique did not follow a particular pattern, as genotypes from different sources were grouped together, while some from same source were also separated into eight different groups. The effect of genetic divergence on the choice of parental stock in improvement breeding programme was discussed.

Key words: Groundnut, factor analysis, principal component analysis, canonical discriminant analysis, single linkage cluster analysis.

INTRODUCTION

Groundnut (*Arachis hypogaea* L.), a member of the family *Fabaceae* is a major source of vegetable oil and plant protein in Africa. It is the World's thirteenth most important food crop, the fourth most important source of edible oil and the third most important source of vegetable protein (Encyclopedia of Agricultural Science, 1994). Multivariate statistical methods and numerical taxonomy has been used extensively in summarizing and describing variation pattern in a population of crop genotypes (Ram and Panwar, 1970; Bartual et al., 1985; Rezai and Frey, 1990; Ariyo, 1990b; Ariyo and Odulaja, 1991; Ariyo, 1993; Flores *et al.*, 1997; Cardi, 1998). The Mahalanobis D^2 statistic has been used to quantify the

degree of divergence in different crops (Ram and Panwar, 1970; Das and DasGupa, 1984; Ariyo, 1987a; Nair et al., 1998; Pintu et al., 2007). The technique gave insight into the most genetically divergent parents that could be used for hybridization purpose. Das and DasGupa (1984) and Ariyo (1987a) noted earlier that, geographical diversity was not always related to genetic diversity and therefore not an adequate index of genetic diversity. Genotypes within clusters often showed great geographical diversity.

Successful establishment of germplasm collections and plant introduction for crop improvement as well as for germplasm conservation require studies in genetic varia-bility within plant populations. Jain and Workman (1966) stated that such genetic variability and heterozygosity within populations existed in both natural and cultivated populations. Wright and Debzhonsky (1970) emphasized that the maintenance of this variability depended on complex interactions among a number of genetic and

---
*Corresponding author. E-mail: scmakinde@yahoo.com, bunmi.makinde@lasunigeria.org.

Table 1. Code names and source/ origin of groundnut genotypes.

Number	Genotype	Source/ Origin
1	ICG – 4998	ICRISAT India
2	ICG – 862	ICRISAT India
3	ICG – 6402	ICRISAT India
4	ICG – 8490	ICRISAT India
5	ICG – 4412	ICRISAT India
6	ICG – 156	ICRISAT India
7	ICG – 14466	ICRISAT India
8	ICG – 12370	ICRISAT India
9	ICG – 2106	ICRISAT India
10	ICG – 4343	ICRISAT India
11	ICG – 12189	ICRISAT India
12	ICG – 442	ICRISAT India
13	ICG – 4598	ICRISAT India
14	ICG – 7000	ICRISAT India
15	ICG – 1399	ICRISAT India
16	ICGY-6M- 5236	Zaria, Nigeria
17	ICG-IS- 11687	Zaria, Nigeria
18	ICGY-5M- 4746	Zaria, Nigeria
19	ICG-IS- 6646	UNILORIN, Nigeria
20	ICG- IS- 3584	UNILORIN, Nigeria
21	ICG49- 85A	UNAAB, Nigeria
22	UGA-7- M	UNAAB, Nigeria

and environmental factors. Ariyo (1987a and b) buttressed this fact further by stating that progress in breeding for economic characters often depends on the availability of a large germplasm representing a diverse genetic variation. He added that for a long term improvement programme, a large and diverse germplasm collection is an invaluable source of parental strains for hybridization and subsequent development of improved varieties. According to White and Gonzalez (1990), Nassir and Ariyo (2005), Aremu et al. (2007) accurate cultivar evaluations and ability to differentiate between cultivars in respect of genetic parameters associated with adaptedness in cultivated plants and their wild progenitors are critical to any plant breeding programme.

The objectives of this study therefore, are to evaluate and determine the variation pattern in collection of groundnut, identify the characters that sort the genotypes into different groups, suggest potential parents that could be used in improvement programme and appraise the suitability of the various multivariate techniques for classification of variation in groundnut.

MATERIALS AND METHODS

The twenty two genotypes of groundnut used in this study comprised of 15 accessions collected from International Crop Research Institute for the Semi-Arid Tropics (ICRISAT), Patancheru, India. The remaining 7 genotypes were collected from

different research centers within Nigeria, Table 1 presents the genotype coding with their collection centre. Planting was done during the raining season of 2009 (April) in the Department of Botany Nursery, Lagos State University-Ojo Campus, Lagos (6° 36'N, 3° 34'E) Lagos State, Nigeria. Following land preparation, they were grown in double-row plots, replicated 3 times in a randomized complete block design.

Each row was 4 m long with 1 m between rows and plants were spaced 40 cm apart within the row to give ten plants in a row. Each stand was thinned to one plant at two weeks after planting. Manual weeding was done at two weeks after planting and subsequently at three weeks intervals to ensure minimal crop-weeds competition. There was no application of inorganic fertilizers and chemicals (herbicides and pesticides).The rainfall, relative humidity and temperature data of the study sites are presented in Table 2.

Data collection

Agronomic and yield data were collected on each genotype. Five internal plants were sampled in each row (that is ten plants in each plot). At maturity, pods were harvested on plant basis to obtain some characteristics. Altogether, data were collected on 33 characters. Table 3 presented the 33 characters and their methods of scoring. Mean values of the characters were computed for the ten sampled plants in each plot. The means of the characters were subjected to analysis of variance and covariance (SAS 2000). The Principal Component Analysis (PCA) and Canonical Analysis were also done. The PCA analysis reduces the dimensions of a multivariate data to a few principal axes, generates an Eigen vector for each axis and produces component scores for the characters (Sneath and Sokal, 1973; Ariyo and Odulaja, 1991). Canonical analysis also measures the axis along which variation between

Table 2. Mean monthly temperature, T (°C), relative humidity, RH (%) and rainfall R (mm) for the study months.

Months	Environmental variable		
	T (°C)	RH (%)	R (mm)
April	29.0	74	157.4
May	28.5	78	320.7
June	26.9	83	69.5
July	26.3	83	18.5
August	26.0	84	85.2

Table 3. Characters used in the analysis and their methods of measurement/ scoring.

S/No	Character	Measurement/ Scoring (s)
1	Days to 50% flowering	Estimated using calendar
2	Height at flowering	Measured (cm)
3	Number of leaves/ plant at flowering	Counted
4	Final height/ plant	Measured (cm)
5	Days to maturity	Estimated using calendar
6	Number of branches/ plant at maturity	Counted
7	Nodes on the main stem/ plant at maturity	Counted
8	Stem girth/ plant at maturity	Measured (cm)
9	Leaflet length	Measured (cm)
10	Leaflet width	Measured (cm)
11	Leaflet length/ width ratio	Estimated
12	Pod width	Measured (cm)
13	Pod length	Measured (cm)
14	Seed length	Measured (mm)
15	Seed width	Measured (mm)
16	Shelling %age	Estimated (%)
17	Number of pods/ plant	Counted
18	Sample seed weight (100 seeds)	Measured(g)
19	Yield/ plant	Measured(g)
20	Growth habit	1 (procumbent); 2 (procumbent 2); 3 (decumbent 1); 4 (decumbent 2); 5 (decumbent 3); 6 (erect); 7 (others)
21	Stem branching pattern	1 (alternate); 2 (sequential); 3 (irregular with flowers on the main stem); 4 (irregular without flowers on the main stem); 5 (others)
22	Stem pigmentation	1 (absent); 2 (present)
23	Stem hairiness	3 (scarce); 7(abundant)
24	Lateral branch habit	1 (non-distichous); 2(distichous)
25	Peg colour	1 (absent); 2 (present)
26	Leaflet shape	1 (cuneate); 2 (obcuneate); 3 (elliptic); 4 (lanceolate); 5 (others)
27	Hairiness of young leaflets	1 (almost glabrous); 2 (sparse and short); 3 (sparse and long); 4 (profuse and short); 5 (profuse and long); 6 (others)
28	Hairiness of mature leaflets	1 (almost glabrous); 2 (sparse and short); 3 (sparse and long); 4 (profuse and short); 5 (profuse and long); 6 (others)
29	Pod beak	1 (absent); 3 (slight); 5 (moderate); 7 (prominent); 9 (others)
30	Pod constriction	0 (none); 3 (slight); 5(moderate); 7 (deep); 9 (very deep)
31	Pod reticulation	0 (smooth); 3 (slight); 5 (moderate); 7 (prominent); 9 (others)
32	Seed colour	1 (one colour); 2 (variegated)
33	Number of seeds/ pod	1 (2-1); 2 (2-1-3); 3 (2-3-1); 4 (2-3-4-1); 5 (2-4-3-1) 6 (3-2-4-1); 7 (3-4-2-1); 8 (others)

Source: IBPGR/ ICRISAT groundnut Descriptors (1981).

Table 4. Eigen values, % and cumulative variance, factor scores and communality of the ten most important characters from factor analysis

Eigen value	Proportion of variation accounted for (%)	Cumulative percentage
7.162	28.170	28.170
5.047	14.262	42.432
4.258	12.690	55.122
3.826	10.951	66.073
3.234	9.423	75.496

Table 5. Eigen values, percent and cumulative variance, factor scores and communality of the ten most important characters from factor analysis.

Character	Factor I	Factor II	Factor III	Factor IV	Communality
Yield per plant	0.281	0.196	-0.655	0.125	0.963
Seed colour	0.033	0.736	-0.325	0.075	0.935
Number of pods per plant	-0.195	-0.210	-0.412	0.438	0.924
Matures leaflet length	-0.701	0.531	-0.108	0.037	0.918
Sample seed (100 seeds) weight	0.352	0.370	0.299	-0.456	0.913
Pod width	0.697	0.538	-0.223	0.208	0.909
Height per plant at flowering	0.144	-0.315	-0.502	-0.499	0.905
Final height per plant	-0.744	0.213	0.165	0.356	0.899
Pod length	0.341	0.803	0.195	-0.144	0.897
Pod beak	0.424	0.105	-0.364	0.375	0.887
Eigen values	6.062	4,047	3.859	3.318	
Percent Variance	18.37	12.26	11.69	4.232	
Cumulative variance	18.37	30.63	42.32	52.58	

entries were maximum (Rezai and Frey, 1990; Ariyo, 1993). Factors and discriminant canonical analysis were also performed using the SPSS (Version 10.0) package. Factor analysis used the covariance matrix of characters (Harman, 1967; Ariyo, 1992) to generate factor loadings and communalities using the method of principal component extraction.

The discriminant canonical analysis summarizes the multivariate data in the same way as the canonical correlation. The analysis uses the Wilks' lambda as the statistics for entering or removing new variables and thereby identifies the variables that provide the best discrimination among the entries. Single Linkage Clusters Analysis (SLCA) was performed to obtain dendrogram and sort genotypes into clusters using the FASTCLUS technique of SAS.

RESULTS

Factor analysis

The results obtained from the factor analysis of the characters are presented in Table 4. The analysis identified 33 factors out of which only four were extracted which together explained 53% of the variance among the entries. The first factor with Eigen value of 6.062 accounted for only 18.37% of the variance and is primarily related to final plant height, pod width, leaflet length, stem pigmentation, number of nodes on the main

stem at maturity and number of leaves at flowering. The factor that accounted for 12.26 % of the total variance is mainly loaded by pod length. The third factor that accounted for 11.69 % of the total variance is mainly described by days to maturity and hairiness of young leaflet. The fourth factor is loaded by pod constriction, plant height at flowering, weight of 100 seeds, stem hairiness and it accounted for just 4.23 % of the total variance. The communality values ranged from 0.963 for yield/ plant to 0.680 for stem branching pattern.

Principal component analysis (PCA)

Results from the PCA presented in Table 5, revealed that only five of the thirty three principal components had Eigen values greater than 3.0 while, the first four axes with Eigen values of 7.162, 5.047, 4.258 and 3.826 respectively, jointly accounted for 66.07% of the total variation among the genotypes. The first five principal axes together explained above 70% of the total variation among the 33 characters that described the 22 genotypes. The major characters described by the first four principal axes are presented in Table 6. The first principal component axis was mainly loaded by vegetative characters.

Table 6. Eigen vectors for major traits of the first four principal components used in the ordination.

Axis 1		Axis 2		Axis 3		Axis 4	
Trait	Score	Trait	Score	Trait	Score	Trait	Score
Final plant height	-0.302	Pod length	0.399	Days to maturity	0.413	Leaflet shape	-0.318
Leaflet length	-0.285	Seed colour	0.366	Hairiness of young leaflet	0.336	Pod constriction	0.274
Stem pigmentation	-0.268	Pod width	0.268	Yield/ plant	-0.333	Plant height at flowering	-0.273
Number of nodes on the main stem at maturity	-0.259	Leaflet length	0.264	Days to 50% flowering	0.304	Stem hairiness	0.271
Number of leaves at flowering	0.251	Seed length	0.263	Plant height at flowering	-0.255	Weight of 100 seeds	-0.250
Leaflet width	-0.251	Leaflet length/ leaflet width ratio	0.253	Stem branching pattern	-0.219	Seed width	-0.244
Number of seeds/ pod	0.241	Seed width	0.234	Number of pods/ plant	-0.209	Growth habit	-0.243
Peg colour	-0.231	Hairiness of mature leaflet	-0.228	Pod beak	-0.185	Number of pods/ plant	0.241
Seed length	0.219	Weight of 100 seeds	0.184	Seed colour	-0.166	Number of seed/ pod	0.225
Pod beak	0.172	Pod reticulation	0.180	Leaflet shape	-0.162	Hairiness of young leaflet	0.220

These were final plant height leaflet length, stem pigmentation, nodes on the main stem at maturity, number of leaves at flowering, leaflet width, seeds per pod, peg colour, seed length and pod beak in that order. Axes two and three were described largely by pod and seed characteristics like pod length, pod width, pod reticulation, pod beak, seed colour, seed length, weight of 100 seeds and number of pods per plant. The fourth axis is loaded by leaflet shape, pod reticulation, plant height at flowering, stem hairiness, weight of 100 seeds, seed width, number of pods per plant, seeds per pod and hairiness of young leaflet.

The configuration of the twenty two groundnut genotypes, along the first three principal component axes are shown in Figures 1, 2 and 3. The ordination of the genotypes on axes 1 and 2 (Figure 1) revealed that ICG 6402 (genotype 3), ICG 12370 (genotype 8), ICG 1399 (genotype 15) and ICG-IS-6646 (genotype 19) were the most distinct genotypes. ICGY-5M-4746 (genotype 18), ICG49-85A (genotype 21) and UGA-7-M (genotype 22) from local sources were most distinct from others in Figures 2 and 3. The remaining genotypes from ICRISAT-India and local sources (Nigeria) did not show any specific pattern in their distribution

Single linkage cluster analysis (SLCA)

The dendrogram from the Single Linkage Cluster Analysis is presented in Figure 4. All genotypes were distinct at 100% level of similarity while at 25% they could no longer be discriminated. ICGY-5M-4746 (G18) and ICG-IS-11687 (G17), both collected locally (from Zaria), were most similar to each other and different from others above 85% level of similarity. ICGY-6M-5236 (genotype 16)

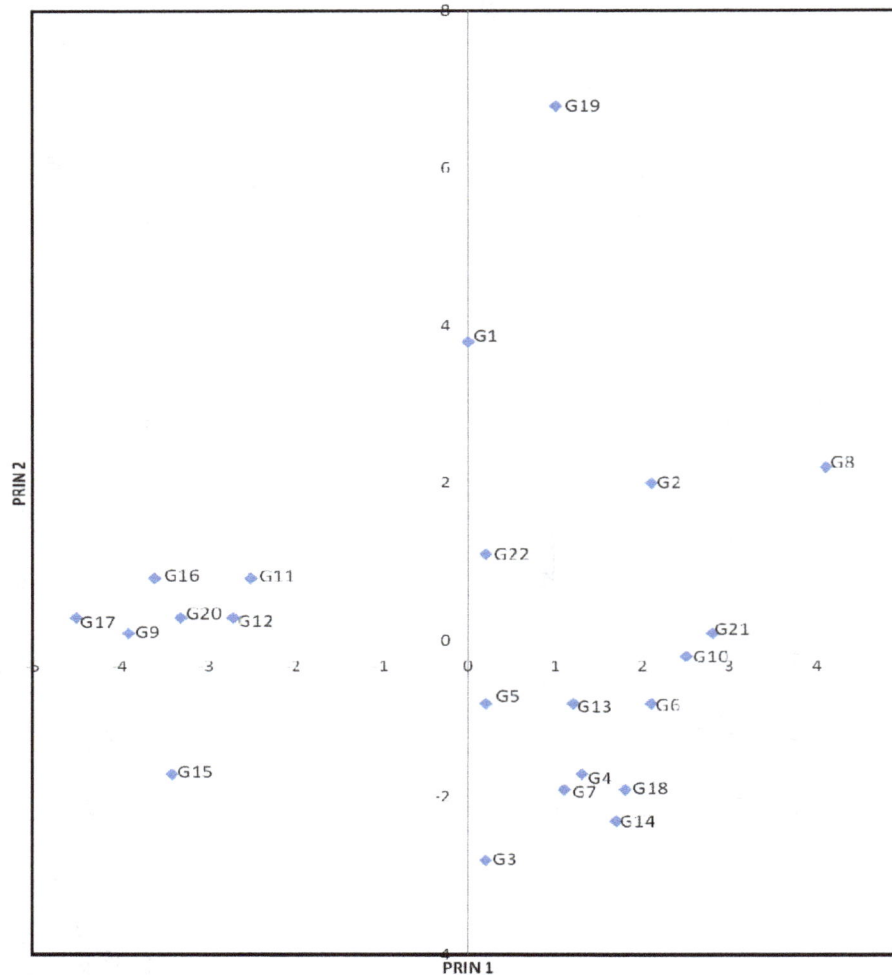

Figure 1. Configuration cf the 22 groundnut genotypes under principal component axes 1 and 2.

formed cluster with others from Zaria collection at 65% level of similarity. At 64% genotype ICG49-85A (G20) and UGA-7-M (G22) formed a cluster, while ICG- 4998 (G1) and ICG- 862 (G2) from ICRISAT (India) formed cluster at 63% level of similarity and they where the most similar genotypes with the local collections. The last two sub-clusters cannot be distinguished from each other at 50% level of similarity. ICG-IS-6646 (G19) and ICG-IS-3584 (G20) joined the cluster at 49 and 48% levels of similarity respectively. Above 45% ICG-4412 (G5) and ICG-156 (G6) cannot be distinguished from each other, ICG-12189 (G11) and ICG-8490 (G4) had joined them to form a cluster at 49 and 35% levels of similarity respectively. Above 33% all the entries had formed eight sub-clusters and by 29% the last three entries ICG-12370 (G8), ICG-2106 (G9) and ICG-4343 (G10) had formed a single cluster with the others.

Table 7 presents the eight clusters, obtained with the FASTCLUS procedure of SAS, showing the pattern of association with characters. Clusters I and VII contained 8 and 2 genotypes respectively. Four genotypes each were grouped into clusters II and III, while the other clusters contained one entry each. Genotype in cluster V was the tallest at flowering and had the largest days to maturity, number of branches at maturity and weight of 100 seeds. Entries 8 and 10 in cluster VII are late flowering with highest yield, while entry 9 that made up cluster VIII had the tallest plants at maturity, highest number of nodes on the main stem at maturity with thickest stems and produced the highest number of pods per plant.

Canonical analysis (CA)

The Eigen values, total variances and correlations between original variables and canonical variables that described the variation in the characters measured are

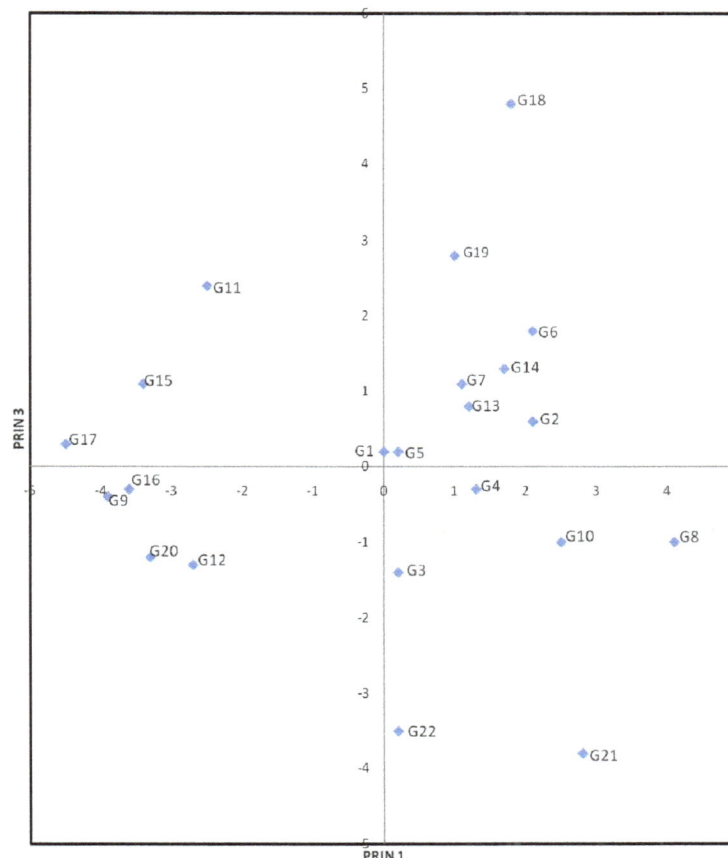

Figure 2. Configuration of the 22 groundnut genotypes under principal component axes 1 and 3.

presented in Table 8. The first five canonical variables had Eigen values greater than 2.0 and accounted for 46.14, 27.72, 11.67, 7.54 and 6.92% of the total variance, respectively. The first four canonical variables however, recorded 93.07% of the variation. Number of leaves at flowering and lateral branch habit, were among the important characters in the first canonical variable while the second canonical variable comprised of number of pods per plant, yield per plant and stem pigmentation. The third canonical variable comprised of number of pods per plant, yield per plant, pod constriction and peg colour while number of leaves per plant, leaflet length and leaflet width were important for the fourth variable.

Discriminant analysis

Table 9 presents the Eigen values, variance and pooled within group correlation between discriminant variable and the canonical discriminant functions. The first four functions had Eigen values that are above 2 and jointly accounted for 99.34% of the total variance. The first two functions accounted for about 84% of the total variance

within the genotypes whereas the third and the fourth functions explained 14.77 and 1.08% of the total variance respectively. The first discriminant function, which accounted for 60.82% of the variance, was highly negatively correlated with number of leaves per plant at flowering (-0.799) but positively correlated with leaflet length (0.392). Number of pods per plant (-0.872) and yield per plant (-0.641) had high negative correlations with the second function while number of seeds per pod had the highest positive correlation (0.368) with the second function. Leaflet shape had the highest positive correlation (0.319) with the third function while hairiness of mature leaflet had the least correlation (0.111) with the third function. The fourth discriminant function correlated negatively with lateral branch habit (-0.495) while number of pods per plant had the least correlation (-0.230). The step-wise order of inclusion of the ten most important variables in the discriminant analysis is shown in Table 10. The order in which the variables were included in the discriminant analysis indicates their relative importance in classifying entries. Number of pods per plant was ranked first in the order of relative importance for discriminating the genotypes. It was followed by number of leaves per

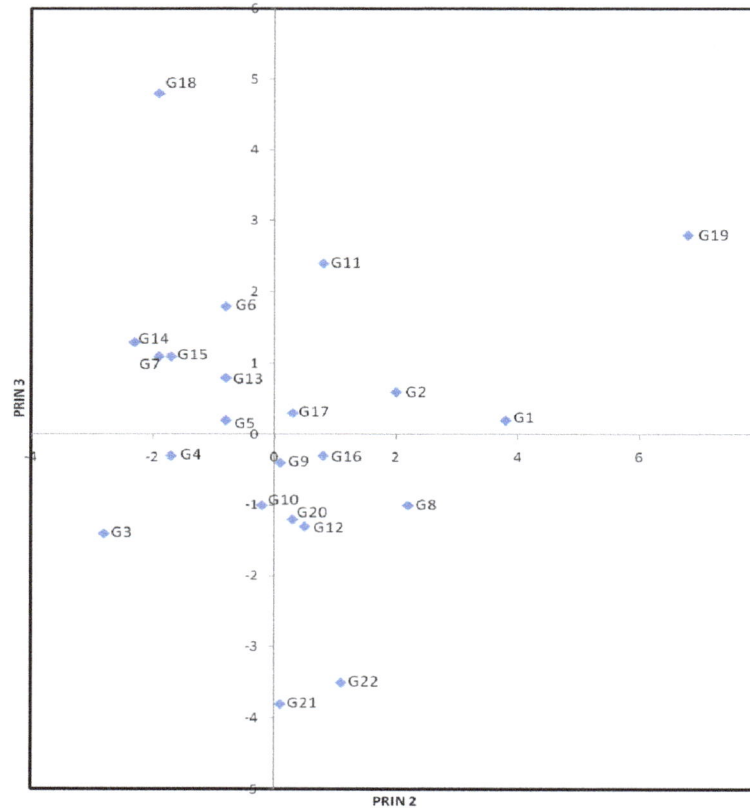

Figure 3. Configuration of the 22 groundnut genotypes under principal component axes 2 and 3.

plant at flowering and hairiness of young leaflet respectively, while the least ranked variable among the top ten was pod reticulation.

DISCUSSION

When dissimilarity between a pair of variety is defined on a multivariate criterion, it is useful to be able to determine the plant characters which cause the dissimilarity to arise and the relative contributions that the various characters make to the total variability in the germplasm (Ariyo, 1993). Factor analysis and principal component analysis identified some similar characters as the most important for classifying the variation among groundnut genotypes. These included; final plant height, leaflet parameters, pod parameters, stem pigmentation, number of nodes on the main stem at maturity and number of leaves at flowering. The similarity between the two techniques had been reported earlier in okra by Ariyo (1993) and rice by Nassir and Ariyo (2007). Although, the two techniques produced similar results, their underlying principles are substantially different from each other. While PCA does not rely on any statistical model and assumptions, factors analysis does. It is also imperative to note that factor analysis

suffers from other drawbacks, such as absence of 'error' structure and the dependence upon scale used to measure the variables (Bartual et al., 1985).

The canonical analysis gave a different picture of the relative importance of the various characters within the entries when compared to the principal component and factor analyses. The analysis considered number of leaves per plant at flowering as the character that best discriminated the groundnut genotypes. Other important variables included, lateral branch habit, pod beak, hairiness of mature leaflet and peg colour. The discriminant analysis also identified number of leaves at flowering as the most important discriminatory trait among the entries. Pod beak, leaflet length, leaflet width, pod constriction and stem branching pattern were other important characters identified by discriminant analysis. Factor analysis captured more of the variation within the entries in higher number of axes compared to other techniques used in this study. However, the techniques showed considerable differences in the characters considered most important for describing the variation among the entries. Differences in results of multivariate techniques, with respect to characters which best summarized the within population variance, had earlier been reported by Ariyo (1993) and Nair et al. (1998).

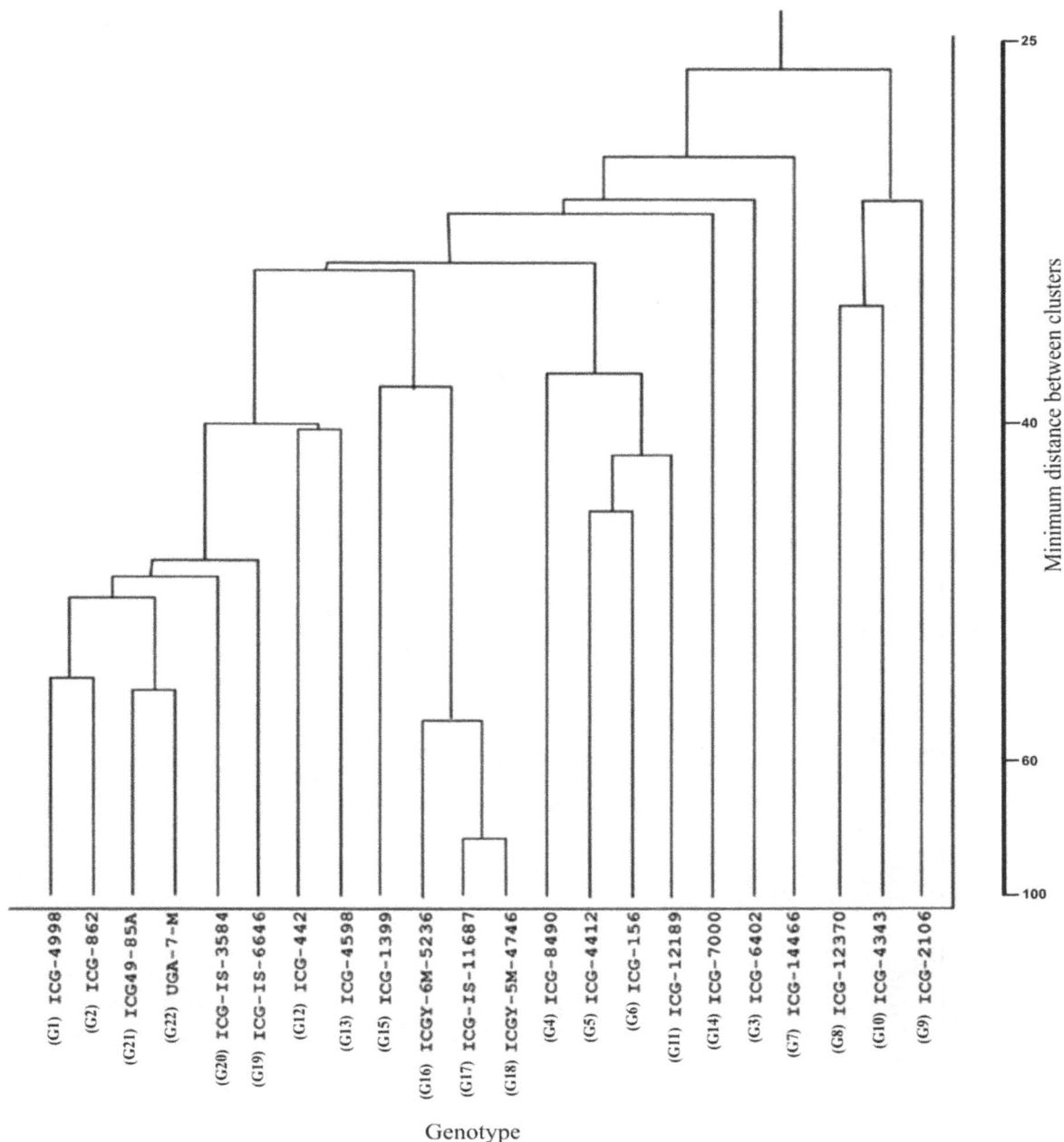

Figure 4. Dendrogram representing relationships of 22 genotypes of groundnut derived from nearest neighbour sorting using Single Linkage Cluster Analysis (SLCA).

Compared to PCA analysis (55%), the canonical correlation analysis accounted for 85.53% of the within entries variance in the same number of axes while the discriminant analysis explained a high figure of 90.16%. The three techniques were, however, better than the factor analysis, which accounted for just 40% of the total variance within entries in the same number of axes. The factor analysis identified final plant height, pod length, days to maturity and pod width as important characters while the discriminant analysis identified number of leaves per plant at flowering, number of pods per plant, leaflet shape and pod beak as the most discriminatory characters. Thus, a combination of factor analysis and any of the PCA, canonical correlation or discriminant analyses would be appropriate for describing the variation in groundnut germplasm.

The grouping of the genotypes by clustering technique did not follow a particular pattern. Some genotypes from the same source were grouped together while others from different sources were clustered together. This

Table 7. Major characteristic pattern of eight clusters (using single linkage cluster method) of groundnut genotypes with their mean values and the standard deviation in parenthesis.

Character	Genotype clusters								Grand mean
	I 1,2,12,13,19,20,21,22,	II 15,16,17,18	III 4,5,6,11	IV 14	V 3	VI 7	VII 8,10	VIII 9	
Days to 50% flowering	29.42 (5.51)	25.92 (0.66)	27.65 (2.63)	28.93	24.13	26.73	32.99 (0.84)	25.80	27.70 (2.75)
Plant height at flowering	20.62 (3.36)	19.71 (4.15)	20.44 (1.00)	22.47	25.02	20.27	22.26 (1.27)	21.65	21.56 (1.71)
Number of leaves at flowering	46.41 (13.99)	40.71 (24.19)	44.32 (15.98)	33.72	36.39	57.84	75.88 (4.14)	33.08	46.04 (14.51)
Days to maturity	126.93 (18.23)	118.05 (6.54)	123.73 (10.83)	126.27	134.93	118.00	133.33 (0.66)	116.40	124.71 (7.03)
Final plant height	47.71 (6.66)	54.58 (15.30)	47.53 (11.96)	49.76	55.21	49.36	47.66 (11.52)	70.11	52.74 (7.65)
Number of branches at maturity	5.07 (0.33)	5.05 (0.24)	4.73 (0.24)	4.61	5.62	4.95	4.92 (0.04)	4.96	4.99 (0.30)
Number of nodes on the main stem at maturity	32.13 (4.28)	31.24 (1.58)	28.85 (3.55)	28.77	32.32	30.13	28.28 (3.29)	35.19	30.86 (2.34)
Stem girth at maturity	2.01 (0.13)	1.82 (0.14)	1.94 (0.24)	2.12	2.17	1.90	2.00 (0.27)	2.20	2.02 (0.13)
Number of pods/ plant	116.90 (38.29)	105.9 (36.41)	134.43 (73.89)	113.49	115.12	130.95	138.63 (49.96)	139.60	124.38 (12.99)
Weight of 100 seeds	42.15 (13.02)	35.31 (3.61)	46.86 (10.94)	50.00	52.64	37.82	39.44 (12.51)	34.04	42.28 (6.89)
Yield/ plant	20.11 (6.14)	20.83 (6.83)	24.06 (7.57)	21.22	17.29	22.95	24.42 (6.26)	19.59	21.31 (2.41)

Table 8. Eigen values, total variance, cumulative variance and correlation between original and canonical variables that describe the variation in 33 traits measured on 22 groundnut genotypes

Canonical variable	Eigen value	Proportion of variance accounted for (%)	Percentage cumulative	Correlation of canonical variable with					
1	14.925	46.14	46.14	Number of leaves/ plant at flowering (-0.318)	Lateral branch habit (0.311)	Pod beak (-0.215)	Hairiness of mature leaflet (0.202)	Pod constriction (-0.184)	Peg colour (-0.178)
2	8.969	27.72	73.86	Number of pods/ plant (-0.671)	Yield/ plant (-0.501)	Stem pigmentation (-0.475)	Lateral branch habit (0.398)	Number of leaves at flowering (0.372)	Number of seeds/ pod (0.309)
3	3.775	11.67	85.53	Number of pods/ plant (-0.588)	Yield/ plant (-0.429)	Pod constriction (-0.324)	Peg colour (0.279)	Leaflet shape (0.258)	Number of seeds/ pod (0.236)
4	2.440	7.54	93.07	Number of leaves at flowering (0.728)	Leaflet length (-0.487)	Leaflet width (-0.452)	Final plant height (-0.385)	Stem branching pattern (-0.362)	Hairiness of young leaflet (0.341)
5	2.240	6.92	100.00	Leaflet shape (0.319)	Stem hairiness (-0.235)	Peg colour (0.194)	Pod constriction (0.187)	Pod beak (0.113)	Hairiness of mature leaflet (0.105)

implies that geographical diversity is not a measure of genotypic diversity in groundnut as reported in okra by Ariyo (1987a). Mean values of characters were more or less continuous across clusters, hence, no sharp distinction between clusters was observed. This was an indication that the characters were under polygenic control. Therefore, improvement programme in groundnut through varietal selection will require painstaking and continuous hybridization and selection efforts for appreciable success (Nassir, 2002). However, the clusters showed some character distinctions that could be employed for hybridization purpose. Cluster III for instance, recorded highest yield per plant but fewer number of pods per plant when compared with cluster VIII, hence genotypes in cluster VIII may give even higher yield if the number of fruits and the number of nodes on the main stem are increased through a careful hybridization with any genotypes in cluster VIII. A high yielding progeny which will have a better combination of height, number of pods per plant and seed weight could be selected from a cross between suitable entries in clusters V and VII. The large amount of genetic variability observed among the genotypes supported the earlier observation by Rao (1985), Siddiquey et al. (2006) and Pintu et al. (2007) that abundant genetic divergence existed in groundnut germplasm. In addition, the pattern of genetic variation would be of great importance to germplasm collectors and plant breeders. The categorization of the diversity among the genotypes into groups with similar characteristics can be used to design a collection strategy (Ariyo, 1993; El-Nasir et al., 2006). Furthermore, the high level of variability exhibited by this population

Table 9. Eigen values, total variance, cumulative variance and pooled within group correlation between discriminant variables and the canonical discriminant functions.

Discriminant Canonical variable	Eigen value	Proportion of variance accounted for (%)	Percentage cumulative	Pooled within group correlation * with					
1	13.698	60.815	60.815	Number of leaves/ plant at flowering (-0.799)	Pod beak (-0.393)	Leaflet length (0.392)	Leaflet width (0.379)	Pod constriction (-0.304)	Stem branching pattern (0.280)
2	3.566	15.829	76.644	Pods/plant (-0.872)	Yield/ plant (-0.641)	Stem pigmentation (-0.499)	Number of seeds/ pod (0.368)	Leaflet shape (0.314)	Pod reticulation (-0.256)
3	3.045	13.517	90.159	Leaflet shape (0.319)	Stem hairiness (-0.217)	Pod constriction (0.216)	Peg colour (0.173)	Pod beak (0.145)	Hairiness of mature leaflet (0.111)
4	2.217	9.841	100.000	Lateral branch habit (-0.495)	Peg colour (0.413)	Number of leaves at flowering (-0.347)	Pod constriction (-0.302)	Hairiness of mature leaflet (-0.288)	Pods/ plant (-0.230)

* Largest absolute correlation between each variable and any discriminant function.

Table 10. Stepwise order of inclusion of the ten most important variables from discriminant analysis.

Variable	Wilks' Lambda	F- value
Number of pods per plant	0.247	95.779
Number of leaves at flowering	0.075	82.068
Hairiness of young leaflet	0.054	67.044
Seed length (cm)	0.041	58.879
Peg colour	0.033	53.040
Seed width (cm)	0.024	52.893
Stem girth at maturity (cm)	0.019	50.424
Stem hairiness	0.015	49.719
Pod length (cm)	0.012	49.181
Pod reticulation	0.009	48.206

*= All F- values are significant at P ≤ 0.01.

indicates that heterosis could be utilized to produce superior hybrid which can be used to enhance crop production. Development of such genotype, however involves the understanding of the variance components in the population (Lukhele, 1981; Makinde, 1988).

Conclusion

Factor analysis captured more of the variation within the entries in higher number of axes compared to other techniques used in this study. However, the techniques showed considerable differences in the characters considered most important for describing the variation among the entries. Thus, a combination of factor analysis and any of the PCA, canonical correlation or discriminant analyses would be appropriate for describing the variation in groundnut germplasm. Genotypes ICG-2106, ICG49-85A and UGA-7-M could serve as a source of genes for earliness. ICG-4998, ICG-12370, ICG-4598, ICG-12189 and ICG-IS-6646 could be exploited for increase in pod yield.

REFERENCES

Abou ETHS, Ibrahim MM, Aboud KA (2006). Stability parameters in yield of White Mustard (*Brassica alba* L.) in different environments. World J. Agric. Sci. 2(1): 47-55.

Aremu CO, Ariyo OJ, Adewole BD (2007). Assessment of selection techniques in genotype x environment interaction in cowpea (Vigna unguiculata (L.) Walp). African J. Agricul. Res. 2(8): 352-355.

Ariyo OJ (1987a). Multivariate analysis and the choice of parents for hybridization in okra (*Abelmoschus esculentus* (L.) Moench). Theor. Appl. Genet. 74: 361–363.

Ariyo OJ (1987b). Stability of performance of okra as influenced by planting date. Theor. Appl. Genet. 74:83-86.

Ariyo OJ (1990b). Measurement and classification of genetic diversity in okra (Abelmoschus esculentus). Ann. Appl. Biol. 116: 335-341.

Ariyo OJ (1992). Factor analysis of vegetative and yield traits in okra (*Hibiscus esculentus*). Indian J. Agric. Sci. 60 (12):793-795.

Ariyo OJ (1993). Genetic diversity in West African okra (*Abelmoschus caillei* L. (Chev.) Stevels- Multivariate analysis of morphological and agronomical characteristics. Genetic Res. Crop Evol. 40: 25-32.

Ariyo OJ, Odulaja A (1991). Numerical analysis of variation among accessions of okra (*Abelmoschus esculentus (L.) Moench*). Ann. Bot. 67:527- 531.

Bartual R, Cabonell EA, Green DE (1985). Multivariate analysis of a collection of soybean cultivars from south-eastern Spain. Euphytica. 34: 113-123.

Cardi T (1998). Multivariate analysis of variation among *Solanum commersoni* (+) *Solanum tuberosum* somatic hybrids with different ploidy levels. Euphytica 99: 35-41.

Das PK and Das Gupta T (1984). Multivariate analysis in black grain (*Vigna mango* (L.) Hepper. Indian J. Genet. 44(2): 243-247.

Encyclopedia of Agricultural Science (1994) Groundnut (*Arachis hypogaea* L.). Acad. press. 3:112.

Flores F, Gutrerrez JC, Lopez J, Moreno MT, Cubero JI (1997). Multivariate analysis approach to evaluate a germplasm collection of *Heydsarum coronarium* L. Genet. Res. and Crop Evol. 44: 545-555.

Harman HH (1967). Modern factor analysis. 2nd ed. University of Chicago Press. Chicago. 124 pp.

International Board for Plant Genetic Resources (IBPGR) and International Crop Research Institute for the Semi - Arid Tropics (ICRISAT) (1981). Groundnut Descriptors. IBPGR Secretariat, Rome. 23pp.

Jain SK, Workman PL (1966). The genetic of inbreeding species. *Advanced genetics.* 14: 55–13.

Lukhele PE (1981). Estimation of genetic variability in sorghum (*Sorghum bicolor* L.) Moenel). Unpublished M. Sc. Thesis, Faculty of Agriculture, Ahmadu Bello University. Nigeria.

Makinde SC (1988). Genetic characterization of okra (*Abelmoschus esculentus* L.) Moench.) cultivars. Unpublished M. Sc. Thesis, Faculty of Science, University of Ilorin, Ilorin. Nigeria.

Nair NV, Ballakrishnan R, Screenivasan TV (1998). Variability for quantitative traits in exotic hybrid germplasm of sugar cane. Gen. Res. and Crop Evol. 45:459-464.

Nassir AL (2002). Studies on genotype x environment interaction, variability and plant character correlations in rice (Oriza sativa L.). Unpublished PhD Thesis, submitted to the post graduate School, University of Agriculture, Abeokuta. Ogun state Nigeria. 123pp.

Nassir AL, Ariyo OJ (2005). Genotype X Environment stability analysis of grain yield of rice (*Oryza sativa* L.). Trop. Agric. (Trinidad) 2005: 1-8.

Nassir AL, Ariyo OJ (2007). Multivariate analysis of variatiob of field-planted upland rice (*Oryza sativa*) in a tropical habitat. Malays. Appl. Biol. 36(1): 47-57.

Pintu B, Shamistha M, Sudhansu SM, Nirmalya B (2007). Influence of genotype on in vitro multiplication potential of *Arachis hypogaea* L. Acta. Bot. Croat. 66(1):15-23.

Ram J, Panwar DVS (1970). Intraspecific divergence in rice. Int. J. Genet. Plant Breed. 30 (1):1-10.

Rao VR (1985). Genetic resources and their use in enhancement of peanut at ICRISAT APRES Procedings. 17:27-30

Rezai A, Frey KJ (1990). Multivariate analysis of variation among wild oat accessions-seed traits. Euphytica. 49:111-119.

Statistcal Analysis System (SAS) (2000). SAS Online Doc. Version 8. Cary, NC: SAS Institute Inc.

Siddiquey MNH, Haque MM, Ara MJF, Ahmed MR, Roknuzzaman M (2006). Correlation and path analysis of groundnut (*Arachis hypogaea* L.). Int. J. Sustain. Agril. Tech. 2(7): 6-10.

Sneath PHA, Sokal R (1973). Numerical taxonomy. W. H. Freeman. San Francisco. 537pp.

White JW, Gonzalez A (1990). Characterization of negative association between seed yield and seed size among genotypes of common bean. Field crops Research. 23: 159-175.

Wright F, Debzhansky TH (1970). Genetics of evolution process. Columbia University press New York. 234pp.

19

Agronomic and molecular evaluation of recombinant inbred lines (RILs) of lentil

C. Bermejo[2], V. P. Cravero[2], F. S. López Anido[1] and E. L. Cointry[1]*

[1]Department of Plant Breeding, Rosario National University, (UNR), CC 14, Zavalla S2125ZAA, Argentine.
[2]CONICET, Zavalla, Argentina.

Sequence-related amplified polymorphism (SRAP) and morphological markers were studied to compare the efficiency of both marker systems in the evaluation of twenty five lentil (*Lens culinaris* Medik.) recombinant inbred lines (RILs) and four testers. Data on 13 morphological traits were collected and analyzed. A total of 240 polymorphic SRAP's bands (76.7%) were scored using four combinations of primers. Cluster analysis and both principal component and principal coordinate analysis were carried out. The entries were grouped in five clusters through procrustes generalized analysis. Relationships among lines revealed by molecular markers were significantly correlated with those based on the agronomic traits (r = 0.75), suggesting that the two systems give similar estimates of genetic relations among the RILs. In future breeding programs parent selections could be based on these traits information in order to broaden the genetic.

Key words: Genetic distance, molecular markers, morphological markers, recombinant inbred lines, SRAP marker.

INTRODUCTION

The genus *Lens* is a member of the legume tribe *Vicieae* which includes the major legume crops of the classical Mediterranean civilizations, faba bean, pea and lentil. *Lens* is a small Mediterranean genus that comprises the cultivated lentil (*Lens culinaris* Medikus subsp. *culinaris*) and 6 related taxa (Ferguson et al., 2000). Lentil (*L. culinaris* Medik.) is a diploid (2n = 2x = 14), autogamous species which is one of the oldest crops in the world, originated in the Near East (Zohary, 1972). The cultivated species, *L. culinaris* has been divided into two sub-species (Barulina, 1930) namely macrosperma (seed diameter, 6 to 9 mm) and microsperma (seed diameter, 2 to 6 mm). The macrosperma type has yellow cotyledons and very light or no pigmentation in their flowers and other plant parts, whereas the microsperma type has red, orange or yellow cotyledons with pigmented flowers and other plant parts. The small size of the seed has some advantages, for example, small seeded varieties in comparison with large seeded varieties were found to be better adapted to dry environments (Erskine, 1996). Genetic variation also exists in lentil for low temperature tolerance under drought condition. Photothermally sensitive genotypes are more tolerant to low temperatures (Keatinge et al., 1996). Large seeded varieties in comparison with small seeded varieties have greater cold tolerance (Erskine, 1996). For culinary use the small seeded varieties are preferred due to less cooking time they demand. However, ssp macrosperma has significantly higher seed saponin content than ssp microsperma (Ruiz et al., 1997). Saponins are a class of bioactive compounds with diverse good properties such as the inhibition of growth and sporulation of a wide range of fungi (Gestetner et al., 1971), the reduction of plasma cholesterol levels in humans (Sidhu and Oakenfull, 1986) and the exhibition of anticancer activity (Konoshima et al., 1992).

One of the major problems that face Argentinian lentil breeders is the narrow genetic base of the current cultivated germplasm (derived from four varieties). This must be broadened from other sources and isolate superior recombinant inbred lines (RILs). While the parental phenotypes can be defined in terms of seed diameter, cotyledon colour and the presence or absence of pigmentation in the flowers, the RILs obtained present mixed phenotypes due to recombination process and these traits lose their properties as agronomic markers.

*Corresponding author. E-mail: ecointry@unr.edu.ar.

On the other hand, the seed diameter might be affected by the environmental conditions in which the crop is grown, and the cultural practices used for production as well (i.e. soil conditions, nutrient deficiency, water stress, extreme temperature, pest infestation, handling operations) (Bishaw et al., 2007). In this sense, varieties with intermediate values for seed diameter are difficult to classify or assign to either ssp. To avoid this problem, when dealing with genetically diverse lentil lines, others morpho-agronomic traits and molecular markers should be considered to provide a relatively unbiased method for their classification and the evaluation of the genetic diversity. Several molecular markers systems were used to determine allelic diversity in lentil collections (Datta et al., 2007). In this work, we propose the use of SRAP (Sequence-Related Amplified Polymorphism) (Li and Quirós, 2001), because they are simpler than AFLP and more reliable than RAPD markers. The objective of the present study was to evaluate morphological traits associated with macrosperma-microsperma seeded type and in combination with SRAP markers, access the efficiency of their use in the classification of recombinant inbreed lines derived from a current lentil breeding programmed.

MATERIALS AND METHODS

The present investigation was conducted at the research station of the Faculty of Agriculture of Rosario University, Argentina (33° 1' S and 60° 53' W), in 2008. The experimental material consisted of 25 F_5 recombinant inbred lines (RILs) (obtained from crosses between lines developed by ICARDA) and four tester (two microsperma type, T_1 and T_2 and two macrosperma type, T_3 and T_4). For each material, 30 seeds were sowed in a randomized block design using 5 dm^3 pots (Kgsoil/pot) filled with a mixture of sterile soil, peat and perlite (1:1:1) as substrate. Plants were grown in greenhouse with natural light. Observations were recorded on number of branch (NB), length (LF) and width (W) of folioles, length (LP) and width (WP) of pods, number of pods per plant (NP), number of folioles (NF), plant height (PH), height of the first pod (HFP), number of nodes at the first pod (NNFP), days to flowering (DF), days to maturity (DM), number of total nodes (NTN) and seed diameter (SD). With the collected data Euclidean distances between RILs were calculated and a cluster analysis was carried out. A dendrogram was generated using "Ward's minimum variance" through the InfoGen software (Balzarini and Di Renzo, 2003).

For DNA extraction and SRAP procedure about 100 mg of fresh leaf was ground in liquid nitrogen and the total genomic DNA was extracted using the Murray and Thompson (1980) protocol based on CTAB. The amplifications were carried out in a thermo-cycler MyCyclerTM (BIO-RAD). At the beginning of the PCR reaction, the annealing temperature was set at 35°C and run for five cycles. Annealing temperature was raised to 50°C for another 35 cycles. Denaturing was done at 94°C for 1 min, while extension was carried out at 72°C for 1 min in all cycles. Four primers forward and two primer reverse were used originating eight primers combinations.

Primers "Forward"

me1, 5'-TGAGTCCAAACCGGATA-3'
me2, 5'-TGAGTCCAAACCGGAGC-3'

me4, 5'-TGAGTCCAAACCGGACC-3'
me5, 5'-TGAGTCCAAACCGGAAG-3'

Primer "Reverse"

em1, 5'-GACTGCGTACGAATTAAT-3'
em2, 5´-GACTGCGTACGAATTTGC-3´

The amplified fragments were separated in denaturing acrylamide sequencing gels and revealed with silver (Li and Quirós, 2001). SRAP fragments were scored for presence or absence as 0 and 1, respectively. Genetic distances were calculated with SRAP data according to Dice's similarity index and a dendrogram was performed. A comparison between morphological and molecular data was carried out through the procrustes generalized analysis using the InfoGen program (Balzarini and Di Renzo, 2003).

RESULTS AND DISCUSSION

Morphological analysis

The variable seed diameter (SD) showed mean values that ranged between 0.4 and 0.62 cm, so, lines could not be classified clearly as belonging to microsperma or macrosperma type. The tester lines showed an average value of 0.4 cm (T1 and T2) and 0.7 cm (T3 and T4). In the cluster analysis this variable was excluded to avoid interference. Relationships among the 25 (RILs) and the four testers revealed by cluster analyses are presented in Figure 1. Three main clusters or groups were formed. Except for LF, HFP and NNFP, significant differences were obtained between clusters for the rest of the evaluated traits. The traits DF, DM and NP showed the highest discriminating values (F = 47.7; p < 0.001; F = 45.6; p < 0.001 and F = 36.4 p < 0.001, respectively).

The RILs included in clusters 1 presented the lowest values for number of branch, number of pods per plant, days to flowering, plant height and number of total nodes; and the highest values for the traits related to the size of leaflets and pods. The testers T3 and T4 were also included in this cluster. This fact implies that the RILs of cluster 1 could be considered of the macrosperma type. The RILs included in groups 2 and 3 differed only by the number of pods and the days to flowering. Group 3 had the highest values for these variables (Table 1); T1 and T2 were included in this group, suggesting that the RILs associated with this group would be microsperma type. The lines associated with cluster 2 correspond to recombinant lines with intermediate characteristics of both subspecies.

Barulina (1930) classified lentil landraces from diverse world areas into macrosperma and microsperma subspecies according to seeds, pods and leaflets characters. The RILs included in Clusters 2 and 3 should be classified like microsperma types. However, the RILs included in Cluster 2 have a lower number of pods per plant. Solh and Erskine (1984) found that seed size and other morphological characters form a continuum between macrosperma and microsperma types. Biçer

Ward
Distancia: (Euclidea)

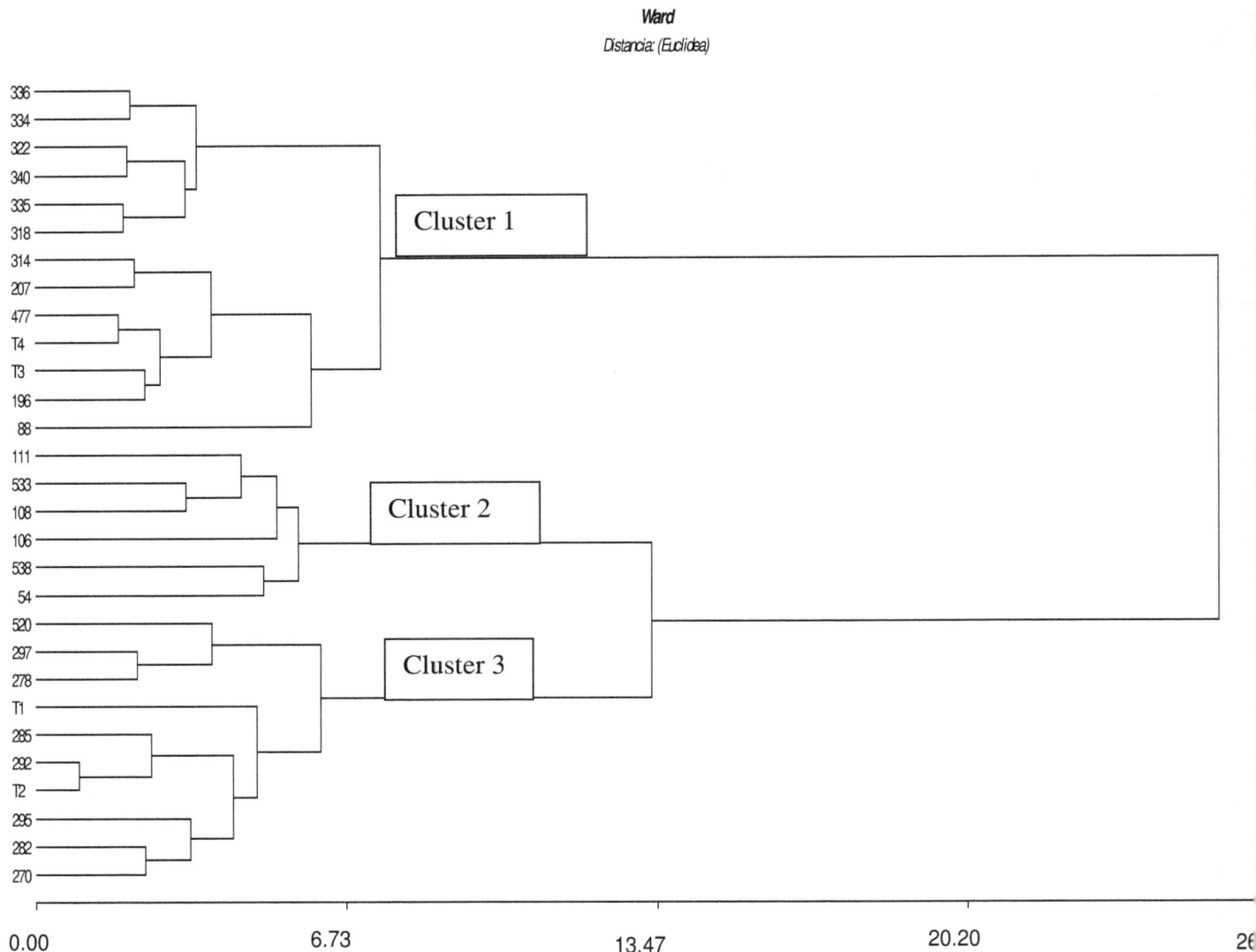

Figure 1. Dendrogram compiled by Ward's method showing the grouping of 25 lentils RILs and four testers based on morphologic traits.

and Sakar (2008) found that the correlation between 1000 seed weight and grain yield was positive, but low and insignificant. Thus, recombinant could have been generated with the morphological characteristics of type microsperma but with less number of seeds per plant.

Molecular data

A total of 8 primers combinations were assayed on the 25 accessions and the four testers. Primer banding patterns that were difficult to score and those that failed to amplify consistently in all genotypes was excluded. Consequently, only four combinations were selected. One RIL was excluded from the trial because the DNA obtained had low quality and it did not allow good

amplification patterns. Pejic et al. (1998) reported that with 150 polymorphic bands it is possible for a researcher to reliably estimate genetic similarities among genotypes within a species. From a total of 317 bands, we found 240 polymorphic fragments (76.7%) with an average of 60 polymorphic bands per combination. This percentage of polymorphism is consistent with those obtained with SRAP markers by Ferriol et al. (2004) in squash, Ahmad et al. (2005) in pistachio, Smutkupt et al. (2006) in highland legumes, and Esposito et al. (2007) in a pea collection.

The relationships between the 24 RILs and the testers revealed by cluster analyses based on Dice distance are shown in Figure 2. Two main clusters can be observed with 5 and 23 accessions, respectively. One of the clusters (cluster 1) included only RILs with a high number

Table 1. Mean values (MV) and standard error (SE) for the four clusters performed with all RILs considering morphologic data.

	Cluster 1 MV±S.E	Cluster 2 MV±S.E.	Cluster 3 MV±S.E.
Number of branch	3.1 ± 0.25^b	4.0 ± 0.45^a	4.1 ± 0.28^a
Width of folioles (cm)	0.3 ± 0.01^a	0.2 ± 0.02^b	0.2 ± 0.01^b
Length of foliole (cm)	1.1 ± 0.06^a	0.9 ± 0.13^a	0.9 ± 0.05^a
Number of folioles	11.4 ± 0.36^a	9.6 ± 0.40^b	10.2 ± 0.20^b
Width of pods (cm)	0.6 ± 0.01^a	0.5 ± 0.04^b	0.5 ± 0.02^b
Length of pods (cm)	1.3 ± 0.03^a	0.9 ± 0.06^b	0.9 ± 0.03^b
Number of pods per plant	20.9 ± 0.94^c	38.6 ± 7.10^b	55.1 ± 3.79^a
Days to flowering	87.4 ± 1.18^b	83.4 ± 2.80^b	101.3 ± 0.21^a
Days to maturity	99.4 ± 1.09^b	90.2 ± 2.37^c	110.0 ± 0.93^a
Plant height (cm)	27.2 ± 1.11^b	37.0 ± 3.27^a	34.9 ± 1.12^a
Number of total nodes	16.6 ± 0.67^b	22.4 ± 1.75^a	21.1 ± 0.81^a
Height of the first pod (cm)	15.8 ± 0.88^a	13.8 ± 1.24^a	12.2 ± 1.40^a
Number of nodes at the first pod	8.7 ± 0.32^a	8.8 ± 0.97^a	7.9 ± 0.74^a

The values followed the same letter are not different at the 5% level.

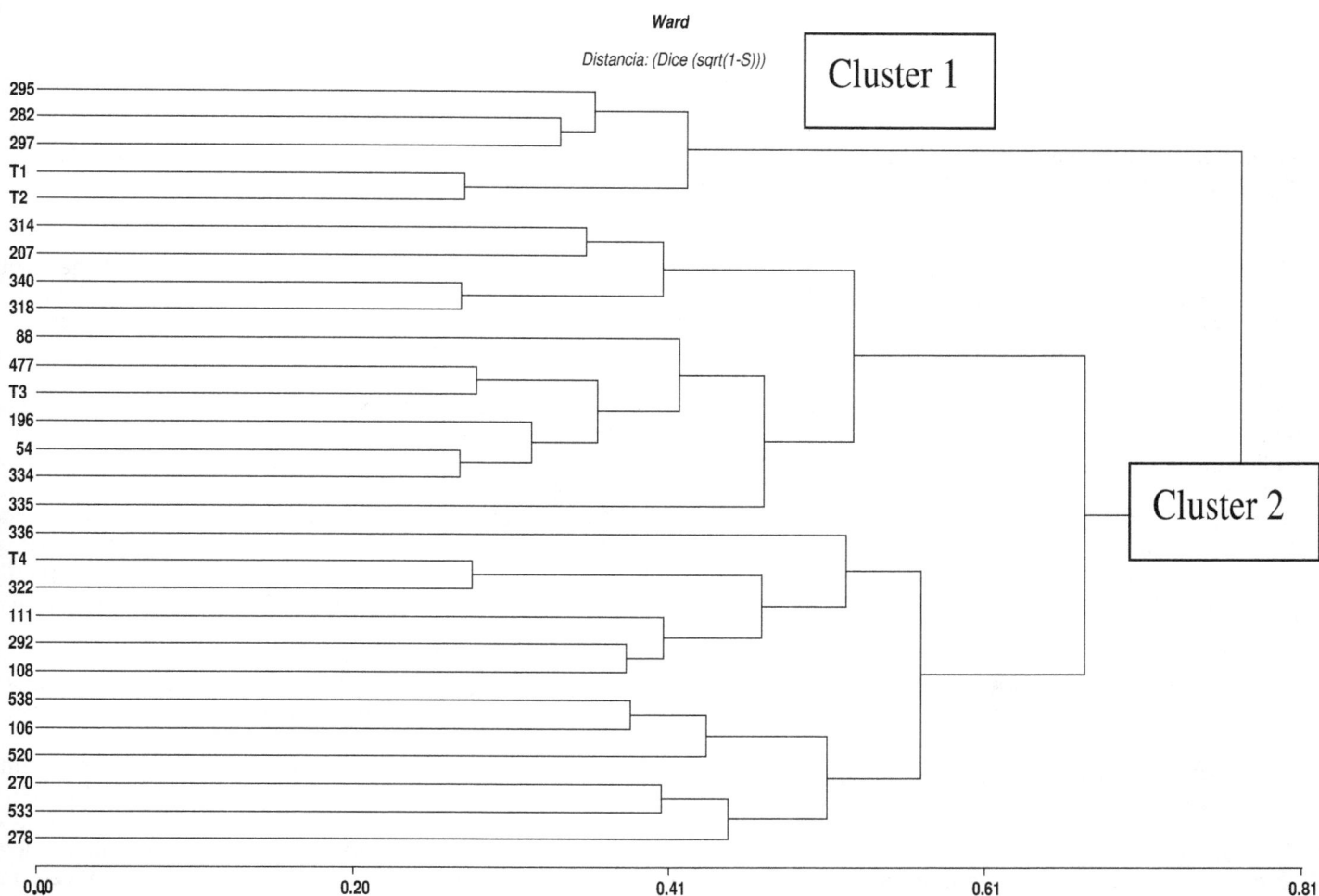

Figure 2. Dendrogram compiled by Ward's method showing the grouping of 24 lentil RILs and four testers based on Dice's distances.

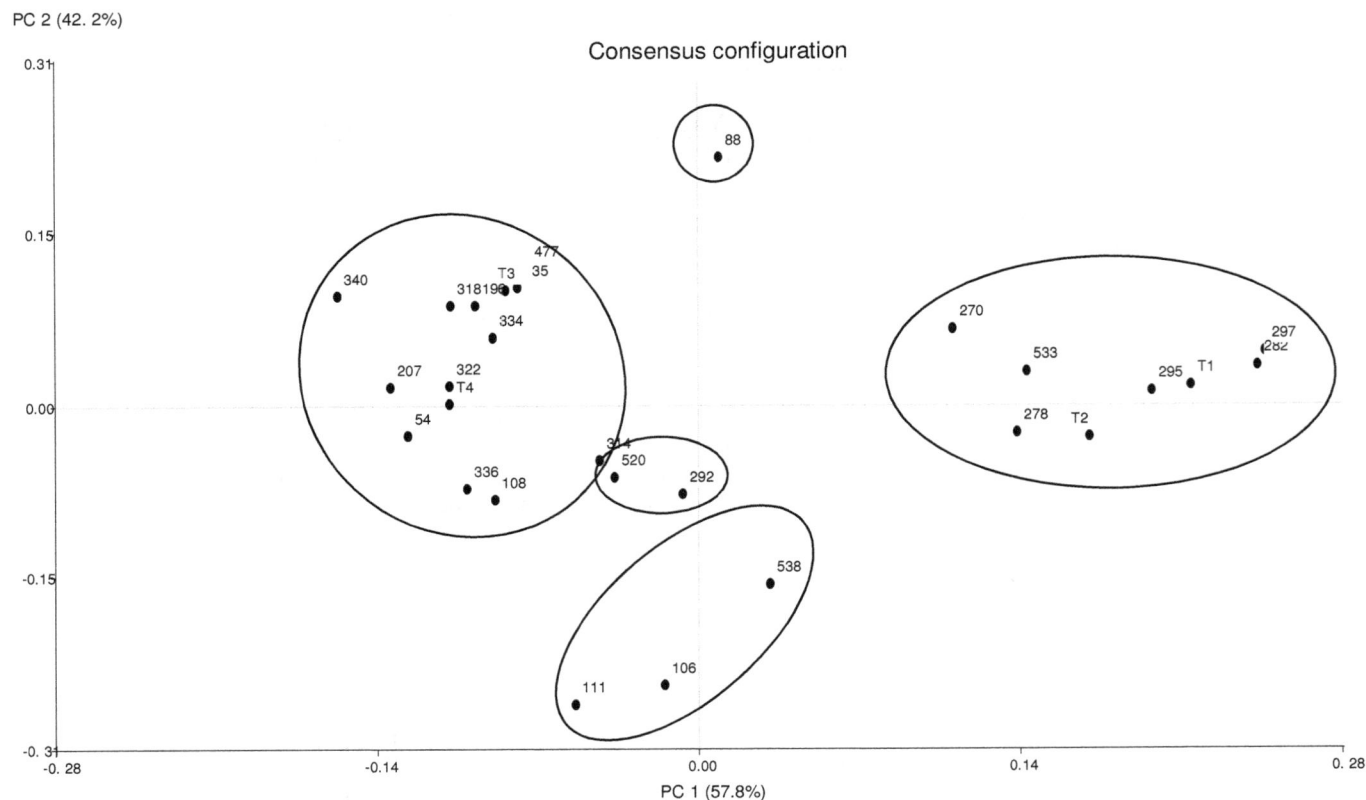

Figure 3. Scatter diagram of the first two principal components (PC$_1$ and PC$_2$) from procrustes generalized of 28 accessions based on 240 SRAP fragments and morphological characters.

of pods. The second cluster comprised RILs with a reduced number of pods. The largest DDI (Dice Distance Index) were observed between RILs included in cluster 1 (DDI = 0.77), meanwhile the lowest DDI were found between the pairs of RILs 54-334 (DDI = 0.27), 54-196 (DDI = 0.29) and 54-461 (DDI = 0.29). Different types of molecular markers were used to determine diversity in lentil collections (Datta et al., 2007). Pioneering works of Abo-elwafa et al. (1995); Ford (1997) with RAPD´s gave important information for germplasm diversity, but also showed that nonspecific PCR based markers could not provide repeatable results in differentiating lentil genotypes. Babayeva et al. (2009) using SSR markers, found a high diversity in Azerbaijan lentil germplasm, as revealed by the low mean pairwise genetic similarities. In our case, the distance index ranged from 0.24 to 0.77 revealing a high genetic variability.

Comparisons between morphological traits and molecular data

Comparison between SRAP and morphological data was carry out using the procrustes generalized analysis. The correlation between Dice similarity index (SRAP data) and Euclidean distance (morphological data) matrices

was 0.75, indicating good correspondence between both data set (Tatineni et al., 1996). RILs distribution for this analysis is showed in the Figure 3. Five big clusters of RILs can be observed and the tester's distribution could be associated with the seed size and the microsperma-macrosperma classification (Barulina, 1930). In several studies carried out in lentil with RAPD markers, the microsperma varieties clustered together whereas the macrosperma varieties conformed another cluster (Sharma et al., 1995, Alvarez et al., 1997, Duran and Pérez de la Vegal, 2004) however, Williams et al. (1974) study did not support the separation of the two types of lentil based on seed size. Our results confirm that this classification could be valid.

The first group included 8 RILs, all of them corresponding to the microsperma type. The second group included fourteen RILs of the macrosperma type. The other three clusters were conformed by RILs with intermediate values between macro and microsperma groups. Abo-elwafa et al. (1995) found identical patterns between accessions belonging to different types. This is an indication that, even when there are differences between the two types of lentils based on grain size and cotyledons colours, this classification is not so suitable, since there is a continuum between the two types, which share the same genetic background.

In breeding programmes for self-fertilizing crops, a large number of crosses are made every year for the introgression of complementary useful genes from one type to other and to isolate transgressive segregants RILs. Therefore, in order to select parents to be involved in a cross for isolating transgressive segregants, greater emphasis needs to be given on the dispersion of genes in the parents. Chahota et al. (2007) founded that, the crosses displaying transgressive segregants having practical utility were observed in the macrosperma × microsperma crosses. Thus, the identification of both types of lentils is one of the key steps in the lentil breeding programmed. The information provided by our study allowed to establish a correlation between morphological and molecular data of r = 0.77. This value indicates an excellent consis-tency between both types of markers and suggests that both types of traits (morphological – molecular) would provide similar estimates on the variability between RILs. This implies that, SRAP markers are an efficient tool in the differentiation of genetic variability. In turn, both morphological and molecular markers are useful to differentiate itself homogeneous groups of lentil.

REFERENCES

Abo-elwafa A, Murai K, Shimada T (1995). Intra-specific and inter-specific variations in Lens, revealed by RAPD markers. Theor. Appl. Genet. 90: 335–340.

Ahmad R, Ferguson L, Southwick SM (2005). Molecular marker analyses of pistachio rootstocks by Simple Sequence Repeats and Sequence-Related Amplified Polymorphisms. J. Hort. Sci. Biotech. 80: 382-386.

Alvarez MT, García P, Pérez de la Vega M (1997). RAPD polymorphism in Spanish lentil landraces and cultivars. J. Genet. Breed. 51: 91-96.

Babayeva S, Akparov Z, Abbasov M, Mammadov A, Zaifizadeh M, Street K (2009). Diversity analysis of Central Asia and Caucasian lentil (Lens culinaris Medik.) germplasm using SSR fingerprinting. Genet. Res. Crop Evol. 56: 293–298.

Balzarini M, Di Renzo J (2003). Infogen. Software para análisis estadísticos de marcadores genéticos. Facultad de Ciencias Agropecuarias. Universidad Nacional de Córdoba. Argentina.

Barulina E (1930). The lentils of the USSR and other countries. Bull. Appl. Bot. Plant Breed. 40: 1–319.

Biçer BT, Sakar D (2008). Heritability and path analysis of some economical characteristics in Lentil. J. Cent. Eur. Agric. 9: 191-196.

Bishaw Z, Niane AA, Van Gastel AJG (2007). Quality seed production. In: Lentil: An Ancient Crop for Modern Time. Chapter 21. (Yadav SS, McNeil D, Philip C, Stevenson PC, Eds.). Springer. The Netherland. pp. 349–383.

Chahota RK, Kishore N, Dhiman KC, Sharma TR, Sharma SK (2007). Predicting transgressive segregants in early generation using single seed descent method-derived micromacrosperma genepool of lentil (Lens culinaris Medikus). Euphytica. 156: 305–310.

Datta S, Sourabh B, Kumari J, Kumar SH (2007). Molecular diversity analysis of lentil genotypes. Pulses news. 18: 2-4.

Durán Y, Pérez de la Vega M (2004). Assessment of genetic variation and species relationships in a collection of Lens using RAPD and ISSR. Span. J. Agric. Res. 2: 538-544.

Erskine W (1996). Seed-size effects on lentil (Lens culinaris) yield potential and adaptation to temperature and rainfall in West Asia. J. Agric. Sci. 126: 335–341.

Esposito MA, Martin EA, Cravero VP, Cointry E (2007). Characterization of pea accessions by SRAP's markers. Scientia Horticulturae. 113: 329–335.

Ferguson ME, Maxted N, Van Slageren M, Robertson LD (2000). A re-assessment of the taxonomy of Lens Mill. Bot. J. Linn. Soc. 133: 41–59.

Ferriol M, Pico B, Fernandez de Córdova P, Nuez F (2004). Molecular diversity of a germplasm collection of squash (Cucurbita moschata) determined by SRAP and AFLP markers. Crop Sci. 4: 653-664.

Ford R (1997). Diversity analysis and species identification in Lens using PCR generated markers. Euphytica. 96: 247–255.

Gestetner B, Assa Y, Henis Y, Birk Y, Bondi A (1971). Lucerne saponins IV. Relationship between their chemical constitution, and haemolitic and antifungal activities. J. Sci. Food Agric., 22: 168–172.

Keatinge JD, Qi A, Kusmenoglu I, Ellis R, Summerfield RJ, Erskine W, Beniwal SPS (1996). Using genotypic variation in flowering responses to temperature and photoperiod to select lentil for the west Asian highlands. Agric. For. Meteorol. 78: 53-65.

Konoshima T, Kokumai M, Kozuka M (1992). Antitumor promoting activities of afromosin and soyasaponin I isolated from Wistaria brachybotrys. J. Nat. Prod. 55: 1776–1778.

Li G, Quirós C (2001). Sequence-Related Amplified Polymorphism (SRAP), a new marker system based on a simple PCR reaction: Its application to mapping and gene tagging in Brassica. Theor. Appl. Genet. 103: 455-461.

Murray JM, Thompson W (1980). Rapid isolation of high-molecular weight plant DNA. Nucl. Acids Res. 8: 4321-4325.

Pejic IP, Ajmone-Marsan M, Morgante V, Kozumplick P, Castiglioni G, Taramino M, Motto M (1998). Comparative analysis of genetic similarity among maize inbred lines detected by RFLPs, RAPDs, SSRs and AFLPs. Theor. Appl. Genet. 97: 1248-1255.

Ruiz RG, Price KR, Rose ME, Fenwick GR (1997). Effect of seed size and testa colour on saponin content of Spanish lentil seed. Food Chem. 58: 223–226.

Sharma SK, Dawson IK, Waugh R (1995). Relationships among cultivated and wild lentils revealed by RAPD analysis. Theor. Appl. Genet. 91: 647–654.

Sidhu GS, Oakenfull DG (1986). A mechanism for the hypocholesterolaemic activity of saponins. Br. J. Nutr. 55: 643–649.

Smutkupt S, Peyachoknagul S, Kowitwanich K, Onto S, Thanananta N, Julsrigival S, Kunkaew W, Punsupa V (2006). Varietal determination and genetic relationship analysis of highland legumes using SRAP markers. J. Genet. Breed. 38: 19-27.

Solh M, Erskine W (1984). Genetic resources of lentil. In: Witcombe, J.R. and W. Erskine (Eds), Genetic resources and their exploitation – Chickpeas, faba beans and lentils, pp. 205-224. Martinus Nijhoff/Dr. W. Junk Publishers for ICARDA and IBPGR, The Hague.

Tatineni V, Cantrell RG, Davis DD (1996). Genetic diversity in elite cotton germplasm determined by morphological characteristics and RAPDs. Crop Sci. 36: 186-192.

Williams JT, Sanchez AMC, Jackson MT (1974). Studies on lentils and their variation I. The taxonomy of the species. SABRAO J. 6: 133-145.

Zohary D (1972). The wild progenitor and place of the cultivated lentil. Lens culinaris. Econ. Bot. 26: 326–332.

20

Correlation and path coefficient analyses in sunflower

Abrar B. Yasin* and Shubhra Singh

Department of Genetics and Plant Breeding, College of Agriculture, Allahabad Agricultural Institute, Deemed University, Naini Allahabad- 211007 (U. P.), India.

Correlation and path coefficient analyses were studied in twenty four diverse genotypes of sunflower in order to understand the relationship and contribution on eight characters towards the grain yield. The yield kg per plant exhibits highly significant and positive correlation with number of seeds per head, head length in diameter and 1000-seed weight at both genotypic and phenotypic level. Path coefficient analysis revealed that number of seeds per head, 1000-seed weight and head length in diameter had the highest and positive direct effect with yield kg per plant. Hence, the study revealed the importance of number of seeds per head, head length in diameter and 1000-seed weight as selection criteria for improvement of yield in sunflower.

Key words: Correlation, path coefficient, genotypes, *Helianthus annuus*, genetic interaction.

INTRODUCTION

Yield is the most economic character in almost all of the crops. Yield is a complex entity and inheritance of yield depends upon a number of characters which are often polygenic in nature and are highly affected by environmental factors (Nadarajan and Gunasekaran, 2005). Knowledge of genetic system controlling yield and its components is useful in understanding the prepotency of the parents and thus help to select parents possessing in-built genetic potential. For efficient selection, programme interrelationship between yield and its components is inevitable and mutual association of plant characters, which is determined by correlation coefficient and is used to find out the degree (strength), mutual relationship between various plant characters and the component character on which selection can be relied upon the genetic improvement of yield. But information on the relative importance of direct and indirect effects of each component characters towards yield is not provided by such studies. Path coefficient is helpful in partitioning the correlation into direct and indirect effects so that relative contribution of each component character to the yield could be assessed (Singh and Narayanam, 2007). In other words, path analysis measures the direct and

indirect contribution of various independent characters on a dependent character. Therefore, the present investigation was undertaken to determine the mutual association among eight selected traits in sunflower and their direct and indirect effects on yield by using path coefficient analysis.

MATERIALS AND METHODS

The materials for the present study consisted of 24 genotypes of sunflower which were introduced from Uttar Pradesh Council of Agriculture Research (UPCAR). Table 1 reveals detailed information about the 24 genotypes used in the present study. The field experiment (3 × 2.4 m) was laid out in randomized block design with three replications at Experimental Center, College of Agriculture, Allahabad Agricultural Institute (Deemed University) Allahabad. Geographically Allahabad Agricultural Institute-Deemed University, Allahabad is situated at 25.35°N latitude, 82.25°E longitude and at an altitude of 78 m above sea level. This region has sub-tropical climate with extreme summer and winter both. During winter season especially in the month of December and January, the temperature drops down as low as 1 - 2°C, while during summer the temperature reaches up to 44 - 46°C. Out of the recommended dose of fertilizers (40:50:40 kg of NPK ha^{-1}), 50% of N and entire dose of P and K were applied at the time of sowing and 2 - 3 seeds were dibbed per hill following a spacing of 60 x 30 cm. After 14 days of sowing, the excess seedlings were thinned out, retaining only one healthy seedling/hill. Five competitive plants were tagged at random in each treatment and in each replication for recording

*Corresponding author. E-mail: abrar_0011@rediffmail.com.

Table 1. List of 24 genotypes of sunflower.

S. no	Accession number	Status	Introduced from	Genetic characteristics
1	ASF-1	Germplasm	UPCAR	Tall, large head, more number of seed per head and high yielding
2	ASF-2	Germplasm	UPCAR	Tall, high 1000-seed weight and large leaf area
3	ASF-3	Germplasm	UPCAR	Semi-dwarf, large head, more no. of seeds per head, large leaf area and high yielding
4	ASF-4	Germplasm	UPCAR	Tall, large head, high 1000-seed weight, Large leaf area, more number of seed per head and high yielding
5	ASF-5	Germplasm	UPCAR	Semi-Tall, Large head, more number of seeds per head and high yielding
6	ASF-6	Germplasm	UPCAR	Semi-dwarf, large head and disease resistant
7	ASF-7	Germplasm	UPCAR	Semi-dwarf and high 1000-seed weight
8	ASF-8	Germplasm	UPCAR	Semi-dwarf, large head, more no. of seeds per head, high 1000-seed weight and high yielding
9	ASF-9	Germplasm	UPCAR	Semi-tall, large head, high 1000-seed weight and large leaf area
10	ASF-10	Germplasm	UPCAR	Tall, more number of leaves and large leaf area
11	ASF-11	Germplasm	UPCAR	Tall, more number of leaves, high 1000-seed weight
12	ASF-12	Germplasm	UPCAR	Dwarf, large leaf area, large head and more number of seeds per head
13	ASF-13	Germplasm	UPCAR	Semi-tall, large head, more number of seeds per head and high yielding
14	ASF-14	Germplasm	UPCAR	Semi-dwarf and large head
15	ASF-15	Germplasm	UPCAR	Tall, more number of leaves and large leaf area
16	ASF-16	Germplasm	UPCAR	Semi-tall, large head, high 1000-seed weight and large leaf area
17	ASF-17	Germplasm	UPCAR	Semi-dwarf, large head, more number of seed per head and high yielding
18	ASF-18	Germplasm	UPCAR	Semi-tall, more no. of Leaves, large leaf area and high 1000-seed weight
19	ASF-19	Germplasm	UPCAR	Tall, large head, high 1000-seed weight, large leaf area, more number of seed per head and high yielding
20	ASF-20	Germplasm	UPCAR	Semi-tall, large head, more number of seeds per head and high yielding
21	ASF-21	Germplasm	UPCAR	Tall, more number of leaves and large leaf area
22	ASF-22	Germplasm	UPCAR	Semi-dwarf and large head
23	ASF-23	Germplasm	UPCAR	Semi-dwarf and high 1000-seed weight
24	ASF-24	Germplasm	UPCAR	Semi-dwarf and large head

UPCAR = Uttar Pradesh Council of Agricultural Research, Lucknow.

Table 2. Genotypic and phenotypic correlation coefficient for nine characters in sunflower.

Characters	r	Plant height (cm)	Leaf length (cm)	Leaf breadth	Leaf area (cm²)	No. of leaves per plant	Head length in diameter	No. of seeds per head	1000-seed weight	Yield kg/Plant
Plant height	rg	1.000						-0.001	0.097	0.004
	rp	1.000						0.053	-0.059	0.009
Leaf length (cm)	rg		0.032					-0.199	-0.462**	-0.110
	rp		0.091					0.051	-0.173	-0.038
Leaf breadth (cm)	rg			0.482*	0.435*			-0.585**	0.009	-0.311
	rp			0.278	0.178			-0.321	0.101	-0.154
Leaf area (cm²)	rg				0.418*	0.765**	0.968**	-0.460*	-0.352	0.245
	rp				0.252	0.641**	0.750**	-0.141	-0.003	0.109
No. of leaves per plant	rg					0.117	0.303	0.424*	0.629**	0.297
	rp					-0.102	0.111	0.266	0.252	-0.195
Head length in diameter	rg						0.495*	-0.068	-0.343	0.192
	rp						0.286	-0.078	-0.010	0.054
No. of seeds per head	rg							0.068	-0.311	-0.251
	rp							0.064	-0.209	0.170
1000-seed weight	rg								-0.640**	-0.031
	rp								-0.010	0.071

*Significant at 5% probability level; **Significant at probability 1% level. rg = genotypic correlation; rp = phenotypic correlation.

detailed observations. The observations were recorded on nine quantitative characters viz., plant height, leaf length, leaf breadth, leaf area (leaf area was worked by using the factor 0.695, Llie and Tonev, 2000), number of leaves per plant, head diameter, number of seeds per head, test weight and yield per plant. Mean values were subjected to different statistical and biometrical analysis. Correlation coefficient at genotypic and phenotypic levels were estimated from the analysis of variance and according to the procedure of Singh and Chaudhary (1977) and path coefficient analysis were done as according to Dewey and

Lu (1959).

RESULTS AND DISCUSSION

In the present investigation, the genotypic correlation coefficients in general were higher than the phenotypic correlation coefficients (Table 2). This could be due to relative stability of the genotypes since the majority of them have been

subjected to a certain amount of selection. Yieldper plant showed significant and positive correlation with number of seeds per head, head length in diameter and 1000-seed weight at both genotypic and phenotypic level. These findings are in close agreement with those of Jaksimovic et al. (1998), Dagustu (2002) in sunflower. The correlation coefficients between the character plant height with leaf breadth, leaf area and head

Table 3. Direct and indirect effect at genotypic and phenotypic levels of different quantitative characters on yield kg per plant of sunflower (*H. annuus* L.).

Characters	Plant height (cm)	Leaf length (cm)	Leaf breadth (cm)	Leaf area (cm)	No. of leaves per plant	Head length in diameter	No. of seeds per head	1000-seed weight (g)	Genotypic/phenotypic correlation coefficient with yield
Plant height	0.009	-0.001	-0.005	0.001	0.002	0.001	0.001	0.001	0.004
	0.025	-0.001	-0.005	0.003	-0.001	-0.003	0.002	0.001	0.009
Leaf length	0.001	-0.034	-0.005	0.021	-0.005	0.001	-0.007	-0.001	-0.110
	0.002	-0.012	-0.003	0.008	0.001	0.001	0.002	-0.003	-0.038
Leaf breadth	0.004	-0.015	-0.011	0.026	-0.007	0.001	-0.022	0.001	-0.311
	0.007	-0.002	-0.017	0.009	0.002	0.001	-0.010	0.017	-0.154
Leaf area	0.004	-0.026	-0.010	0.027	-0.010	0.001	-0.017	-0.001	0.245
	0.006	-0.007	0.013	0.012	0.002	0.001	-0.004	0.001	0.109
No. of leaves per plant	-0.001	-0.010	-0.005	0.017	-0.016	0.001	-0.012	-0.001	0.297
	-0.003	-0.001	0.005	0.003	0.008	-0.001	-0.007	0.012	-0.195
Head length in diameter	0.004	0.002	0.004	0.005	-0.001	0.202	-0.009	0.291	0.488**
	0.007	0.097	0.215	-0.001	0.001	0.221	-0.005	0.214	0.749**
No. of seeds per head	0.001	0.007	0.006	0.013	0.005	0.248	0.337	0.001	0.603**
	0.001	0.007	0.006	-0.002	-0.002	0.171	0.232	0.305	0.718**
1000-seed weight	-0.001	0.016	0.001	0.010	0.010	0.233	-0.007	0.392	0.643**
	-0.001	0.002	-0.002	0.001	0.001	-0.001	0.208	0.215	0.423*

Residual = 0.0011 (genotypic), residual = 0.0046 (phenotypic), bold number = direct effect.

length in diameter were positively significant. Similar relationship was reported by Singh and Labana (1990). A tall plant supporting many leaves could increase total biomass production through increased carbon fixation, which can ultimately be portioned to reproductive organs. Thus, the total number of internodes, which are the sites of leaf initiation, should remain constant during breeding for short plant type. It was found that at genotypic level, leaf length showed positive and significant correlation with leaf breadth but negative and significant correlation with 1000-seed weight. This trait showed positive and significant correlation with leaf area both at genotypic and phenotypic level. At genotypic and phenotypic level, leaf breadth showed positive and significant correlation with leaf area. This trait showed positive and significant correlation with number of leaves per plant but negative and significant correlation with number of seeds per head at genotypic level. Leaf area showed positive and significant correlation with number of leaves per plant and negative significant correlation with number of seeds per head at genotypic level. Similar observation was also confirmed by Rajan (1976). The genotypic and phenotypic correlation coefficient of yield with its contributing components were portioned into direct and indirect effects through path coefficient analysis (Table 3) in order to formulate a sound basis for selection of the important contributing characters to the yield in sunflower. Thus, the correlation in conjugation with path coefficient can give a better insight into the cause and effect relationship between the different pairs of characters. Path analysis revealed that all the characters except number of seeds per head, head length in diameter and 1000-seed weight had small positive/negative direct effect on yield per plant. Plant height and leaf area had positive small direct effect on yield per plant, while leaf length, leaf breadth and number of leaves per plant had negative small direct effect on yield per plant. The characters number of seeds per head, head length in diameter and 1000-seed weight exhibited positive and significant association with yield per plant both at genotypic and phenotypic level and on partitioning the correlation, it was observed that these traits showed positive and high direct effect on yield per plant. These results are in agreement with that of Patil et al. (1996); Dagustu (2002). As expected, the number of seeds per head, head length in diameter and 1000-seed weight showed positive and significant association with yield as well as these traits also exhibit positive and high direct effect on yield. Hence, selection for all the above traits would be more effective to bring about simultaneous improvement for yield and other yield components in sunflower.

Conclusion

It is concluded from the present investigation that, yield has positive and significant association with number of seeds per head, head length in diameter and 1000-seed weight at both genotypic and phenotypic level. Path

coefficient analysis indicates that number of seeds per head, 1000-seed weight and head length in diameter have maximum positive direct effect with yield. The selection of these traits would be more effective to bring about simultaneous improvement for yield in sunflower.

REFERENCES

Dagustu N (2002). Correlation coefficient of seed yield components in sunflower. Turk. J. Field Crops 7: 15-19.
Dewey OR, Lu KH (1959). Correlation and path coefficient analysis of yield components in crested wheat grass seed production. Agron. J. 51: 515-518.
Llie L, Tonev T (2000). Relation between leaf area, leaf dimension and optimizing the equalizing coefficient K (LxBx0.695). Bulgarian J. Agric. Sci. pp. 399-404.
Nadarajan N, Gunasekaran M (2005). Quantitative Genetics and Biometrical Techniques in Plant Breeding. Kalyani Publishers pp. 27-28.
Patil BR, Rudra RM, Vijaykumar CHM, Basappa H, Kulkarni RS (1996). Studies correlation and path analysis in sunflower. J. Oil Seed Res. pp. 162-166.
Rajan SS (1976). Correlation in sunflower with yield components. Division of Genetics Indian Agri. Res. Institute pp. 669-674.
Singh P, Narayanam SS (2007). Biometrical Techniques in Plant Breeding, Kalyani Publishers pp. 56-57.
Singh SB, Labana KS (1990). Correlation and path analysis in sunflower. J. Crop Sci. pp. 49-53.
Singh RK, Chaudhary BD (1977). Biometrical methods in quantitative genetic analysis, Kalyani Publishers, New Delhi pp. 57-58.

21

Identification of coupling- and repulsion-phase markers in rice for brown planthopper resistance genes using F$_2$S of IR 50 X CO 46

Venkateswarlu Yadavalli[1,2]*, Gajendra P. Narwane[2], P. Nagarajan[2] and M. Bharathi[3]

[1]Department of Biochemistry, School of Life Sciences, University of Hyderabad. Hyderabad (AP), India-500 046, India.
[2]Centre for Plant Molecular Biology and Biotechnology, Tamil Nadu Agricultural University, Coimbatore, Tamilnadu, India-641003, India.
[3]Department of Entomology, Tamil Nadu Agricultural University, Coimbatore, Tamilnadu, India-641003, India.

Brown planthopper, *Nilaparvata lugens* Stål (Homoptera: Delphacidae) is one of the most threatening pests, often significantly reducing the rice yield. The breeding of resistant cultivar has been the most effective way of controlling this pest. Recent advancements in DNA marker technology together with the concept of marker-assisted selection (MAS) provide new solutions for selection of more durable brown planthopper (BPH) resistant genotypes in rice. F$_2$s of a cross between IR 50 and CO 46 and their susceptible and resistant parents were used for the present study. In bulk segregant analysis, random amplification of polymorphic DNA (RAPD) primer, OPC 11 (5'AAAGCTGCGG 3') showed co-dominant banding pattern, which amplified a susceptible phenotype specific marker OPC11$_{856}$ associated with repulsion phase. It also amplified resistant phenotype specific markers, OPC 11$_{817}$ which are associated in coupling phase to the resistant allele. The OPA11$_{817}$ RAPD marker could be used in a cost effective way for marker assisted selection of BPH resistant rice genotypes.

Key words: RAPD markers, brown planthopper, bulk segregant analysis, hopper burn, *Oryza sativa*.

INTRODUCTION

Rice (*Oryza sativa* L.) is the most important cereal crop of the world. Outbreaks of the insect pests are closely associated with insecticide misuse, especially during the early crop stages. These insecticide sprays, usually directed at leaf feeding insects, disrupt the natural biological control, which favour the brown planthopper (BPH) development as a secondary pest. Both the nymphs and adults of the brown planthopper insert their sucking mouthparts into the plant tissue to remove plant sap from phloem cells resulting in a severe damage symptom known as 'Hopper burn' besides transmitting rice grassy stunt virus and ragged stunt virus as vectors (Heinrichs, 1979; Rivera et al., 1966).

Incorporating resistance gene(s) from wild species into cultivated species can be an alternative approach to develop BPH resistance in susceptible commercial cultivars (Rahman et al., 2009). Studies conducted by many researchers have investigated the genetics of resistance in rice to brown planthopper. To date, 21 genes have been reported out of which eleven resistance loci reported so far in rice have been identified from wild species. These are *Bph20(t)* on chromosome 4 and *Bph21(t)* on chromosome 12 (Rahman et al., 2009), *Bph10* on the long arm of chromosome 12 from *Ovalipes australiensis* (Ishii et al., 1994), *Bph12(t)* on the short arm of chromosome 4 from *Oryza latifolia* (Yang et al., 2002), *Bph13(t)* on the long arm of chromosome 2 from *Otto eichingeri* (Liu et al., 2001), another *Bph13(t)* against BPH biotype 4 on the short arm of chromosome 3 from *Oryza officinalis* (Renganayaki et al., 2002), *Bph14(Qbp1)* and *Bph15(Qbp2)* on the long arm of chromosome 3 and the short arm of chromosome 4, respectively, from *O. offcinalis* (Huang et al., 2001),

*Corresponding author. Email: venkibiotech@gmail.com, lb06ph08@uohyd.ernet.in.

Bph18(t) on the long arm of chromosome 12 from *O. australiensis* (Jena et al., 2006), and *bph11(t)* and *bph12(t)* on the long arm of chromosome 3 and chromosome 4 respectively, from *O. officinalis* (Hirabayashi et al., 1999).

Dominant gene, *Bph₁*, governing resistance in 'Mudgo', 'MTU15', 'Co22', and 'MGL2' (Athwal et al., 1971), where as a single recessive gene, *bph₂*, conveys resistance in 'ASD7' and 'Ptb18'. Sri Lankan cultivar Rathu Heenati has a dominant gene for resistance, which is non-allelic to, and independent of, *Bph₁* and was designated as *Bph₃* (Lakshminarayana and Khush, 1977). Another Sri Lankan cultivar, Babawee, has a recessive gene for resistance (Lakshminarayana and Khush, 1977), this gene is independent of *bph₂* and is designated as *bph₄*. Later *bph₅* in 'ARC10550', *Bph₆* in 'Swarnalatha' and *bph₇* in 'T12', by means of BPH biotypes from Bangladesh (Kabir and Khush, 1988; Bharathi and Chelliah, 1991). Another recessive gene, *bph₈,* identified in 'Thai Col.5', 'Thai Col 11' and 'Chin Saba'. In the Sri Lankan local cultivars, Pokkali, Balamawee, and Kaharamana a dominant gene, *Bph₉*, identified by Nemato et al. (1989).

Populations of BPH were categorized into five biotypes on the basis of their differential reactions to a set of reference cultivars (Chelliah and Bharathi, 1993). In most of the map construction, F_2 segregating populations are the result of selfing F_1s of two homozygous inbred lines. Most of the molecular maps to date are based on segregation data from F_2 progenies (Jena and Khush, 1992; Wu et al., 2002). A novel BPH resistance gene has been introduced into cultivated rice lines from a distantly related species of *Oryza* and the gene has been mapped with a DNA marker by RAPD and bulked segregant analysis method (Jena and Khush, 1992; Jeon et al., 1999). Bulk segregant analysis (BSA) is a rapid procedure for identifying markers in specific regions of the genome, in which two pools contrasting for a trait are analyzed to identify markers that distinguish them (Michelmore et al., 1991). Two types of molecular markers have been used to develop detailed genetic maps in, RFLPs and RAPDs in rice (Huang et al., 1997; McCouch et al., 1997; Jeon et al., 1999). PCR based RAPD markers have been used for tagging agronomic traits in several crops (Martin et al., 1991; Nair et al., 1995; Ford et al., 1999; Manninen, 2000). RFLPs are often co-dominant but are restricted to regions with low or single copy sequences, moreover it requires large amount of highly pure DNA, specific probes and time consuming. RAPD relies on the differential enzyme amplification of small DNA fragments using PCR with arbitrary oligonucleotide primers (usually 10 mers). Polymorphisms result from either chromosomal changes in the amplified regions or base changes that alter primer binding. The procedure is rapid, requires only small quantity of DNA, which need not be of high quality, and involves no radioactivity. As no southern hybridization is involved, polymorphisms can be detected in fragments containing highly repeated sequences; this provides

markers in regions of the genome previously inaccessible to analysis. Due to its simplicity, inexpensiveness still using for marker assisted selection in *Pisum sativum* for nematode resistance (Burrow et al., 1996; Garcia et al., 1996), scab resistance in *Malus domestica* (Yang et al., 1997), sclerotinia rot resistance in *Trifolium pratense* (Page et al., 1997), brown planthopper resistance in F_3 population of interspecific cross between *O. sativa* and *O. officinalis* (Jena et al., 2002), anthracnose resistance and angular leaf spot resistance in *Phaseolus vulgaris* (Caixeta et al., 2003; Martin and Menarim, 2000), soybean mosaic virus resistance in *Glycine max* (Zheng et al., 2003), rust resistance in *Vicia faba* (Avila et al., 2003), rust resistance in *P. sativum* (Vijayalakshmi et al., 2005), Fusarium wilt resistance in *Cajanus cajan* (Kotresh et al., 2006), karnal bunt resistance in *Triticum aestivum* (Kumar et al., 2006), anthracnose resistance gene in sorghum (Singh et al., 2006), *Arachis hypogaea* for rust resistance (Mondal et al., 2007; He and Prakash, 1997). *Bph 1* has been tagged in rice by using bulked segregant analyses, with 520 RAPD primers to identify markers linked to the BPH resistance gene (Kim and Sohn, 2005). In the present study 170 RAPD primers were used for tagging of bph resistance genes in the F_2s of IR 50 x CO 46.

MATERIALS AND METHODS

Screening for BPH resistance

Screening for BPH resistance was done in the green house, paddy breeding station (PBS), coimbatore by 'standard seed box screening' test (Heinrichs and Mochida, 1984). The F_2s of the cross IR 50 X CO 46 were tested along with the susceptible check (TN1) and the resistant check (CO 46) for their resistance to BPH infestation. Seven days after sowing, when the seedlings were at three-leaf stage they were infected with second and third instar nymphs (8 nymphs of BPH per seedling). When the seedlings of the susceptible check were almost completely dead, the test entries were rated according to the damage rating of the standard evaluation system for rice (International Rice Research Institute, 1996) (Table 1).

Isolation of genomic DNA

Isolation of genomic DNA was done following the method recommended by Jena et al. (2002) with slight modifications. To extract DNA from the parents, the F_2s from the scored plants, the surviving plants immediately after screening were freed from the insects and were planted in separate clay pots in order to grow those to grown up stage for about 15 to 20 days so as to extract sufficient quantity of DNA.

RAPD and bulk segregant analysis

The RAPD analysis was performed following the method recommended by Saiki et al. (1988) with required modifications. A total of seventy-three decamer primers obtained from Operon Technologies Inc., California, USA were used in this study. 10 ng of template DNA was used for PCR amplification, which is carried out

Figure 1. Artificial screening for BPH resistance in the F_2s. Each lane consists of 17 seedlings. First 10 lanes from left to right constitutes a total of 170 seedlings of the cross IR 50 x CO 46. 11th lane is the susceptible check (IR 50) and 12th lane is Resistant check (CO 46).

in a total volume of 20 µL. The final concentrations are 100 pmol primers, 0.5 mM each of dGTP, dATP, dTTP, dCTP, one unit of Taq polymerase (Invitrogen) and 1X PCR buffer containing 1.5 mM $MgCl_2$. The programme for amplification was setup for initial denaturation at 95°C for 3 min, and then thirty amplification cycles each amplification cycle contained one denaturation step at 94°C for one minute, annealing step at 36°C for 40 s and one extension step at 72°C. Final extension was set for three min at 72°C in a thermal cycler (Eppendorf, USA). PCR amplification products were run on 1.2% agarose gels containing 0.2 µg/mL ethidium bromide in a standard horizontal gel electrophoresis unit (Broviga, Chennai, India) having TBE buffer (90 mM Tris-borate, 1 mM EDTA pH 8.0). The DNA bands were photographed in a gel documentation system.

Initially IR 50 (Susceptible) and Co 46 (resistant) parents were screened with all 73 decamer primers. Polymorphic primers were tested on two DNA bulks, as well as parents (Michelmore et al., 1991). Those primers, which show polymorphism between the parents were used to test for polymorphism in the F_2 population of resistant and susceptible populations each comprising ten samples. Resistant bulk comprises of the genomic DNA from all ten resistant populations in the same way susceptible bulk comprises of genomic DNA from all 10 susceptible populations which are used in this study.

EXPERIMENTAL RESULTS

Inheritance pattern of BPH resistance gene in the F_2 progeny

A total of 170 F_2s were categorized as resistant and susceptible. The plants showing a damage score of 3 and 5 were grouped as resistant (124) and plants showing a damage score of 7 and 9 were grouped as susceptible (46). This data fitting well to the expected 3:1 ratio ($x^2 = 0.38$, P 0.50 to 0.75) (Figures 1, 2 and Table 1).

Identification of RAPD marker linked to BPH resistance

Out of the total 73 RAPD primers initially screened on the parental lines of IR 50 and CO 46, 18 primers (24.6%) showed amplification in both parents. Among these 18 primers, 7 primers did not show any polymorphism and the remaining 11 primers showed reproducible polymorphism between parents. Among these 11 primers, 4 primers showed co-dominant banding pattern of polymorphism between parents. This could be useful for distinguishing heterozygotes from homozygotes. Three primers produced dominant amplicons specific to resistant parent and four primers shown susceptible parent specific amplicons (Table 2). Out of four co-dominant RAPD primers, one primer that is, OPC 11 shown distinct, repeatable and high degree of polymorphism in the resistant parent, resistant bulk, susceptible parent and susceptible bulks. OPC 11 generated polymorphic DNA fragments of OPC 11_{817} (817 bp) and OPC 11_{856} (856 bp). Out of these two markers identified, OPC 11_{817} was associated in coupling

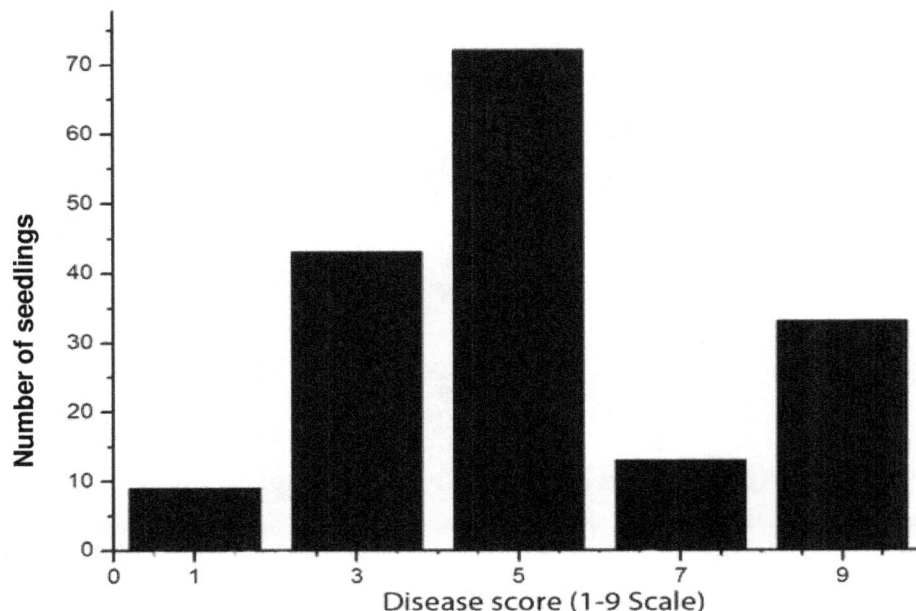

Figure 2. Hopper burn severity score of 170 F_2 seedlings of the cross IR 50 x CO 46 based on standard evaluation system.

Table 1. Standard evaluation system for rice brown planthopper damage.

Scale	Criteria	Category
0	None	-
1	Very slight damage	Highly resistant
3	First and second leaves with orange tips; slight stunting	Resistant
5	More than half the leaves with yellow - orange tips; pronounced stunting	Moderately resistant
7	More than half of the plants wilting or dead and remaining plants severely stunted or drying	Moderately susceptible
9	All plants dead	Susceptible.

Table 2. Pattern of polymorphism between parents (IR 50 and CO 46) detected by RAPD analysis using Operon primers.

Polymorphic type	Primers (No)	Polymorphism (%)
Co-dominant	4	5.4
Dominant (Resistant specific band)	3	3.8
Recessive (Susceptible specific band)	4	5.19
Monomophic	7	9.09
No amplification	55	71.42
Total	73	100

phase to the resistant allele, while OPC 11_{856} was linked in repulsion phase (Figure 3).

DISCUSSION

The performance of the parents IR 50, CO 46 and the F_2s

during screening reveals the consistency of the screening protocols for BPH resistance in this study. DNA amplified products obtained from PCR analysis using random primers have been proposed as an alternative method in targeting DNA sequences for genetic characterization and mapping (Williams et al., 1990).

Relatively higher number of amplified products per

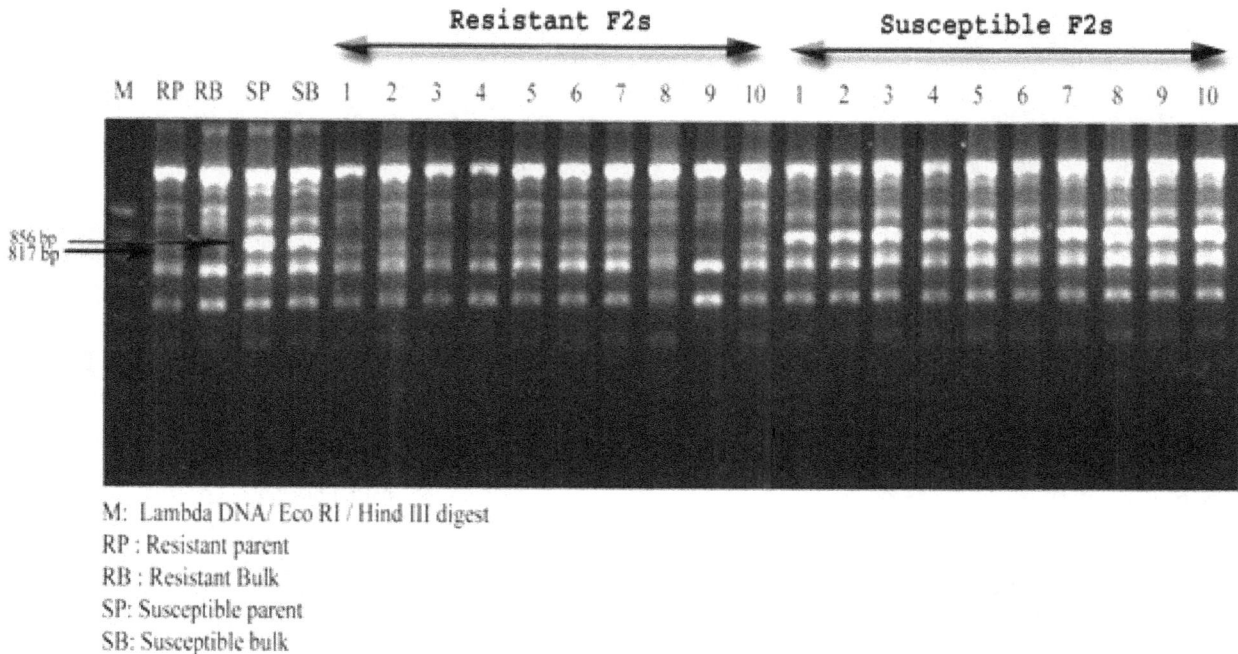

M: Lambda DNA/ Eco RI / Hind III digest
RP : Resistant parent
RB : Resistant Bulk
SP: Susceptible parent
SB: Susceptible bulk

Figure 3. Co-segregation banding pattern of the RAPD primer OPC 11. OPC 11_{817} was associated in coupling phase to the resistant allele, while OPC 11_{856} was linked in repulsion phase.

primer were found in rice, when compared to other plants, like maize (Welsh and McClelland, 1990). One of the most time consuming requirements of DNA marker development, is the need to screen entire mapping populations, with every probe or primer and this has been removed by the bulk segregant analysis (BSA). The minimum size of the bulk is determined by the frequency with which linked loci might be detected as polymorphic between the bulked samples. For a dominant RAPD marker, the probability of a bulk of 'n' individuals having band and a second bulk of equal number of individuals not having a band will be $2(1-1/4)^n(1/4)^n$, when a locus is linked to the target gene (Michlemore et al., 1991). Of the four primers (OPC 10, OPC 11, OPE 14 and OPM 13) that were tested in the BSA with the resistant and susceptible bulks along with the resistant parent and susceptible parent and their F_2s, only one primer, that is, OPC 11 showed co dominant phenotype specific banding pattern. Two pools contrasting for a trait, that is, resistant and susceptible to BPH were analyzed to identify markers that distinguish them. Markers that are polymorphic between the bulks were genetically linked to the loci that determine the trait was used to construct the pools (Michelmore et al., 1991). Results obtained in F_2 seedlings indicate that RAPDs are co-dominant, highly polymorphic and informative in nature. These co-dominant RAPD markers are comparatively rare. Similar to other kinds of co-dominant markers, these co-dominant RAPDs can be of particular value for the purpose of linkage analysis because they provide maximum linkage information per individual in the segregating populations. Co-dominant markers provides easy discrimination between recombinant homozygotes to recombinant heterozygote's (Williams et al., 1990; Mohan et al., 1997; Semagn et al., 2006). RAPD markers which show co-dominant nature were successfully employed in marker assisted selection (MAS) in various crops (Jena et al., 2002; Mondal et al., 2007).

Phenotypic evaluation should be performed with more reliable methods to avoid false positives in further MAS (Mackill and Ni, 2001). A clear polymorphism between the bulks comparable to that between the parents was observed. Poulson et al. (1995) suggested that when bulks are constructed from enough individuals, the BSA is sufficiently robust to cope with the low level of phenotypic misclassification. Bulk segregant analysis by using RAPD markers were successfully used in the development of linked molecular markers. Thus, $OPA11_{817}$ RAPD marker could be used in a cost effective way for marker assisted selection of BPH resistant rice genotypes

ACKNOWLEDGEMENTS

The authors sincerely acknowledge DBT and the Government of India, for providing financial assistance to the study. Also, they acknowledge Mr. K. Kalyana Babu, Scientist, VKPAS, Almora (UK), India and Mr. M.S.R Krishna for their critical suggestions on the preparation of the manuscript.

REFERENCES

Athwal DS, Pathak MD, Bacalangco EH, Pura CD (1971). Genetics of resistance to brown planthopper and green leafhopper in *Oryza sativa* L. Crop. Sci., 1: 747-750.

Avila CM, Sillero JC, Rubiales D, Moreno MT, Torres AM (2003). Identification of rapd markers linked to the uvf-1 gene conferring hypersensitive resistance against rust (*Uromyces viciae*-fabae) in vicia faba l. Theor. Appl. Genet., 107(2): 353-358.

Bharathi M, Chelliah S (1991). Genetics of rice resistance to brown planthopper and relative contribution of genes to resistance mechanisms. Rice genetics ii, IRRI, Rice genet. II, pp 255-261.

Burrow MD, Simpson CE, Paterson AH, Starr JL (1996). Identification of peanut (*Arachis hypogaea* L.) RAPD markers diagnostic of root-knot nematode (*Meloidigyne arenaria* (neal) chitwood) resistance. Mol. Breed., 2: 368-379.

Caixeta ET, Bor´em A, Fagundes SA, Niestche S, Barros EG, Moreira MA (2003). Inheritance of angular leaf spot resistance in common bean line bat 332 and identification of RAPD markers linked to the resistance gene. Euphytica, 134: 297-303.

Chelliah S, Bharathi M (1993). Biotypes of the brown planthopper, *Nilaparvata lugens* (homoptera: Delphacidae)—host inxuenced biology and behavior. Chemical ecology of phytopathogous insects. International Science Publishers New York, pp 133-148.

Ford R, Pang ECK, Taylor PWJ (1999). Genetics of resistance to ascochyta blight (*Ascochyta lentis*) of lentil and the identification of closely linked rapd markers. Theor. Appl. Genet., 98: 93-98.

Garcia GM, Stalker HT, Shroeder E, Kochert GA (1996). Identification of RAPD, SCAR and RFLP markers tightly linked to nematode resistance genes introgressed from *Arachis cardenasii* to *A.hypogaea*. Genome, 39: 836-845.

He G, Prakash CS (1997). Identification of polymorphic DNA markers in cultivated peanut (*Arachis hypogaea* L.). Euphytica. 97:143–149.

Heinrichs EA (1979). Control of leafhopper and planthopper vectors of rice viruses. In: Moramorosch K, arris KF (eds) leafhopper vectors and planthopper disease agents. Academic Press, New York.

Heinrichs EA, Mochida O (1984). From secondary to major pest status: The case of insectcticide-induced rice brown planthopper, *Nilaparvata lugens* resurgence. Protection Ecol., 7:201-218.

Hirabayashi H, Kaji R, Angeles ER, Ogawa T, Brar DS, Khush GS (1999). RFLP analysis of a new gene for resistance to brown planthopper derived from *O. officinalis* on rice chromosome 4. Breed. Sci., 48: 48.

Huang N, Arnold P, Mew T, Magpantay G, McCouch S, Guiderdoni E, Xu J, Subudhi P, Angeles R, Khush GS (1997). RFLP mapping of isozymes, RAPD and QTLs for grain shape, brown planthopper resistance in a doubled haploid rice population. Mol. Breed., 1: 1-8.

Huang Z, He G, Shu L, Li X, Zhang Q (2001). Identification and mapping of two brown planthopper resistance genes in rice. Theor. Appl. Genet., 102: 929-934.

International Rice Research Institute (ed) (1996). Standard evaluation system for rice, 4th edn. International Rice Research Institute, Manila, Philippines.

Ishii T, Brar DS, Multani DS, Khush GS (1994). Molecular tagging of genes for brown planthopper resistance and earliness introgressed from oryza australiensis into cultivated rice, *O. Sativa*. Genome, 37 (2): 217-221.

Jena KK, Jeung JU, Lee JH, Choi HC, Brar DS (2006). High-resolution mapping of a new brown planthopper (bph) resistance gene, bph18(t), and marker-assisted selection for BPH resistance in rice (*Oryza sativa* L.). Theor. Appl. Genet., 288: 288-297.

Jena KK, Khush GS (1992). Introgression of genes from *Oryza officinalis* well ex watt to cultivated rice, *O. Sativa* L. Theor. Appl. Genet., 80: 737-745.

Jena KK, Pasalu IC, Rao YK, Varalaxmi Y, Krishnaiah K, Khush GS, Kochert G (2002). Molecular tagging of a gene for resistance to brown planthopper in rice (*Oryza sativa* L.). Euphytica. 129:81–88.

Jeon YH, Ahn SN, Choi HC, Hahn TR, Moon HP (1999). A RAPD marker for the gene conferring resistance to indian biotype of bph. Rice Genet. Newsl., 15: 133-134.

Kabir MA, Khush GS (1988). Genetic analysis of resistance to brown planthopper in rice, (*Oryza sativa* L.). Plant Breed., 100: 54-58.

Kim SM, Sohn JK (2005). Identification of a rice gene (bph 1) conferring resistance to brown planthopper (*Nilaparvata lugens* Stål) using STS markers. Mol. Cells, 20(1): 30-34.

Kotresh H, Fakrudin B, Punnuri SM, Rajkumar BK, Thudi M, Paramesh H, Lohithaswa H, Kuruvinashetti MS (2006). Identification of two RAPD markers genetically linked to a recessive allele of a fusarium wilt resistance gene in pigeonpea (*Cajanus cajan* L. Mill sp.). Euphytica, 149: 113-120.

Kumar M, Luthra OP, Chawla V, Chaudhary L, Saini N, Poonia A, Kumar R, Singh AP (2006). Identification of rapd markers linked to the karnal bunt resistance genes in wheat. Biol. Plant. 50(4): 755-758.

Lakshminarayana A, Khush GS (1977). New genes for resistance to brown planthopper in rice. Crop Sci., 17: 96-100.

Liu GQ, Yan H, Fu Q, Qian Q, Zhang ZT, Zhai WX, Zhu LH (2001). Mappping of a new gene for brown planthopper resistance in cultivated rice introgressed from *Oryza eichingeri*. Chin. Sci. Bull., 46: 738-742.

Mackill DJ, Ni J (2001). Molecular mapping and marker assisted selection for major gene traits in rice. Rice genetics iv. Science Publishers, Inc., IRRI, Philippines, pp 137-151.

Manninen OM (2000). Associations between anther-culture response and molecular markers on chromosomes 2h, 3h and 4h of barley (*Hordeum vulgare* L.). Theor. Appl. Genet., 100:57-62.

Martin ALA, Menarim H (2000). Identification of RAPD marker linked to the co-6 anthracnose resistant gene in common bean cultivar ab 136. Genet. Mol. Biol., 23 (3): 633-637.

Martin GB, Williams JGK, Tanksley SD (1991). Rapid identification of markers linked to a pseudomonas resistance gene in tomato by using random primers and near isogenic lines. Proc. Natl. Acad. Sci. USA, 88: 2336-2340.

McCouch SR, Chen X, Panaud O, Temnykh S, Xu Y, Cho YG, Huang N, Ishii T, Blair M (1997). Microsatellite marker development, mapping and applications in rice genetics and breeding. Plant. Mol. Biol., 35: 89-99.

Michelmore R, Paran WI, Kesseli RV (1991). Identification of markers linked to disease resistance genes by bulked-segregant analysis : A rapid method to detect markers in specific genome regions by using segregating populations. Proc. Natl. Acad. Sci. USA, 88: 9828-9832.

Mohan M, Nair S, Baghwat A, Krishna TG, Yano M, Bhatia CR, Sasak T (1997). Genome mapping, molecular markers and marker-assisted selection in crop plants. Mol. Breed., 3: 87-103.

Mondal S, Badigannavar AM, Murty GSS (2007). RAPD markers linked to a rust resistance gene in cultivated groundnut (*Arachis hypogaea* L.). Euphytica, 159: 233-239.

Nair S, Bentur JS, PrasadaRao U, Mohan M (1995). DNA markers tightly linked to a gall midge resistance gene (gm2) are potentially useful for marker-aided selection in rice breeding. Theor. Appl. Genet., 91: 68-73.

Nemato HR, Ikeda, Kaneda C (1989). New genes for resistance to brown planthopper, in rice. Japan J. Breed., 39: 23-28.

Page D, Delclos B, Aubert G, Bonavent JF, Mousset-Declas C (1997). Sclerotinia rot resistance in red clover: Identification of rapd markers using bulked segregant analysis. Plant Breed., 116: 73-78.

Poulson DME, Henry RJ, Johnston RP, Irwin JAG, Rees RG (1995). The use of bulk segregant analysis to identify a RAPD marker linked to leaf rust resistance in barley. Theor. Appl. Genet., 91: 270-273.

Rahman ML, Jiang W, Chu SH, Qiao Y, Ham TH, Woo MO, Lee J, Khanam MS, Chin JH, Jeung JU, Brar DS, Jena KK, Koh HJ (2009). High-resolution mapping of two rice brown planthopper resistance genes, bph20(t) and bph21(t), originating from *Oryza minuta*. Theor. Appl. Genet., 119: 1237-1246.

Renganayaki K, Fritz AK, Sadasivam S, Pammi S, Harrington SE, McCouch SR, Kumar SM, Sam RA (2002). Mapping and progress toward map-based cloning of brown planthopper biotype-4 resistance gene introgressed from *Oryza officinalis* into cultivated rice, *O. sativa*. Crop Sci., 42: 2112-2117.

Rivera CT, Ou SH, Lida TT (1966). Grassy stunt disease of rice and its transmission by *Nilaparvata lugens* (Stål). Plant Dis. Rep., 50: 453-456.

Saiki RK, Gelfand DH, Stoffel S, Scarf SJ, Higuchi R, Horn GT, Mullis KB, Erlich HA (1988). Primer directed enzymatic amplification of DNA with a thermostable DNA polymerase. Science, 239: 487-491.

Semagn K, Bjørnstad A, Ndjiondjop MN (2006). Principles, requirements and prospects of genetic mapping in plants. Afr. J. Biotech., 5(25): 2569-2587.

Singh M, Chaudhary K, Singal HR, Magill CW, Boora KS (2006). Identification and characterization of RAPD and SCAR markers linked to anthracnose resistance gene in sorghum [sorghum bicolor (L.) moench]. Euphytica, 149: 179-187.

Vijayalakshmi S, Yadav K, Kushwaha C, Sarode SB, Srivastava CP, Chand R, Singh BD (2005). Identification of RAPD markers linked to the rust (Uromyces fabae) resistance gene in pea (Pisum sativum). Euphytica, 144: 265-274.

Welsh J, McClelland M (1990). Fingerprinting of genomes using PCR with arbitrary primers. Nucleic Acid Res., 18: 7213-7218.

Williams JGK, Kubelik AR, Livak KJ, Rafalski JA, Tingey SV (1990). DNA polymorphisms amplified by arbitrary primers are useful as genetic markers. Nucleic Acids Res., 18: 6531-6535.

Wu JY, Wu HK, Chung MC (2002). Co-dominant RAPD markers closely linked with two morphological genes in rice (Oryza sativa L.) Bot. Bull. Acad. Sin., 43: 171-180.

Yang H, Ren X, Weng Q, Zhu L, He G (2002). Molecular mapping and genetic analysis of a rice brown planthopper (Nilaparvata lugens Stål) resistance gene. Hereditas, 136(1): 39-43.

Yang HY, Korban SS, Kruger J, Schmidt H (1997). The use of a modified bulk segregant analysis to identify a molecular marker linked to a scab resistance gene in apple. Euphytica, 94: 175-182.

Zheng C, Chang R, Qiu L, Chen P, Wu X, Chen S (2003). Identification and characterization of a RAPD/SCAR marker linked to a resistance gene for soybean mosaic virus in soybean. Euphytica, 132: 199-210.

22

Combining ability studies over environments in Rajmash (*Phaseolus Vulgaris* L.) in Jammu and Kashmir, India

Asif M. Iqbal, F. A. Nehvi, Shafiq A. Wani, H. Qadri, Z. A. Dar and Aijaz A. Lone*

Division of Plant Breeding and Genetics, Sher-e-Kashmir University of Agricultural Sciences and Technology of Kashmir, Shalimar, 191121, J and K, India.

The experimental material comprised 45 F1's generated by crossing ten diverse lines of common bean (*Phaseolus vulgaris* L.) in a half diallel fashion and their parents were evaluated in RBD in two replications at three locations during Kharif, 2008. The pooled analysis revealed that both *gca* and *sca* were influenced by environments, which suggested that studies are being conducted over the environments to get unbiased estimates. The sca x e interaction was greater than gca x e interaction for most of the traits. The compassion of relative magnitude of gca and sca variances indicated greater magnitude of sca variances for all the traits, indicating greater importance of non additive gene action for the inheritance of these traits. SKUA-R-607 and SKUA-R-608 showed high combining ability for seed yield and yield attributing traits. The most promising crosses, in order of merit for seed yield, were Shalimar rajmash 1 x SKUA-R-607, SKUA-R-608 x SKUA-R-106 and Shalimar rajmash -1 x SKUA- R-612.

Key words: Common bean, diallel, combining ability over environments, gene action.

INTRODUCTION

The common bean (*Phaseolus vulgaris* L.) is the world's most important grain legume for direct human consumption (Goncalves et al., 2008). It is grown in subtropical or temperate regions throughout the world and during the cool, dry season in tropical areas (Barelli et al., 2000). Like other legumes, they supply proteins, carbohydrates, vitamins and minerals and complement cereals, roots and tubers that compose of the bulk of diets in most developing countries. Brazil is the world's greatest common bean producer, producing more than 2.2 million tons, which represents 17.3% of the world's production (Goncalves et al., 2008). India, China, Myanmar, Mexico and the United States are the next highest producers and they produce XX of the world's production. In India, Rajmash (*P. vulgaris* L.) is a minor pulse crop confined to Himachal Pradesh, Uttar Pradesh, Jammu and Kashmir and North Eastern States, because

it requires cool temperatures for growth. The genetic improvement of *P. vulgaris* L. in India has been accomplished primarily by targeting breeding strategies based on consumer preference for seed size, shape and colour. The North Western Himalayan regions, comprising parts of Himachal Pradesh and Jammu and Kashmir, are rich sources of rajmash diversity for seed, colour, shape and plant type.

The choice of parents in a breeding programme for hybridization is one of the most critical considerations, since the selection on the basis of performance does not provide clear information. Genetic parameters, such as combining ability, play a significant role in crop improvement, since they help in characterizing the nature and magnitude of genetic effects governing yield and component traits, besides pinpointing the promising parents to be used in the creation of genetic variability for eventual use in development of suitable varieties. Diallel analysis developed by Griffing (1956) offers an excellent means of obtaining information on differential parental combinations in terms of general and specific combination ability and nature and extent of gene action.

*Corresponding author. E-mail: ajazalone@rediffmail.com.

Table 1. Analysis of variance for morphological, yield, yield attributing and quality traits in Rajmash (*P. vulgaris* L.) pooled-over environments.

Source of variation	d.f	Primary branches plant⁻¹	Secondary branches plant⁻¹	No. of pods plant⁻¹	Pod length (cm)	No. of seeds pod⁻¹	100-seed weight (g)	Seed yield plant⁻¹	Protein content (%)
Environments	2	0.14*	6.44**	10.49**	6.62**	0.39**	74.61	4.68	59.07
Replications	1	-	-	-	-	-	16.33	-	4.75**
Replications within the environments	3	0.20**	0.48	0.31	8.12**	0.27**	16.33	1.45**	4.75**
Genotypes	54	0.55**	5.14**	5.25**	4.02**	0.73**	88.6**	9.36**	23.41**
Parents	9	1.49**	2.00**	3.09**	2.70**	0.68**	42.92**	2.50**	13.64**
Crosses	44	0.31**	3.39**	4.09**	2.72**	0.61**	79.30**	7.19**	23.28**
Parents versus crosses	1	2.82**	110.34**	75.84**	72.88**	6.31**	885.26**	166.43**	117.34**
Genotypes x environments	108	0.16*	2.05**	2.11**	2.31**	0.55**	30.99**	2.14**	5.36**
Parents x environments	18	0.02	1.66**	0.39*	0.59	0.41**	11.16**	0.53	4.43**
Crosses x environments	88	0.19**	2.14**	2.47**	2.55**	0.58**	35.59**	2.50**	5.65**
Parents x crosses x environments	2	0.04	1.85**	1.44**	6.79**	0.64**	7.22	0.82	0.88
Pooled error	0.03	0.21	0.22	0.38	0.06	2.38	0.33	1.18	

* and ** Significant at 5 and 1% levels, respectively.

However, the scope of such studies is limited if the environmental studies are not conducted in the different environments, as the combining ability and inheritance of quantitative traits vary with environments. Increasing the number of environments reduces the contribution of pooled error and additive wit environment variances (Eberhart et al., 1995). The use of diallel analysis procedures for choosing parents in *P. vulgaris* breeding programme has recently received higher emphasis from bean breeders (Barelli et al., 2000; Goncalves et al., 2008). Therefore, the present study was undertaken to study the combining ability estimates, combining ability x environment interaction and nature and magnitude of gene action in common bean.

MATERIALS AND METHODS

The present study was carried out at three locations, namely: Pulse Research Sub Station, SKUAST-K Habak Srinagar, Jammu and Kashmir, India; Krishi Vigyan Kendra Farm Pombai, SKUAST-K Anantnag, Jammu and Kashmir, India and Faculty of Agriculture, SKUAST-K Wadoora, Jammu and Kashmir, India. Ten diverse genotypes of rajmash namely SKUA-R-602, Shalimar rajmash-1, SKUA-R-607, SKUA-R-608, SKUA-R-609, SKUA-R-612, SKUA-R-91, SKUA-R-106, SKUA-R-23 and SKUA-R-153 were crossed in a diallel fashion of Griffing (1956) in all possible combinations without reciprocals. The 45 F1's and 10 parents were evaluated in RBD with two replications at each location during Kharif, 2008. Each progeny was grown with inter and intra spacing of 30 and 10 cm, respectively. The observations were recorded on five competent plants for different traits namely: primary branches plant⁻¹, secondary branches plant⁻¹, number of pods plant⁻¹, pod length (cm), number of seeds pod⁻¹, 100 seed weight (g), seed yield plant⁻¹ (g) and protein content (%). The protein was estimated according to the modified Kjeldhal's method of Piper (1966). The estimates of variance for both the general and specific combining abilities and their effects were computed according to Models I (fixed effect model) and II (parents and crosses, excluding reciprocals) as given by Griffing (1956).

RESULTS AND DISCUSSION

The pooled analysis over the environments (Table 1) revealed significant differences between genotypes, the parents and crosses interacted differentially for the traits under study in different environments. The crosses interacted more

Table 2. Analysis of variance of combining ability and estimates of components of variance for morphological, yield, yield attributing and quality traits in Rajmash (*P. vulgaris* L.) pooled-over environments.

Source of variation	d.f	Primary branches plant⁻¹	Secondary branches plant⁻¹	No. of pods plant⁻¹	Pod length (cm)	No. of seeds pod⁻¹	100-seed weight (g)	Seed yield plant⁻¹ (g)	Protein content (%)
Gca	9	0.72**	1.81**	1.42**	2.90**	0.31**	96.45**	4.13**	9.65**
Sca	45	0.19**	2.72**	2.86**	1.83**	0.37**	33.60**	4.78**	12.12**
gca x environments	18	0.04**	1.60**	0.65**	22.48**	0.32**	16.87**	0.77**	1.76**
sca x environments	90	0.09**	0.91**	1.13**	102.28**	0.26**	15.22**	1.13**	2.86**
$\hat{\sigma}^2 g$	-	0.02**	0.05	0.03	0.07	0.01	2.64	0.11	0.25
$\hat{\sigma}^2 s$	-	0.05**	0.87**	0.91**	0.54**	0.11**	10.80**	1.54**	3.84*
$\hat{\sigma}^2 g$ x environment	-	0.001	0.12	0.04	0.08	0.02	1.30	0.05	0.09
$\hat{\sigma}^2 s$ x environment	-	0.07	0.80	1.02	0.94	0.23	14.03	0.96	2.27
$[\hat{\sigma}^2 D / \hat{\sigma}^2 A]^{1/2}$	-	1.22	2.94	3.60	1.44	3.31	1.42	2.64	2.77

* and ** significant at 5 and 1% levels, respectively.

markedly with environments, thereby suggesting that hybrids did not have the same relative performance across locations. Dethe et al. (2008) reported significant difference among parents and crosses for numbers of pods plant⁻¹, 100 seed weight, primary branches plant⁻¹ and seed yield plant⁻¹. The analysis of variance for the combining ability (Table 2) analysis revealed significant mean squares for both the general and specific combining abilities of all the traits under study in the data pooled over the environments. The significance of the interaction arising from *gca* and *sca* with the environments, revealed that the alleles controlling the *gca* and *sca* behaved differently in the different environments.

The presence of significant *gca* or *sca* x environment interaction have been reported by Matzinger et al. (1959) and Kunkaew et al. (2006) for seed yield plant⁻¹, with contradictory reports revealing non significant interaction of *gca* and *sca* x environment for number of pods plant⁻¹ and

100 seed weight. Comparison of relative magnitude of *gca* and *sca* variances indicated greater magnitude of *sca* variance for all the traits studied in the data pooled over the environments, thereby indicating greater importance of non-additive gene action for the inheritance of these traits. Importance of non-additive gene action for number of seeds pod⁻¹, seed yield plant⁻¹, primary branches plant⁻¹, secondary branches plant⁻¹ and seed weight have been reported by Barelli et al. (2000), Sofi et al. (2006) and Saleem (2009). However, the average degree of dominance was in the range of over-dominance for all the traits studied (Table 2), which was revealed by Sofi et al., 2008.

The perusal of *gca* effects of parents (Table 3) revealed that none of the parents was a good general combiner for all the traits studied. For bold seeds, high combining ability as indicated by SKUA-R-612 and SKUA-R-90 was associated with average or high combining ability for seed

yield plant⁻¹ and protein content (except SKUA-R-91). The high combining ability for seed yield plant⁻¹ as indicated by SKUA-R-607 and SKUA-R-608 was accompanied with high combining ability for primary branches plants⁻¹, pod length and number of seeds pod⁻¹ (SKUA-R-607); and for secondary branches plant⁻¹, number of pods plant⁻¹, 100 seed weight and protein content (SKUA-R-608) in pooled analysis. Further study revealed that there has been no correlation between seed yield and desirable *sca* effect for quality, as most of the cross combinations exhibiting desirable *sca* effect for quality revealed non-significant effect of *sca* for seed yield plant⁻¹. The best five cross combinations for yield and quality traits (Table 4) combinations for yield and quality traits on the basis of *sca* and *per se* performance, revealed that SKUA-R-608 was involved in the maximum number of cross combination (16) revealing desirable *sca* effects, followed by Shalimar rajmash-1 (14) for yield and yield attributing traits in pooled environments. While

Table 3. Effects of general combining ability on morphological, yield and yield attributing quality traits in Rajmash (*P. vulgaris* L.) pooled-over environments.

Parents	Primary branches plant⁻¹	Secondary branches plant⁻¹	No. of pods plant⁻¹	Pod length (cm)	No. of seeds pod⁻¹	100-seed weight (g)	Seed yield plant⁻¹ (g)	Protein content (%)
SKUA-R-602	-0.09** (2.86)	0.01 (6.53)	-0.20* (5.50)	-0.07 (7.76)	0.04* (4.31)	1.44** (34.15)	0.07 (7.06)	0.09 (19.89)
Shalimar rajmash 1	-0.02 (2.80)	0.07 (5.96)	-0.26** (5.26)	0.33** (8.66)	-0.03 (3.60)	1.16** (33.74)	-0.01 (5.03)	0.10 (20.45)
SKUA-R-607	0.24** (3.96)	-0.18** (4.96)	0.04 (5.10)	0.47** (9.60)	0.20** (4.00)	-3.47** (29.64)	0.61** (5.19)	-1.01** (17.99)
SKUA-R-608	0.27** (4.06)	0.45** (6.56)	0.13* (5.66)	0.25** (8.20)	0.03 (3.40)	2.08** (38.01)	0.43** (6.11)	0.34** (21.08)
SKUA-R-609	-0.06** (2.83)	-0.05 (5.76)	-0.01 (5.46)	0.06 (7.63)	-0.05* (3.36)	0.31 (37.77)	0.03 (6.26)	-0.21 (16.67)
SKUA-R-612	-0.05* (2.96)	-0.01 (5.63)	0.22** (6.00)	-0.24** (7.83)	-0.12* (3.56)	0.60* (36.36)	-0.10 (6.45)	0.45** (18.59)
SKUA-R-91	0.02 (2.80)	-0.04 (6.16)	-0.01 (6.16)	0.06 (8.26)	0.04* (3.70)	0.40 (36.53)	0.01 (6.62)	-0.70** (17.74)
SKUA-R-106	-0.14** (2.66)	-0.31** (5.63)	-0.07 (5.90)	-0.28** (7.26)	-0.07* (3.30)	-1.83** (31.93)	-0.57* (5.97)	0.34** (20.94)
SKUA-R-23	-0.04 (3.06)	0.26** (6.76)	0.36** (7.66)	-0.27* (7.56)	0.02 (3.66)	-0.39 (36.58)	-0.27* (5.63)	0.61** (20.97)
SKUA-R-153	-0.05* (2.90)	-0.18** (6.66)	-0.19* (5.73)	-0.31* (7.80)	-0.06* (3.20)	-0.301 (34.22)	-0.21* (5.53)	-0.02 (18.38)
SE(g)±	0.02	0.05	0.05	0.06	0.02	0.17	0.06	0.12
SE(gᵢ-gⱼ)±	0.03	0.07	0.07	0.10	0.04	0.25	0.09	0.18

* and ** Significant at 5 and 1% levels, respectively; Values in parentheses denote mean.

Table 4. Top ranking specific cross combinations on the basis of sca and *per se* performance in pooled-over environments.

Parameter	Per se performance	Sca	
Primary branches plant⁻¹	SKUA-R-607 x SKUA-R-609, SKUA-R-607 x SKUA-R-91, SKUA-R-608 x SKUA-R-153, Shalimar-Rajmash-1 x SKUA-R-608, SKUA-R-607 x SKUA-R-23	SKUA-R-602 x SKUA-R-106, Shalimar-Rajmash-1 x SKUA-R-91, SKUA-R-608 x SKUA-R-23, SKUA-R-612 x SKUA-R-106, SKUA-R-607 x SKUA-R-609	L x L A x A H x L L x L H x L
Secondary branches plant⁻¹	Shalimar-Rajmash-1 x SKUA-R-608, SKUA-R-608 x SKUA-R-106, SKUA-R-607 x SKUA-R-23, SKUA-R-608 x SKUA-R-609, Shalimar-Rajmash-1 x SKUA-R-23	Shalimar-Rajmash-1 x SKUA-R-608, SKUA-R-607 x SKUA-R-23, SKUA-R-608 x SKUA-R-106, SKUA-R-607 x SKUA-R-609, Shalimar-Rajmash-1 x SKUA-R-612	A x H L x H H x L L x A A x A
Number of pods plant⁻¹	Shalimar-Rajmash-1x SKUA-R-23, SKUA-R-612 x SKUA-R-106, SKUA-R-607 x SKUA-R-91, SKUA-R-602 x SKUA-R-106, SKUA-R-607 x SKUA-R-609	SKUA-R-602 x SKUA-R-106, Shalimar-Rajmash-1 x SKUA-R-23, Shalimar-Rajmash-1 x SKUA-R-608, SKUA-R-608, SKUA-R-607 x SKUA-R-91, SKUA-R-608 x SKUA-R-153	L x A L x H L x H A x A H x L

Table 4. Contd.

Pod length (cm)	SKUA-R-608 x SKUA-R-609, SKUA-R-607 x SKUA-R-91, SKUA-R-602 x SKUA-R-607, Shalimar-Rajmash-1 x SKUA-R-608, SKUA-R-91 x SKUA-R-23	SKUA-R-608 x SKUA-R-609, SKUA-R-602 x SKUA-R-106, SKUA-R-91 x SKUA-R-23, SKUA-R-612 x SKUA-R-106, SKUA-R-608 x SKUA-R-153	H x A A x L A x L L x L H x L
Number of seeds pod^{-1}	SKUA-R-607 x SKUA-R-23, SKUA-R-608 x SKUA-R-609, SKUA-R-608 x SKUA-R-106, Shalimar-Rajmash-1 x SKUA-R-23, SKUA-R-607 x SKUA-R-153	SKUA-R-608 x SKUA-R-609, SKUA-R-608 x SKUA-R-106, SKUA-R-607 x SKUA-R-23, Shalimar-Rajmash-1 x SKUA-R-23, SKUA-R-607 x SKUA-R-153	A x L A x L H x A A x A H x L
100 seed weight (g)	Shalimar-Rajmash-1 x SKUA-R-612, SKUA-R-602 x SKUA-R-608, SKUA-R-608 x SKUA-R-106, SKUA-R-602 x SKUA-R-612, Shalimar-Rajmash-1 x SKUA-R-607	Shalimar-Rajmash-1 x SKUA-R-607, SKUA-R-608 x SKUA-R-106, Shalimar-Rajmash-1 x SKUA-R-612, SKUA-R-23 x SKUA-R-153, Shalimar-Rajmash-1 x SKUA-R-153	H x L H x L H x H A x A H x A
Seed yield plant^{-1} (g)	SKUA-R-607 x SKUA-R-609, SKUA-R-608 x SKUA-R-91, SKUA-R-607 x SKUA-R-612, SKUA-R-607 x SKUA-R-91, SKUA-R-609 x SKUA-R-612	SKUA-R-607 x SKUA-R-609, SKUA-R-608 x SKUA-R-91, SKUA-R-609 x SKUA-R-612, Shalimar-Rajmash-1 x SKUA-R-153, Shalimar-Rajmash-1 x SKUA-R-23	H x A H x A A x A A x L A x L
Protein content (%)	SKUA-R-612 x SKUA-R-23, SKUA-R-602 x SKUA-R-23, SKUA-R-608 x SKUA-R-612, SKUA-R-608 x SKUA-R-153, SKUA-R-612 x SKUA-R-153	Shalimar-Rajmash-1 x SKUA-R-91, SKUA-R-608 x SKUA-R-153, SKUA-R-602 x SKUA-R-23, SKUA-R-609 x SKUA-R-153, SKUA-R-612 x SKUA-R-153	A x L H x A A x H A x A H x A

assessing the performance of parents on the basis of general combining ability, it was observed that most of the desirable cross combinations involved high x low, average x low, high x average and average x average general combiners, which have also been reported by several workers in most of the crop species (Ram and Rajput, 1999; Ganesamurthy and Seshadri, 2002).

The overall ranking of genotypes studied revealed that genotypes like SKUA-R-602, Shalimar rajmash-1 SKUA-R-607, SKUA-R-608, SKUA-R-23 and SKUA-R-153 could be a useful source of elite allelic resources based on their general combining ability effects, specific combining ability effects and *per se* performance of both the parents and crosses. These parents are expected to give heterotic combinations resulting from the non-additive gene action and also results in the evolution of transgressive segregants for high yielding genotypes ensuing from complementary epitasis, additive type of gene action and recombination of latent genetic variability hidden in the heterogenetic blocks of such crosses. Normally, *sca* alone would not contribute tangibly in the improvement of self pollinated crops, except where commercial heterosis is feasible. However, *sca* resulting from the heterozygosity of polygenes governing yield and yield components may result in the evolution of recombinants possessing desirable gene aggregates in a homozygous line. The promising crosses, showing high yield (Shalimar rajmash 1 x SKUA-R-607, SKUA-R-608 x SKUA-R-106 and Shalimar rajmash -1 x SKUA- R-612) may be improved through conventional breeding methods such as bi-parental mating and diallel selective mating. Thereafter, it is followed by the pedigree method of selection, so that the tight linkage of any broken and transgressive segregants may be isolated.

REFERENCES

Barelli MAA, Vidigal MCG, Amaral ATJ, Filho PSV, Scapim CA (2000). Diallel analysis of the combining ability of common bean (*Phaseolus vulgaris* L.) cultivars. Braz. Arch. Biol. Technol., 43: 409-414.
Dethe AM, Patil JV, Misal AM (2008). Combining ability analysis in mungbean. J. Food Legumes, 21(3): 200-201.

Eberhart SA, Salhuana W, Taba S (1995). Principles of Tropical maize breeding. Maydica, 40: 339-335.

Ganesamurthy K, Seshadri P (2002). Diallel analysis in soybean (*Glycine max* L.). Madras Agric. J., 89: 14-17

Goncalves VMC, Silverio L, Elias HT, Filho PSV, Kvitschal MV, Retuci VS, Silva CR (2008). Combining ability and heterosis in common bean (*Phaseolus vulgaris* L.) cultivars. Pesq. Agrpec. Bras. Brasilia, 43(9): 1143-1150.

Griffing B (1956). Concept of general and specific combining ability in relation to diallel crossing systems. Aus. J. Biol. Sci., 9: 463-493.

Kunkaew W, Julsrigival S, Senthong C, Kariadee D (2006). Estimation of heterosis and combining ability in azukibean (*Vigna angularis*) under highland growing conditions in Thailand. CMU J., 5(2): 163-168.

Matzinger DF, Sprague GF, Cockerham CC (1959). Diallel crosses of maize in experiments repeated over locations and years. Agron. J., 51: 346-350.

Piper CS (1966). Soil and Plant Analysis. University of Adelaide, Australia, Hans Publishers, Bombay, India, pp. 233-237.

Ram D, Rajput CBS (1999). Line x Tester analysis for green pod yield and its components in french bean (*Phaseolus vulgaris* L.). Horticult. J., 14(2): 155-161.

Saleem SA (2009). Heterosis and combining ability in a diallel cross of eight faba bean (*Vicia faba* L.) genotypes. Asian J. Crop Sci., 1(2): 66-76.

Sofi P, Rather AG, Wani SA (2006). Combining ability and gene action studies over environments in field pea (*Pisum sativum* L.). Pak. J. Biol. Sci., 9(14): 2689-2692.

23

Genetic diversity for sustainability of rice crop in Indian Punjab and its implications

Joginder Singh

Department of Economics and Sociology, Punjab Agricultural University, Ludhiana, India.
E-mail: drjogindersingh@hotmail.com.

The varietal diversity in rice crop in Indian Punjab has been increasing at a fast rate due to variable soil and climatic conditions and variable objective functions of the farmers. In spite of the fact that productivity of rice increased over time, as a result of improvement in technology and its fast adoption, the declining resource base, particularly water, soil health and pest resistance, the genetic diversity of rice crop has been able to sustain the yield improvement as compared to wheat crop which witnessed a reverse trend of decline in genetic diversity.

Key words: Indian Punjab, genetic diversity, sustainability, green revolution, basmati rice.

INTRODUCTION

Due to highly diverse agro-climatic conditions in India and the high importance of rice in the food basket of the consumers, the crop is grown under varied conditions and thus has high degree of genetic diversity. The M. S. Swaminathan Research Foundation in India claims that India alone has 100,000 traditional varieties still in use by the farmers around the country and another 300,000 that have become extinct (Hamilton, 2006). Historically, evidence exists about rice being a staple food crop of Indian Punjab as early as 2000 - 1500 BC due to which landraces cultivated in the state possessed good cooking and eating qualities, most of which belonged to basmati rice (Sidhu and Singh, 2000). Due to various reasons, later the emphasis shifted from rice to wheat in dietary pattern and subsequently in production pattern too. Therefore, rice ceased to be an important crop of Punjab state till the initiation of 'Green Revolution' in mid-sixties when it again stormed in crop pattern of the state, now covering about two-thirds of the cultivated area of the state.

The organized efforts of genetic improvement of rice in Punjab were started in the plains in 1926 and in hills in 1936. Till 1940, a number of varieties such as Jhona 349, Mushkan 7, Mushkan 41, Basmati 370, Sathra 278, Palman Suffaid 246 and Mahlar 346 with typical characteristics of high yielding, resistant to lodging, tolerant to salts, good cooking and eating qualities were released. Partition of the country in 1947 and political reorganization of Punjab state in 1966 gave serious jolts to rice improvement work which was ultimately shifted to

Kapurthala Regional Centre of Punjab Agricultural University.

The remarkable progress of food grain production (mainly wheat and rice) in the post-green revolution period in Punjab was a mixed blessing. It was possible to generate huge surpluses apart from meeting the requirements of the fast growing population of the country and improvement in the economic lot of farmers. The sad part of it was the high rate of resource depletion, particularly water and soil health, increasing pest resistance and air pollution due to large scale burning of crop residue and diminishing ecological bio-diversity apart from almost stagnation of yield and thus overall slowing pace of farm economy. The average yield of rice was only 1009 kg/ha in 1960 - 1961 but touched a level of 3715 kg/ha in 2006 - 2007 (Table 1). The area under the crop was hardly 227 thousand hectares in 1960 – 1961 which went up to 2611 thousand hectares in 2000 - 2001 and reached a plateau thereafter. On the other hand, as a result of shift in crop pattern in favour of rice, a number of marginal crops such as tobacco, pulses, oilseeds, millets, sorghum, cluster beans, maize, sun hemp which had significant area have almost vanished from the state agriculture scene. A glance of compound growth rates (CGRs) would show that the average yield increased at 5.29% in 1970's, 1.29% in 1980's but it almost stagnated in 1990's. However, during the past five years, it has somewhat revived due to favourable climatic conditions. The corresponding CGRs in area worked out to 13.36, 5.43, 2.51 and -0.20%, respectively (Table 1).

Table 1. Area, average yield and production of rice in Punjab.

Year	Rice		
	Area (000 ha)	Average yield (Kg/ha)	Production (000 tonnes)
1960-61	227	1009	229
1970-71	390	1765	668
1980-81	1183	2733	3233
1990-91	2015	3229	6506
2000-01	2611	3506	9154
2006-07	2425	3715	9010
Compound growth rates (%)			
1970-71 to 1979-80	13.36	5.29	18.65
1980-81 to 1989-90	5.43	1.29	6.72
1990-91 to 1999-00	2.51	0.02	2.53
2000-01 to 2006-07	-0.20	1.65	1.45

Statistical abstracts of Punjab.

The production of rice in Punjab was 9 million tones in 2006 - 2007 from the area of 2.4 million hectares under the crop. The average yield in terms of rice worked out to 3715 kg/ha. Out of the total rice area, basmati strains covered 122 thousand hectares thus accounting for 5.03% of total area under rice crop. The comparative average yield of basmati rice was almost half of that of non-basmati and was estimated at 2019 kg/ha (Singh, 2007).

The slower rates of variety change dampened the productivity in Punjab and also offset the positive impact on diversifying genetic base through plant breeding (Smale, 2008). The varietal and genetic scenario of wheat and rice crops has undergone metamorphosis in Punjab. Guided by economic forces, a few high yielding varieties quickly replaced most of the low yielding ones. Of course, susceptibility to pests, diseases and lodging and quality parameters do get significant weight age in the choice of suitable varieties. This paper was attempted with the following specific objectives:

(1) To study changes in varietal and genetic diversification of rice crop in Punjab agriculture.
(2) To bring out the impact of such diversification on production.
(3) To suggest policy implications of such developments.

METHODOLOGY

Although, it is well understood that the genetic diversity is not congruent to varietal diversity but by and large, the number of parents in every successively introduced variety are increasing with the genetic improvement of the rice crop in the state (Sidhu and Singh, 2004). Due to non-availability of information about parentage in back tracing of varieties, it was presumed that increase in the number of varieties and thus parentage makes the genetic base much more diverse having implications on sustainability of the crop.

For making production estimates, a survey on 'Prospects of rice crop in Punjab' was carried out by Punjab Agricultural University every year by taking a sample of about 1000 farmers (Singh, 2008). The District Farm Management Specialists were entrusted with the job of collecting data regarding shift in area, sowing time, seed rate, area under different varieties etc. The data on varietal scenario have been used for this analysis. The Theil Entropy Index (E) (Shorrocks, 1984) has been used for measuring the diversification as under:

$$E = \sum_{i=1}^{n} \{Pi \log (1/Pi)\}$$

Where Pi = Proportion of area under ith variety, N = Number of varieties raised by farmers.

Therefore, the value of E should vary from 0 to log n. The secondary data (Economic & Statistical Organization ESO, Punjab (2008), various issues) were also used to construct series of probable explanatory variables, explaining the yield of rice crop in the regression analysis.

RESULTS AND DISCUSSION

Varietal scenario

There was in general, increase in the number of rice varieties adopted by the farmers over time. In 1982 - 1983, only 7 varieties with dominance of PR 106 were in cultivation. The number gradually swelled to 15 in 1999 - 2000. As more and more of genetic material, recommended or unrecommended, started pouring in the area, varietal diversification increased. The varietal diversification index measured by Theil's Entropy index was 0.3591 in 1982 - 1983 which increased to 0.6246 in 1992 - 1993, 0.8008 in 2002 - 2003 and 0.8974 in 2007 - 2008 (Table 3 and Figure 1). In early 1980s, PR106, Jaya and IR8 were the dominant varieties, potentials of which

Table 3. Varietal diversification of rice in Punjab (Area as % of total area under rice crop).

Year	PR106	Indrasan	PR108	Pusa44	IR8/Jaya	PR111	PR110	PR103	PR109	Satha	Basmati	PR113	PR114	PR115	PR116	PR118	Sharbati	Others	DI
1982-1983	71.2				22.0			1.7			3.6							1.5	0.3591
1983-1984	73.4				17.3			2.6			5.0							1.7	0.3667
1984-1985	67.0				24.4			4.1			3.6							0.9	0.3933
1985-1986	60.2				30.7			1.0			3.2							3.2	0.4359
1986-1987	48.7	5.4	6.5		28.7			0.9	2.8		4.8							2.2	0.6150
1987-1988	36.5	7.5	15.9		19.3			1.3	14.9		3.8							0.8	0.7275
1988-1989	53.8	7.7	8.6		17.1			4.1	1.4		5.5							1.8	0.6369
1989-1990	58.4	6.8	7.3		17.1			3.1	0.2		4.9							2.2	0.5828
1990-1991	55.3	7.5	3.6	11.3	13.3			3.1	0		2.8							3.1	0.6392
1991-1992	62.1	6.6	3.0	9.6	10.5		0.1	2.7	0.2	0.6	2.8							1.8	0.5915
1992-1993	61.4	5.5	3.1	9.4	8.8		1.7	3.4	0.1	2.3	3.4							0.9	0.6246
1993-1994	52.1	6.9	4.3	17.6	8.8		2.1	3.6		1.0	2.7							0.9	0.6801
1994-1995	48.2	6.4	7.9	15.7	10.9	3.2	2.0	1.9		0	2.5							1.3	0.7265
1995-1996	40.7	5.6	7.5	17.3	15.1	8.1	2.2	0.6		0	1.9							1	0.7601
1996-1997	33.2	5.2	8.2	20.4	12.5	14.8	1.7	1.3	0.1	0.4	1.6							0.6	0.8006
1997-1998	23.2	4.5	7.5	27.8	10.0	21.6	1.0	2.0	0.2	0.4	1.6							0.2	0.7936
1998-1999	19.1	2.2	2.5	30.5	9.9	22.0	0.3	1.2		0.8	2.5	6.5						2.5	0.8199
1999-2000	20.3	2.0	2.3	26.9	7.7	10.7	0.3	0.7		2.2	2.4	15.3	5.0					4.2	0.9008
2000-2001	8.4	0.0	1.6	10.3	5.2	5.8		0.2		2.7	1.5	16.0	37.4	1.2	4.2			5.5	0.8716
2001-2002	3.3	3.8	2.4	23.4	2.3	4.9				1.1	3.3	2.5	11.0	0.4	37.3			4.3	0.8352
2002-2003	3.1	4.9	2.0	20.8	1.2	4.8				1.1	3.9	2.2	8.2	0.2	43.1	0.1		4.4	0.8008
2003-2004	2.2	5.1	2.4	17.6	0.6	4.6				0.6	12.7	2.0	4.2	0.1	40.0	2.8		5.1	0.8394
2004-2005	2.3	1.2	2.2	30.1	0.3	6.5		0.1		1.8	7.1	3.3	3.3	0.3	25.3	7.6	2.4	6.3	0.9078
2005-2006	1.8	1.1	2.3	34.5	0.4	9.2				3.5	8.9	6.5	3.2	0.1	13.5	8.4	4.9	1.6	0.9311
2006-2007	1.9	0.9	0.7	36.7	0	8.5				2.6	3.8	7.7	6.7	0.2	7.5	8.6	5.0	9.2	0.9183
2007-2008	1.5	0.5	0.3	33.0	0	7.4				1.3	3.8	6.4	5.3	0.5	9.3	8.6	3.5	18.6	0.8974

DI=Diversification index.

dampened over time. These were replaced by some improved varieties, many of which also could not stand the test of time. Some of the new varieties were recommended by Punjab Agricultural University and the others endeavored by the farmers themselves by subsequent testing on their fields. Pusa 44 and Indrasan are the glaring cases which are not recommended for the state owing to its long duration, high water requirements and susceptibility to bacterial leaf blight but due to higher yield and better quality, these have been quite popular with the farmers and have still been occupying a significant part of the rice area in the state (Table 3). Basmati varieties have been occupying 4 - 6% of the total area under rice in the state. The traditional basmati belt of Punjab preferred basmati 370 having long aromatic grains and rice straw as preferred livestock feed. Pakistani basmati was successfully tried by the farmers of the area because it had higher yield but lacked

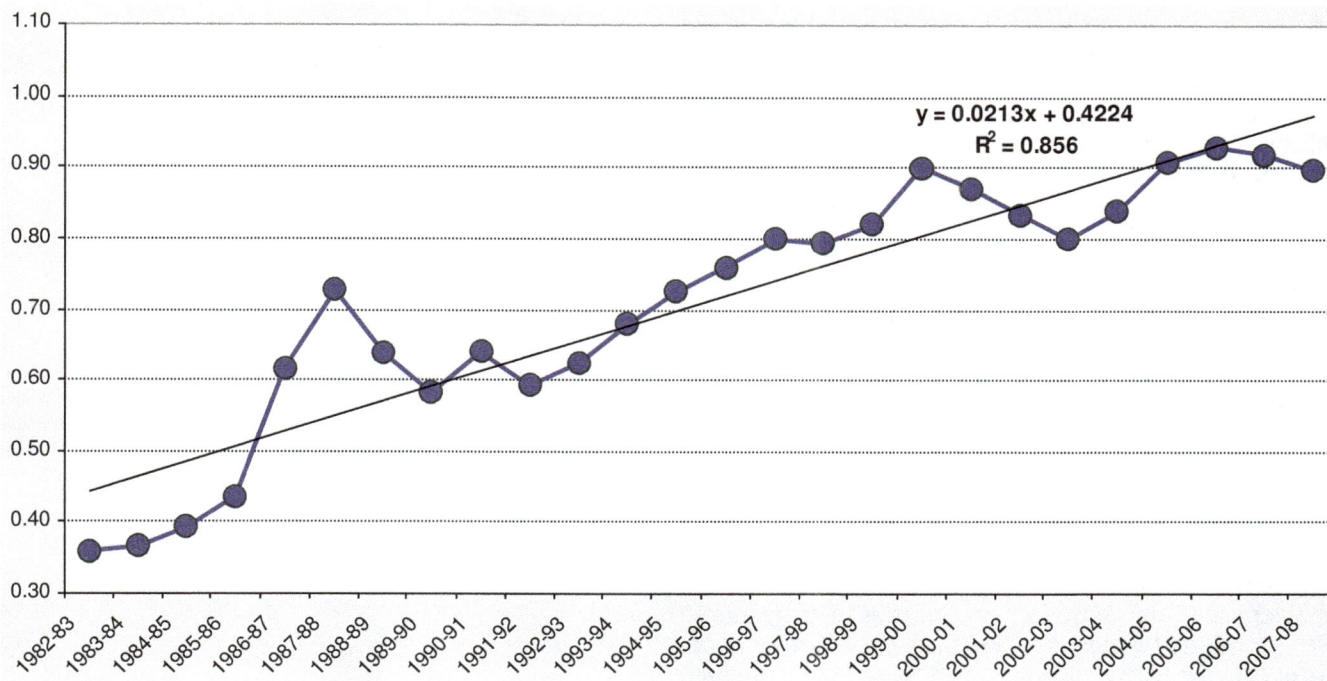

Figure 1. Diversification index of rice in Punjab.

Table 2. Estimated production function with varietal diversity.

Variable	Coefficient	t-value
Intercept	3436.24**	
Varietal diversity	2.6995*	2.2389
Rainfall	-0.5023	-1.0420
Chemical fertilizer use	8.5779*	2.1529
Zinc sulphate	0.0277*	2.1273

R^2 = 0.6899, *Significant at 0.05 probability level; ** Significant at 0.01 probability level.

fragrance. Trade followed the practice of mixing it with Basmati 370. Later on Basmati 370 and Pakistani Basmati were crossed resulting in introduction of Basmati 386. Recently, Sarbati and Basmati 1121 have also been introduced by the farmers but not recommended by the research system due to their susceptibility to Foot Rot and Bacterial Leaf Blight. The basmati strains are considered relatively more environmentally friendly as these are transplanted late and their water requirement goes down sizably. Currently, there is a wide range of rice varieties grown in Punjab. The range varies from short duration of 65 days (Satha) to 165 days (Pusa 44) in the field, apart from 30-45 days in nursery. The quality and thus price based variations can be seen from coarse (IR8) to best quality basmati rice (Basmati 386). The varieties have varying degrees of susceptibility to pests and diseases, lodging and other biotic and abiotic

stresses. Similarly, early and late sown varieties are also adopted by the farmers depending upon the farm situations.

Therefore based on different farm situations with respect to crop rotation, market conditions, suitability of soil and water resources, past experience etc., the rice growers have wider choice of seeds. The breeders do not favour it but the farmers have tailored genetic system to suit their requirements.

Impact of varietal diversity

The results of linear regression analysis are presented in Table 2. It may be seen from this table that natural factors such as rainfall did not exert significant effect on yield of rice crop in Punjab because almost all the entire cultivated area, particularly of rice crop had assured surface and subsurface groundwater supply system. The significant effects of fertilizers and zinc indicated that replenishment of soil fertility with use of chemical fertilizers and even micro-nutrients such as zinc, manganese and ferrous is becoming important with continuously taking rice crop year after year (Appendix 1). The significance of varietal diversity for improving crop yield is sufficient indication of its importance for sustaining the rice crop in the state.

Therefore, it can be inferred that cultivation of rice crop started with the onset of 'green revolution' in mid sixties and is now occupying a prominent place in the crop pattern and research system of the state and food security of the country. The varietal diversity in rice crop in Indian Punjab has been increasing at a fast rate, possibly due to variable agro-climatic conditions under which the crop is grown and variable objective functions of the farmers. In spite of the fact that productivity of rice increased over time as a result of improvement in technology and its fast adoption, the declining resource base, particularly water, soil health and pest resistance, the genetic diversity of rice crop has been able to sustain the yield improvement as compared to wheat crop which witnessed a reverse trend of decline in genetic diversity in the state (Smale, 2008).

REFERENCES

Economic a Statistical Organization (ESO) Punjab (2008). Statistical Abstracts of Punjab. Chandigarh.

Hamilton PR (2006). How many Rice Varieties are there? Rice Today IRRI, Philippines. 3:50.

Sidhu GS, Singh J (2004). Genetic improvement of rice varieties in Punjab. In. Genetic Improvement of Rice Varieties of India by Sharma SD and Prasada Rao U New Delhi Today and Tomorrow's Printers and Publishers, Part II.

Singh J (2008). Prospects of Rice crop in Punjab, Department of Economics & Sociology, PAU, (Mimeographed).

Singh J (2007), Impact Assessment Report on Basmati crop in Punjab, Sir Ratan Tata Trust, Mumbai.

Shorrocks AF (1984). Inequality Decomposition by Population Subgroups. Econometrica. 48.

Smale M, Joginder S, Favo SD, Zambrano P (2008). Wheat breeding, productivity and slow variety change: evidence from Punjab of India after the Green Revolution. Australian J. Agric. Resour. Econ. 52: 419-432.

Appendix 1. A glance of important parameters of Punjab agriculture.

Year	Zinc sulphate (tonnes)	Ferrous sulphate (tonnes)	Manganese sulphate (tonnes)	Fertilizers with major nutrients (000 tonnes)	Pesticides (TGM) (tonnes)	Area irrigated (%)
1980-1981	5042	0	0	762	3200	86
1990-1991	9018	378	0	1220	6500	94
2000-2001	23340	1360	80	1407	7005	97
2005-2006	32000	3500	568	1686	6020	97
2006-2007	NA	NA	NA	1717	6000	97

NA=Not available, Statistical abstracts of Punjab, 2008 and agricultural statistics of Punjab, 2009.

24

Screening of superior chickpea genotypes for various environments of Iran using genotype plus genotype × environment (GGE) biplot analysis

Asghar Ebadi segherloo[1]*, Sayyed Hossain Sabaghpour[2], Hamid Dehghani[1] and Morteza kamrani[3]

[1]Department of Plant Breeding, Faculty of Agriculture, Tarbiat Modarres University, Tehran, Iran.
[2]Dry land Agricultural Research Institute, Kermanshah, Iran.
[3]Department of Agronomy and Plant Breeding, Faculty of Agriculture, University of Tabriz, Tabriz, Iran.

Superior crop cultivars must be identified through multi-environment trials (MET) and on the basis of multiple traits. The objective of this study was to explore the effect of genotype (G) and genotype × environment interaction (GE) on grain yield of 17 chickpea genotypes (*Cicer arietinum* L.) in six different research stations of Iran. GGE (G plus GE) biplot methodology was used to evaluate phenotypic stability in genotypes. A site regression (SREG) analysis to assess G × E interactions and to identify stable genotypes of chickpea was undertaken. These genotypes were developed by various breeders at different research institutes/stations of Iran and International Center for Agricultural Research in Dray Areas (ICARDA). Results indicated that the first two principal components explained 95% of the total GGE variation, with PC1 and PC2 explaining 73 and 22%, respectively. Genotypes Flip 93-93, Flip 94-123C and S 96002 had the highest mean yield and genotype Bivanij had the poorest mean yield. Thus the performance of genotype ILC 6142 was highly variable, whereas genotypes S 96003, Flip 93-48C and S 96027 were highly stable. Collective analysis of the biplots suggests four chickpea mega-environments in Iran. The first mega-environment contained locations Kermanshah, Gorgan and Ghachsaran, with genotype Flip 93-93 being the winner. Genotype Flip 85-57 × 12-071-1005 gave the highest performance in location Ilam and genotypes S 96032 and Bivanij gave the highest performance in locations Urmia. The Lorestan made up the other mega-environment with ILC 6142 as the winner.

Key words: *Cicer arietinum* L., genotype × environment interaction, site regression analysis, GGEBiplot.

INTRODUCTION

Legumes have been considered as a rich source of protein throughout the world and contain approximately three times more proteins than cereals. Chickpea (*Cicer arietinum* L.) is one of the top five important legumes on the basis of whole grain production (FAO, 2000). It is a staple food crop in many tropical and subtropical countries of Asia. Chickpea is the third most important pulse crop in the world, representing 14% of total world pulse production (Kelley et al., 2000). Chickpea is grown on 700,000 ha in Iran and ranks fourth in the world after India, Pakistan and Turkey. It is the most important legume of the country and grown on more than 64% of the total food legume area (FAO, 2001). Iran is currently one of the world's largest net importers of agricultural products, importing about 30% of its requirements. Rapid population growth is expected to increase the demand for food. Iran is working towards increasing its agricultural efficiency. To increase its efficiency, the agricultural sector of Iran is attempting to

---

*Corresponding author. E-mail: asghar_ebadi@yahoo.com.

Abbreviations: AMMI, additive main effect and multiplicative interaction effect; **E,** environment main effect; **G,** genotype main effect; **GE,** genotype × environment interaction; **GGE,** G plus GE; **MET,** multi- environment trials; **PC,** principal component; **SREG,** site regression models.

improve chickpea production with identification and introducing the stable and adaptive cultivars.

Chickpea varieties must show high performance for yield and other essential agronomic traits. Multi-environment trials (MET) play an important role in selecting the best cultivars (or agronomic practices) to be used in future years at different locations and in assessing a cultivar's stability across environments before its commercial release. When the performance of cultivars is compared across sites, several cultivar attributes are considered, of which grain yield is one of the most important. Cultivars grown in MET trials react differently to environmental changes. This differential response of cultivars from one environment to another is called genotype × environment (GE) interaction. GE interactions are an important issue facing plant breeders and agronomists. A significant GE interaction for a quantitative trial such as grain yield can seriously limit progress in selection. The study of the GE interaction may assist understanding of stability concept. Information on the structure and nature of GE interaction is particularly useful to breeders because it can help determine if they need to develop cultivars for all environments of interest or if they should develop specific cultivars for specific target environments.

Phenotypic stability has been extensively studied by biometricians who have developed numerous methods to analyze it (Eberhart and Rusell, 1966; Lin et al., 1988; Huhn, 1979; Kang and Pham, 1991). Usually a large number of genotypes are tested over a number of site and year, and it is often difficult to determine the pattern of genotypic responses across environments without the help of graphical display of the data (Yan et al., 2001). The biplot technique (Gabriel, 1971) provides a powerful solution to this problem. Biplot analysis is a multivariate analytical technique that graphically displays the two-way data and allows visualization of the interrelationship among environments, and the interrelationship between genotypes and environments. Biplots are useful for summarizing and approximating patterns of response that exist in the original data (Gabriel, 1971). Two types of biplots, GE biplot (Zobel et al., 1988) and GGEbiplot (Yan et al., 2000), were used to visualize the genotype × environment two-way data but each had its unique functions. The "GE" biplot refers to graph of the genotype by environment interaction obtained from the additive main effects and multiplicative interactions (AMMI) model. The "GGE" refers to the genotype main effect (G) plus the GE interaction, which are the two sources of variation of the site regression (SREG) model (Burgueno et al., 2001). The measured yield of each cultivar in each test environment is a measure of environment main effect (E), genotype main effect, and GE interaction (Yan and Kang, 2003). Typically, E explains up to 80% or higher of the total yield variation, however it is G and GE that are relevant to cultivar evaluation (Yan, 2002). Yan et al. (2000) presented standard biplots of the site

regression model to enhance its interpretation for selecting the best performing cultivars in subsets of sites. In analyzing Ontario winter wheat performance trial data, Yan and Hunt (2001) used a GGEbiplot constructed from the first two principal components (PC1 and PC2) derived from PC analysis of environment-centered yield data. GGEbiplot can be useful in two major aspects. The first is to display the which-won-where pattern of the data that may lead to identify high-yielding and stable cultivars and discriminating and representative test environments (Yan et al., 2001). A major challenge of plant breeding is finding the useful information within the quantities of data. The GGEbiplot graphically displays G and GE of a MET in a way that facilitates visual cultivar evaluation and mega-environment identification. The GGEbiplot software was chosen to facilitate the application of the GGEbiplot methodology in MET data analysis and the analyses of two-way data.

The objectives of this study were to (1) interpret G main effect and GE interaction obtained by SREG analysis of yield performances of 17 chickpea genotypes over sixteen environments; (2) use the GGEbiplot technique to examine the possible existence of different mega-environments in chickpea-growing regions in Iran; (3) visually assess how to vary yield performances across environments based on the GGEbiplot, and other objectives were to apply this method to determine discriminating ability and representativeness of the environments.

MATERIALS AND METHODS

Experimental design and plant materials

Data analyzed in this study were obtained from sets of chickpea yield trials conducted for three years (2002-2004) at six different research stations in Iran including Ghachsaran (GHA), Gorgan (GOR), Urmia (EUR), Ilam (ILA), Kermanshah (KER) and Lorestan (LOR). The detailed description of these test locations is given in Table 1. In each environment (year X location), 17 genotypes were tested. The genotypes were developed by various breeders at different research institutes/stations of Iran and International Center for Agricultural Research in Dray Areas (ICARDA). The names of genotypes, cods and origin of these genotypes are given in Table 2. At each environment a randomized complete block design with four replications was used. The trial fields were plowed with tractors usually from June to August and disc harrowed few days prior to planting time. The experimental plots consisted of four rows of 4 m length. Row to row and plant-to-plant distances was kept at 30 and 10 cm respectively at all the environments. Weeds were controlled by hand-weeding about two or three times, as required. Neither herbicides nor insecticides were used in any trials, as there was no need for them. Data on seed yield were taken from the middle two rows of each plot, leaving aside the guard rows on either side of a plot. Upon harvested seed yield was determined for each genotype at each test environments, the average was computed in accordance with the experimental design.

Data analysis

Analysis of variance was conducted by SAS (SAS Institute, 1996), to determine the effect of location (L), genotype (G) and GE interaction among these factors, on grain yield. Correlation coefficients between

Table 1. Agro-climatic characteristics of testing environments.

Environments		Mean	Latitude	Altitude m	Temp(°C)[a]		Rainfall (mm)		Soil condition	
Location	Year	(Kg ha⁻¹)	Longitude		Min	Max	PS[b]	GS[c]	Texture	Type[d]
Gorgan	2002	2026.8	36°51′N	13.3	4.4	31.5	100.2	290.3	Sandy-loam	Cambisols
	2003	1998.4	54°16′E		4.1	33.5	178	543		
	2004	2616.5			3.8	34.2	135	425		
Kermanshah	2002	1249.0	34°19′N	1322	3.8	38	121.2	358.6	Silt-loam	Cambisols
	2003	1157.5	47°07′E		3	39.5	45	216		
	2004	1456.8			5.3	37	128.4	398.5		
Lorestan	2002	1115.1	23°26′N	1147.7	5.6	38.2	155.2	499	Silt-loam	Regosols
	2003	957.6	48°17′E		3.4	34.2	119.6	369.5		
	2004	1181.9			4	32				
							140.1	430.8		
Urmia	2002	1214.1	37°27′N	1091	2.8	36	101	300.1	Sandy-loam	Cambisols
	2003	1283.8	57°55′E		3.5	38.7	85.3	254		
	2004	1376.3			4	35	71.4	233.7		
Ghachsaran	2003	2053.3	30°10′N	669.5	6.4	39.1	145.2	487.5	Silt-loam	Regosols
	2004	2011.9	50°50′E		5.3	39.2	180	575		
Ilam	2003	1904.0	33°38′N	1363.4	5	32.1	183	564	Silt-loam	Cambisols
	2004	1834.0	46°25′E		4.9	37.6	150.3	458		

[a] Temp(°C) = Mean seasonal temperature; [b] PS = Preseasonal rainfall; [c] GS = Growing season; [d] Type = According to FAO system of soil classification.

Table 2. Genotype code, name and origin of 17 chickpea genotypes.

Genotype code	Name	Origin	Genotype code	Name	Origin
G1	S 96002	ICARDA	G10	Flip 93-48C	ICARDA
G2	S 95293	ICARDA	G11	Flip 94-60C	ICARDA
G3	S 96003	ICARDA	G12	Flip 94-30C	ICARDA
G4	S 96027	ICARDA	G13	ILC 482-205C	ICARDA
G5	S 96078	ICARDA	G14	Flip 94-123C	ICARDA
G6	S 96032	ICARDA	G15	Flip 85-57 × 12-071-1005	ICARDA
G7	S 96019	ICARDA	G16	Kurosh × 12-071	Iran
G8	Flip 93-93	ICARDA	G17	Bivanij	Iran
G9	ILC 6142	ICARDA			

pairs of environments were computed via SAS (SAS Institute, 1996). In addition, principal component axes (PCAs) were extracted and statistically tested by Gollob's (1968) F-test procedure (Vargas and Crossa, 2000). The first two components were used to obtain a biplot by GGEbiplot software (Yan 2001), which is a windows application that fully automates biplot analysis.

RESULTS AND DISCUSSION

In this investigation, partitioning and interpretation of the G main effect and GE interaction were based on SREG models. The measured yield of each genotype in each test environment is mixture of environmental main effect (E) genotype main effect (G), and interaction term of genotype and environment (GE), however it is G and GE that are relevant to cultivar evaluation (Yan, 2002). Yan et al. (2000) proposed a standard biplot of G + GE based on a SREG model referred to GGEbiplot. It was constructed using the first two principal components (PC1 and PC2) derived from subjecting the environment-centered data to singular-value decomposition.

Table 3. Site regression (SREG) analysis of variance for grain yield (kg ha^{-1}) of the genotypes across locations.

Source	Df	Mean square	F-test	Explained (%)
Model	101	272683.98	36.00	
location (L)	5	11108550.28	426.94	85.03**
Genotype (G)	16	744449.52	63.11	12.57**
G × L	80	189616.27	10.19	2.03**
Interaction PCA 1	21	985640	14.50	72.79**
Interaction PCA 2	19	334022	4.92	22.20**
Interaction PCA 3	17	36234	0.53	2.14
Interaction PCA 4	15	30902	0.45	1.60
Interaction PCA 5	13	22403	0.33	1.00
Residuals	86	96943.39		

** Significant at the 0.01 probability level.

Table 4. Genotype (G), location (L) and genotype by location (GL) variance terms for yield lentil multi-environmental trials, 2002 to 2004.

Year	Source	Df	Sum of squares	Explained (%)
	L	5	67127073	49.27
2002	G	16	31220720	22.92
	GL	80	37888656	27.81
	L	5	72782281	67.00
2003	G	16	8489371	7.81
	GL	80	27364623	25.19
	L	3	86027776	82.58
2004	G	16	6196491	5.95
	GL	48	11956078	11.48

The site regression analysis of variance of grain yield (kgha^{-1}) of the 17 genotypes tested in sixteen environments showed that 92.24% of the total sum of squares was attributable to location effects, only 6.18% to genotypic effects, and 1.57% to GE interaction effects (Table 3). The variance components for the location, genotype and genotype × location based on the yearly data are presented in Table 4 which gives an overall picture of the relative magnitudes of the genotype (G), location (L), and genotype × location interaction (GL) variance terms. Location was always the most important source of yield variation accounting for 49.27 to 82.58% of the total variance. The large yield variation due to L, which is irrelevant to cultivar evaluation and mega environment investigation (Gauch and Zobel, 1996), justifies selection of SREG procedures for analyzing the MET data.

Results from SREG analysis also showed that the first principal component axis (PC1) of the genotype main effect plus interaction captured 72.79% of the sum of squares in 21.87% of degrees of freedom. Similarly, the second principal component axis (PC2) explained a further 22.2% of the GGE sum of squares. Furthermore, PC1 and PC2 had sums of squares greater than that of

genotypes. The mean squares for the PC1 and PC2 were significant at P = 0.01. An F-test at P = 0.01 suggested that two principal component axes of the interaction were significant for the model with 101 degrees of freedom.

The GGEbiplot graphically displays G plus GE of a MET in a way that facilitates visual cultivar evaluation and mega-environment identification (Yan et al., 2000). Only two PC (PC1 and PC2) are retained in the model because such a model tends to be the best model for extracting patterns and rejecting noise from the data. In addition, PC1 and PC2 can be readily displayed in a two-dimensional biplot so that the interaction between each genotype and each environment can be visualized (Yan and Hunt, 2002).

There are numerous ways to use a GGEbiplot, but the polygon view of the biplot is most relevant to the mega-environments identification. For this purpose, the genotypes that are connected with straight lines so that a polygon is formed with all other genotypes contained within the polygon (Figure 1A). The vertex genotypes in this investigation are Flip 93-93, ILC 6142, Bivanij, Kurosh × 12-071 and Flip 85-57 × 12-071-1005. These genotypes are the best or the poorest genotypes in some or all of the locations since they had the longest distance from the origin of biplot. There are five sectors in Figure 1A. The

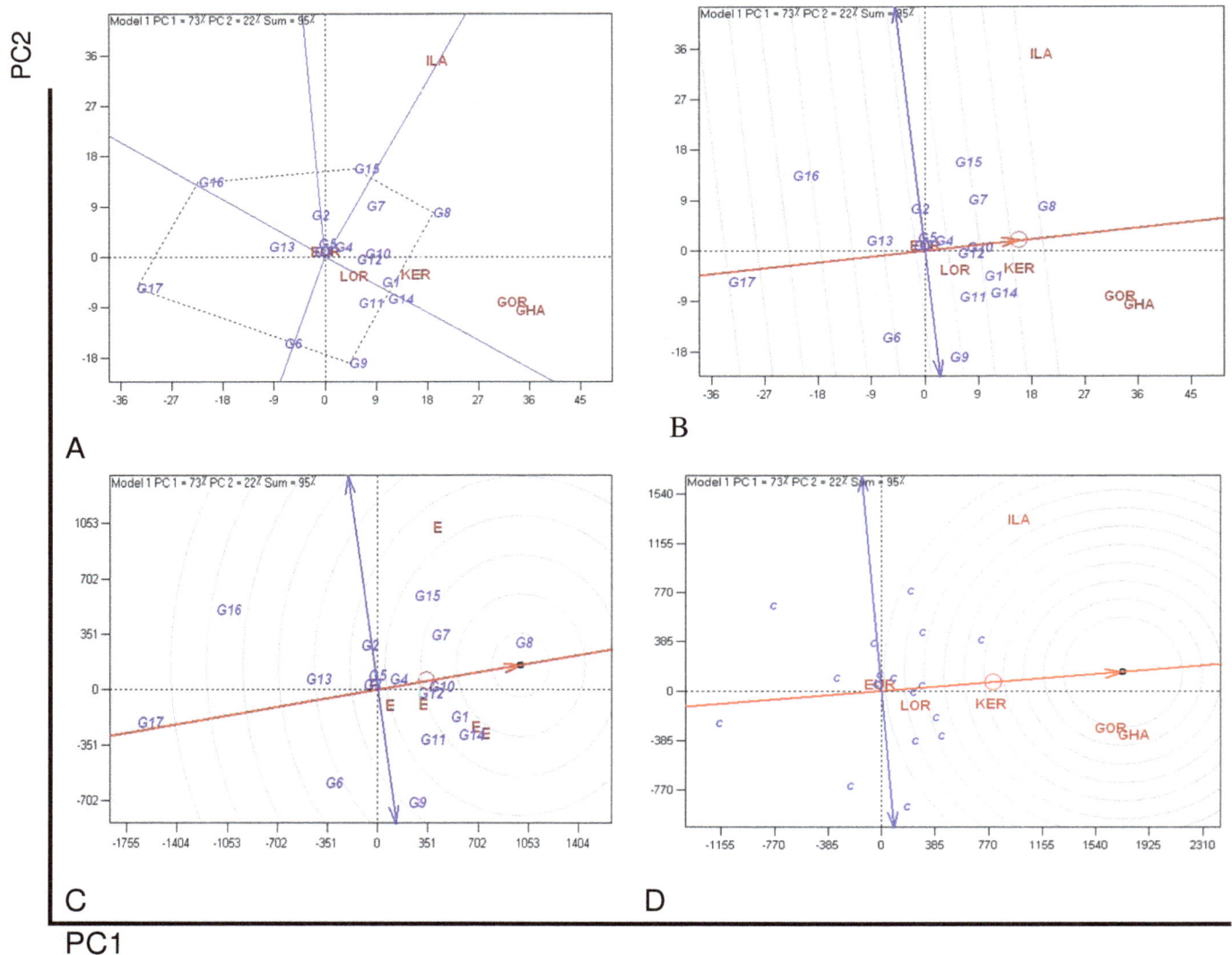

Figure 1. (A) Mega-environment and their winning genotypes, (B) Cultivars ranking based on both average yield and stability (C) Comparison of the locations with the *ideal* location based on both discrimininating ability and representativeness of the target location, (D) Comparison of the genotypes with the *ideal* genotype for both mean yield and stability.

vertex genotype for each sector is the one that gave highest yield for locations that fall within that sector. Therefore, the first mega-environment contained locations KER, GOR and GHA, with genotype Flip 93-93 being the winner. Genotype Flip 85-57 × 12-071-1005 gave the highest performance in location ILA and genotypes S 96032 and Bivanij gave the highest performance in locations EUR. Also genotype ILC 6142 gave the highest performance in locations LOR. Genotype Kurosh × 12-071 did not give the highest yield in any of the locations, that is it was the poorest genotype in all of locations. Another use of Figure 1A is that the locations are grouped based on the best genotypes and we have four groups of locations: ILA as a group, EUR as a group, KER, GOR and GHA and LOR as a group. Another application of the GGEbiplot

geometry is to visually identify the mean performance and stability of genotypes. The mean yield of the genotypes can then be approximated by nominal yields of the genotypes in that mean location. In Figure 1B, genotypes Flip 93-93 and Flip 94-123C had the highest mean yield and genotypes Bivanij and Kurosh × 12-071 had the poorest mean yield. Mean yields of the genotypes were in the following order: Flip 93-93>S 96019>Flip 94-123C>S 96002>Flip 93-48C>Flip 85-57 × 12-071-1005>Flip 94-30C>Flip 94-60C>S 96027> ILC 6142>S 96078>S 95293>S 96003>S 96032>ILC 482-205C>Kurosh × 12-071>Bivanij.

The performance of genotypes ILC 6142, Kurosh × 12-071 and Flip 85-57 × 12-071-1005 is highly variable (less stable), whereas genotypes S 96003, Flip 93-48C and S 96027 are highly stable. An ideal genotype is one that has

PC2

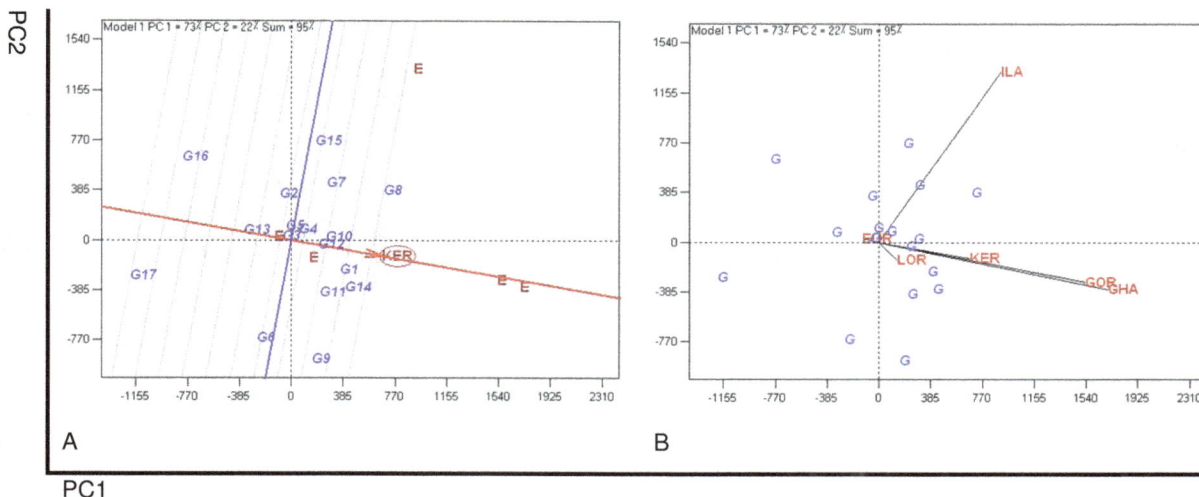

PC1

Figure 2. (A) The performance of different cultivars in a location (KER), (B) correlation between locations.

Table 5. Correlation coefficients among test locations.

Location	Urmia	Gorgan	Kermanshah	Lorestan	Ghachsaran
Gorgan	-0.36				
Kermanshah	-0.27	0.82**			
Lorestan	-0.15	0.47	0.62**		
Ghachsaran	-0.30	0.96**	0.84**	0.58*	
Ilam	-0.13	0.39	0.35	-0.12	0.38

both high mean yield and high stability. The center of the can centric circles in Figure 1C represents the position of an ideal genotype, which is defined by a projection on to the mean-location axis that equals the longest vector of the genotypes that had above-average mean yield and by a zero projection on to the perpendicular line (zero variability across environments). A genotype is more desirable if it is closer to the ideal genotype. Therefore genotypes Flip 93-93, Flip 93-48C and S 96002 are more desirable than other genotypes.

Discriminating ability is an important measure of a test location. A test location, lack of discriminating ability provides no information about the cultivars and, therefore the test location is useless. Another equally important measure of a test location is its representativeness of the target location. If a test location is not representative of the targets location, it is not only useless but also misleading since it may provide biased information about the tested cultivars. An ideal location should be highly differentiating of the genotypes and at the same time representative of the target location. The GGEbiplot way of measuring representativeness is to define an average location and use it as a reference or benchmark. The average location is indicated by small circle (Figure 1D). The ideal location, represented by the small circle with an

arrow pointing to it, is the most discriminating of genotypes and yet representiveness of the other tests locations. Therefore Gorgan, Ghachsaran and Kermanshah were relatively desirable test locations, whereas Urmia, Lorestan and Ilam were relatively undesirable test locations.

Figure 2A illustrates graphic comparison of the relative performance of all genotypes at location Kermanshah. In this figure, genotypes Flip 93-93, Flip 94-123C and S 96002 had the highest at Kermanshah and genotypes Bivanij and Kurosh × 12-071 had the poorest yield.

The vector view of a GGEbiplot provides a succinct summary of the interrelationship among the environments (Yan, 2002). Figure 2B is referred to as the vector view of the GGEbiplot, in which the environments are connected with the biplot origin via lines. This view of the biplot helps understand the interrelationships among the environments. The cosine of the angle between the vectors of two environments approximates the correlation coefficient between them. Therefore, the most prominent relations by Figure 2B are: A near zero correlation among Ilam and Kermanshah, Gorgan, Ghachsaran and Lorestan as indicated by the near perpendicular vectors (r = cos90 = 0); a positive association among Kermanshah, Gorgan, Ghachsaran and Lorestan as indicated by acute angles. The correlation coefficients among the six test locations are presented in Table 5. The correlation coefficients among the locations indicate that the biplot currently shows relationship among the location that had relatively large loading on both PC1 and PC2 (Table 5). The number of correlation coefficients increases quickly to an unmanageable level as more locations are involved. Such a vector view of a biplot can be used to identify different mega-environments, thus that test locations from different mega-environments should have large angles, hence low or negative correlations.

It is clear that the GGEbiplot method is an excellent tool for visual MET data analysis. Analysis of stability and identification of mega-environments on chickpea using this method has not been already reported. In addition, this study indicated the possibility of improving progress from selections under diverse location conditions by applying GGL biplot. Multivariate analysis such as SREG analysis is an important tool for breeders, geneticists, and agronomists for analysis of MET data. We agree that G and GE must be considered simultaneously in genotype evaluation and mega-environment analysis. Compared with conventional univariate methods of the MET data analysis, SREG procedures have some advantages. The most important advantage of these methods is graphical presentation of the MET data, which greatly enhances our ability to understand the patterns of the data. These methods have a usage in selecting superior genotypes and test environments for a given mega-environment. This useful application is available in SREG and AMMI models by aid of GGEbiplot software and AMMIWINS program, respectively, and these can improve the identification of mega-environments and favorable genotypes. These methods are important tools for selecting high yielding, stable genotype. In conclusion, we suggest use of the SREG analysis for identification of favorable genotypes and mega-environments in chickpea.

REFERENCES

Burgueno J, Crossa J, Vargas M (2001). SAS programs for graphing GE and GGEbiplots. Biometrics and Statistics Unit, CIMMYT.

Eberhart SA, Russell WA (1966). Stability parameters for comparing varieties. Crop Sci., 6: 36-40.

FAO (2000). FAO Production Yearbook. Vol. 54, pp. 67, 71, 113.

FAO (2001). UN Food and Agriculture Organization. Rome, Italy.

Gauch HG, Zobel RW (1996). AMMI Analysis of Yield Trials. In: M.S. Kang and H.G. Gauch, Jr. (ed.) Genotype by Environment Interaction. CRC Press, New York. pp. 1–40.

Gabriel KR (1971). The biplot graphic display of matrices with application to principal component analysis. Biometrika, 58: 453-467.

Gollob HF (1968). A statistical model which combines features of factor analytic and analysis of variance techniques. Psychometrika, 33: 73-115.

Huhn M (1979). Beitrage zur Erfassung der phanotypischen stabilitat. I. Vorschlag einiger auf Ranginformationnen beruhenden stabilitatsparameter. EDV in Medizin und Biologie, 10: 112-117 (in German).

Kang MS, Pham HN (1991). Simultaneous selection for high yielding and stable crop genotypes. Agron. J., 83: 161-165.

Kelley TG, Parthasarathy RP, Grisko-kelley H (2000). the pulse economy in the mid 1990s: A review of global and regional development. Page 1-29 in R. Knight, ed. Linking research and marketing opportunities for the pulses in the 21st century. Kluwer Academic Publishers, Pordrecht, the Netherlands.

Lin CS, Binns MR (1988). A superiority measure of cultivar performance for cultivar × location data. Can. J. Plant Sci., 68: 193-198.

SAS Institute Inc (1996). SAS/STAT/IML User's Guide, Version 8, 4th edn. Cary, NC.

Vargas M, Crossa J (2000). The AMMI analysis and the graph of the Biplot in SAS. CIMMYT, Int. Mexico, p. 42.

Yan W, Hunt LA (2001). Interpretation of genotype × environment interaction for winter wheat yield in Ontario. Crop Sci., 41: 19-25.

Yan W, Hunt LA (2002). Biplot analysis of multi-environment trial data. Quant. Genetics Genom. Plant Breeding J., 19, 289-303.

Yan W, Kang MS (2003). Ggebiplot Analysis: A Graphical Tool for Breeders. Geneticists and Agronomists, CRD Press. Boca Raton.

Yan W (2001). GGEbiplot a windows application for graphical analysis of multi-environment trial data and other types of two-way data. Agron. J., 93: 1111-1118.

Yan W (2002). Singular-value partitioning in biplot analysis of multienvironment trial data. Agron. J., 94: 990-996.

Yan W, Paul L, Crossa J, Hunt LA (2001). Two types of GGEbiplot for analyzing multi-environment trial data. Crop Sci., 41: 656-663.

Yan W, Hunt LA, Sheng Q, Sulavnics Z (2000). Cultivar evaluation and mega-environment investigation based on the GGEbiplot. Crop Sci., 40: 597-605.

Zobel RW, Wright MJ, Gauch HG (1988). Statistical analysis of a yield trial. Agron. J., 80: 388–393.

25

Characterization of okra (*Abelmoschus spp.* L.) germplasm based on morphological characters in Ghana

D. Oppong-Sekyere[1], R. Akromah[1], E. Y. Nyamah[2], E. Brenya[3] and S. Yeboah[1]

[1]Department of Crop and Soil Sciences, KNUST-Kwame Nkrumah University of Science and Technology, Kumasi, Ghana.
[2]Department of Horticulture, KNUST-Kwame Nkrumah University of Science and Technology, Kumasi, Ghana.
[3]CRIG-Cocoa Research Institute of Ghana, P. O. Box 8, New Tafo-Akim, Ghana.

Twenty five accessions of okra collected in Ghana were evaluated for phenotypic identity, diversity and quality based on morphological characters. Qualitative and quantitative characteristics were measured and scored as specified by the standard international crop descriptor for okra. A dendrogram was generated for morphological data based on the simple matching coefficient, and four cluster groups were observed. The distribution of the accessions into the groups, based on the morphological traits had no unique geographical relationship. The results of the matrix of similarity among the 25 accessions performed by NTsys pc programme placed two accessions in a tie, suggesting that, they were identical. Eight accessions were placed at above 80% similarity, meaning that, the accession pairs were closely related, and three accessions were 50% similar, which means they matched at half the characters measured. Six pairs of accessions measured were somewhat diverse, which can be exploited by plant breeders for further improvement. The genetic affinity between the accessions from different regions and ethnic groups could however be due to the selection and exchange of okra between farmers from different regions and ethnic groups. Distinct morphotypes exist in the Ghanaian okra germplasm, depicted by variation in petal colour, pubescence of the leaf and stem, fruit shape, anthocyanin pigmentation and number of days to 50% flowering.

Key words: Dendrogram, diversity, morphotypes, pubescence, vegetables.

INTRODUCTION

Okra (*Abelmoschus esculentus* L.) is a warm-season annual herbaceous vegetable crop which can be found in nearly every market in Africa. It is grown primarily for its young immature green fruits and fresh leaves used in salads, soups and stews. The crop, which is generally self-pollinated (Martin, 1983), belongs to the Malvaceae (mallow) family and has its origin in West Africa (Joshi et al., 1974).

The okra provides an important source of vitamins and minerals (Lamont, 1999). Grubben et al. (1977) have also reported significant levels of carbohydrate, potassium and magnesium. The seeds of okra are reported to contain between 15 and 26% protein and over 14% edible oil content (NARP, 1993). The crop is the fourth most popular vegetable in Ghana after tomatoes, capsicum peppers, and garden eggs (Sinnadurai, 1973), and its production is widespread across all the major regions. About 10-15 t /ha of yield can be obtained under good management (NARP, 1993). The world okra production was estimated at 4.8 million tons (Gulsen et al., 2007). In Ghana, okra is found in its fresh state in almost all markets in Ghana during the rainy season and in a dehydrated form during the dry season, particularly in Northern Ghana due to its strong commercial value. Okra has vital importance as food diet among the inhabitants of the cities and villages.

Characterization of crops is a very essential first step in any crop improvement programme (De Vicente et al., 2005). Characterization of genetic resources, therefore, refers to the process by which accessions are identified, differentiated or distinguished according to their character or quality (traits) (Merriam-Webster, 1991). Characterization provides information on diversity, within and between crop collections. This enables the identification of unique accessions essential for curators of gene banks (Ren et al., 1995). Moreover, information obtained on genetic relatedness among genetic resources of crop plants is useful, both for breeding and for the purposes of germplasm conservation (Brown et al., 1990).

Notwithstanding the potential of the crop, there is no improved variety for cultivation in Ghana (Kumar et al., 2010). More so, there has not yet been any previous reported attempt by breeders at improving vegetable in terms of developing core collections for higher yield and quality. The accessions under cultivation, over the years in the various regions across the country are landraces. Nevertheless, these landraces are associated with challenges such as, high susceptibility to diseases and pests, for example, nematodes. In addition, these landraces have long maturity periods yet short harvesting duration. They are of poor nutritional quality, non-standard in shape, colour and size, making them unfit for the Ghanaian vegetable export market. This has a consequential effect of causing a reduction in the per capita income of the nation. It is therefore important that plant breeders developed improved varieties of the okra vegetable, which seems to be the last concern in their research programmes for adoption by Ghanaian vegetable farmers and for the export market. Varieties that are perennial in growth habit and at the same time, combine higher yields and early maturity with longer harvest duration and more so resistant to diseases and pests, would be ideal to the okra vegetable industry in Ghana. Improved varieties in terms of fruit size, shape and colour are also very much desired in the Ghanaian okra export market. It is against this backdrop that this characterization and genetic diversity study was necessitated. More importantly, the Ghanaian ecotypes as an important first step should improve the crop in Ghana by providing the foundation to enhance their potential use. This study sought to afford us the opportunity to assess qualitative and quantitative variations among collections of the Ghanaian okra landraces through morphological evaluation and thus exploit such variations in breeding programmes to develop improved, high yielding varieties.

MATERIALS AND METHODS

Twenty-five accessions of okra were used as experimental materials to assess the differences in morphological traits. Among the accessions, twelve were collected from the Plant Genetic Resources Research Institute (PGRRI) of CSIR, twelve from College of Agricultural Science, University of Education, Winneba-Mampong, and one from the Department of Horticulture, KNUST-Kumasi. The study was conducted in the major and minor cropping seasons of 2008 at the experimental field of the Department of Crop and Soil Sciences, KNUST. The land, which had been allowed to lie uncultivated for one year after a previous harvest of groundnut was slashed, ploughed and harrowed to a fine tilt for the experiment. Randomized complete block design (RCBD) was used as experimental design with four replications. The total experimental area was 14 m x 83 m. Seeds were sown directly in the field at a rate of two seeds per hill. Seedlings were thinned to one plant per stand two weeks after germination. There were a total of 100 plots with each plot measuring 2.4 m by 1.8 m. Each plot had sixteen plants with a spacing of 60 cm by 45 cm. Standard agronomic practices including thinning, weeding and watering were adopted. Compound fertilizer in the form of N.P.K. 15:15:15 at a rate of 250 kg/ha and urea at a rate of 125 kg/ha were applied to the plants at 30 days after sowing.

Data were collected on plant growth habit, general growth appearance/branching, flowering characteristics, pigmentation and pubescence of the various plant parts, fruit characteristics and leaf characteristics. Data were recorded from the four tagged plants in each plot.

A standardized crop descriptor for okra (IBPGR, 1991) was used to measure the various parameters studied (Tables 1 and 2). A dendrogram showing the distinct clusters among the 25 okra accessions was constructed using Numerical Taxonomy and Multivariate Analysis System (NTSYS version 2.11s; Rohlf, 2005) and similarity coefficients were calculated by simple matching produced by UPGMA (Rohlf, 2005).

RESULTS

General observations

In general, all the okra accessions showed relatively wide ranges of variations for all morphological characters observed. Most of the plants showed erect growth habits while leaf and stem colours were predominantly green. Petal or flower colour was mostly golden yellow among the okra, whereas fruit orientation was largely intermediate for all accession studied. Majority of the fruits produced green and smooth fruits. Most of the accessions showed symptoms of okra mosaic virus and okra leaf curl disease.

Vegetative characters

Plant growth habit, general appearance and branching

Branching position-at-main-stem (general growth appearance) of okra accessions were 60% in occurrence for unique orthotrop axis (UOA). Densely branched all over (DBO) and densely branched base (DBB) characters were 20% in frequency (Figure 1). Figure 2 shows the DBB, UOA and DBO growth habits observed in this study.

Fruit characters

Fruit colour

The results showed that fruit colour displayed five distinct variations that ranged from common green, green with

Table 1. Evaluated characteristic of okra collection. Coding of qualitative characters is according to IBPGR, 1991 descriptors for okra.

S/N	Code for character Qualitative	Parameter measured	Character codes
1	SC	Seed colour	1=dark, 2=black, 3=whitish to dark, 4=purple to black
2	SSh	Seed shape	1=Roundness, 2=Kidney, 3=Spherical
3	SS	Seed size	1=Small, 2=medium, 3=large
4	BPMS	Branching position at main stem	1=UOA-unique orthotrop axis, 2=DBO-densely branched all over, 3=DBB-densely branched base
5	MLC	Mature leaf colour	1=Green, 2=Green+red veins
6	LSh	Leaf shape	From types 1 to 11
7	LBr (cm)	Length of branches	0= no branches, 1= branches rarely > 10cm
8	LRC	Leaf rib colour	1=Green, 2=Green+red veins
9	PtC	Petiole colour	1=Green, 2=Green+red veins, 3=purple
10	PC	Petal colour	1=Golden yellow, 2=yellow
11	CDR	Colour of the darkest ridges	1=light, 2=dark, 3=light to dark
12	StC	Stem colour	1=green, 2=green+purple tinge, 3=purple
13	NES	Number of epicalyx segments	1=8 to10, 2=5 to 7, 3=>10
14	FSp	Flowering span	1=Single flowering, 2=grouped flowing
15	NSfS	Number of segments from the stigma	From 5 to 12 segments
16	FC	Fruit colour	1 =Green, 2=green+red spots, 3=dark green to black, 4=green to yellow, 5=purple
17	FP	Fruit pubescene	1=Smooth, 2=little rough, 3=downy+hairs
18	FSh	Fruit shape	From types 1 to 15
19	NR/F	Number of ridges per fruit	1=0, 2=b/n 5 and 12, 3=5ridges
20	PFMS	Position of fruit from the main stem	1=intermediate, 2=slightly falling, 3=horizontal, 4=Erect, 5=Drooping
21	LFP	Length of fruit peduncle	1=1 to 3cm, 2=>3cm
22	SI	Susceptibility to insects	Scale: 1 to 9; (Podagrica spp, Aphids, Cotton stainer): NS=0-1,WS=1-3,IS=3-5, HS=6-9
23	Sdi	Susceptibility to diseases	Scale: 1 to 9; (OMV, OLCV): NS=0-1, WS=1-3, IS=3-5, HS=6-9

Table 2. Qualitative traits that varied among okra collection studied.

S/N	Accession	SC	SSh	SS	BPMS	MLC	LSh	LBr (cm)	LRC	PtC	PC	CDR	StC	NES	FSp	NSfS	FC	FP	FSh	NR/F	PFMS	LFP	SI	Sdi
1	GH 4487 Muomi	1	1	3	1	1	2	0	1	1	1	1	1	1	1	9	1	1	2	1	1	1	2	2
2	GH 4482 Muomi	2	2	2	1	2	1	0	2	2	1	1	2	1	1	9	1	1	2	1	1	1	2	2
3	GH 4499 Fetri	3	1	2	1	1	2	0	1	2	1	1	1	1	1	12	1	1	8	2	1	1	2	2
4	GH 1169 Fetri	3	1	3	1	2	2	0	2	2	2	1	2	2	1	7	1	1	8	2	1	1	2	2
5	GH 4376 Atuogya	2	2	2	1	2	2	0	2	2	2	1	1	2	1	9	1	2	2	2	2	1	2	2
6	GH 4490 Fetri	4	1	1	1	2	7	0	2	3	1	1	3	3	1	6	1	1	8	2	1	1	2	2
7	GH 3801 Pora	3	3	1	1	2	7	0	2	3	2	1	3	1	1	9	5	2	3	2	3	1	9	9

Table 2. Contd.

#	Accession	SC	SSh	SS	BPMS	MLC	LSh	LBr	LRC	PtC	PC	CDR	StC	NES	FSp	NSfS	FC	FP	FSh	NR/F	PFMS	LFP	SI	Sdi
8	GH 6102 Fetri	3	2	1	1	2	0	1	1	1	1	1	1	6	1	1	8	2	1	1	2	2		
9	GH 4964 Muomi	2	2	2	2	1	1	1	1	1	2	1	1	6	1	1	2	1	1	1	2	2		
10	GH 5793 Gyeabatan	2	2	1	2	1	1	2	2	3	1	1	2	5	2	2	4	2	2	1	2	2		
11	GH 5787 Asontem	2	3	2	2	2	0	1	1	1	1	1	1	9	1	2	8	2	4	1	0	0		
12	GH 3736 Fetri	3	3	1	1	2	0	1	1	1	1	1	1	5	1	2	8	2	1	1	5	5		
13	Atuogya-tiatia	4	3	1	3	1	1	2	2	1	2	2	2	6	2	2	4	2	3	2	2	2		
14	DA/08/001Wun mana	3	3	2	2	3	1	2	2	2	1	1	3	9	3	3	6	2	1	2	5	0		
15	DA/08/02Sheo mana	4	1	1	3	3	0	2	2	1	1	3	2	5	1	2	14	2	1	1	0	0		
16	DA/08/02 Ason-Wen	4	3	1	1	2	0	2	2	1	2	1	1	7	1	1	8	2	1	1	2	2		
17	Atuogya-Asante	2	2	2	2	3	1	3	1	2	1	2	2	10	2	2	1	2	1	2	2	2		
18	Asontem	4	3	1	1	4	0	1	1	1	2	1	1	9	1	1	8	2	4	1	2	2		
19	DA/08/04Wun mana	2	2	2	2	2	1	2	2	2	2	2	2	9	2	3	15	2	3	2	2	2		
20	DA/08/004 Agbodro	3	3	2	1	4	1	2	2	1	1	2	2	5	1	1	7	1	5	2	0	0		
21	Gbodro-wild	2	2	2	2	2	0	2	2	2	1	2	2	9	2	3	3	3	1	2	2	2		
22	DA/08/02Dikaba	3	3	1	3	2	0	1	1	2	1	1	1	6	1	2	13	2	4	1	0	0		
23	DA/08/03Sheo mana	4	1	1	3	2	1	1	1	1	3	1	1	10	4	3	15	2	4	2	0	0		
24	Atuogya-tenten	1	2	3	3	2	0	1	1	2	2	2	2	5	1	1	2	1	3	2	5	5		
25	KNUST/SL1/07Nkrumahene	1	3	1	1	2	0	1	1	2	2	2	2	9	1	1	2	1	2	2	0	0		

SC, See colour; SSh, seed shape; SS, seed size; BPMS, branching position at main stem; MLC, mature leaf colour; LSh, leaf shape; LBr, length of branches; LRC, leaf rib colour; PtC, petiole colour; PC, petal colour; CDR, colour of the darkest ridges; StC, stem colour; NES, number of epicalyx segments; FSp, Flowering span; NSfS, number of spines from the stigma; FC, fruit colour; FP, fruit pubescence; FSh, fruit shape; NR/F, number of ridges per fruit; PFMS, position of fruit from the main stem; LFP, length of fruit peduncle; SI, susceptibility to insects; Sdi, susceptibility to diseases.

red spots, dark green to black, green-to-yellow to completely purple (Figure 3). In total, 72% of the accessions produced green fruits while 8% displayed green-with-red-spotted fruits, dark green to black fruits and green to yellow-fruits. A small portion (4%) of the accessions had tinged purple fruits. Variety GH3801 Pora had a unique purplish pigmented fruit colour. Figure 4 gives a pictorial view of variations in fruit colour among the okra studied.

Fruit pubescence

Fruit pubescence showed wide variation among the okra accessions. Sixty-four percentages of okra accessions showed fruits with no hairs (Figure 5), while the rest had rough, downy or little hairs on their fruits representing 4, 12 and 20% occurrences, respectively. From these results, one can conclude that the okra fruit has no hairs.

Fruit shape (form)

Fruit shape showed the greatest diversity among the okra accessions; from short and triangular to long straight or long curved. From the results in Figure 6, 8% of total accessions bore fruits with shape scores of 1, 3, 4, and 15, respectively,

according to the descriptor (IBPGR, 1991; Plate 1). Twenty percent accessions bore fruits with a shape score of 2, while 4% bore fruit with shape scores 6, 7, 13 and 14. Fruit shape scores 5, 9, 10, 11 and 12, did not show any occurrence (zero percentages) while shape score of type 8 recorded the highest occurrence of 32% of the okra accessions. Figure 7 shows variations in fruit shape among accession studied.

Number of fruit ridges

Number of fruit ridges and position of fruits on main stem are useful only for cultivated forms of

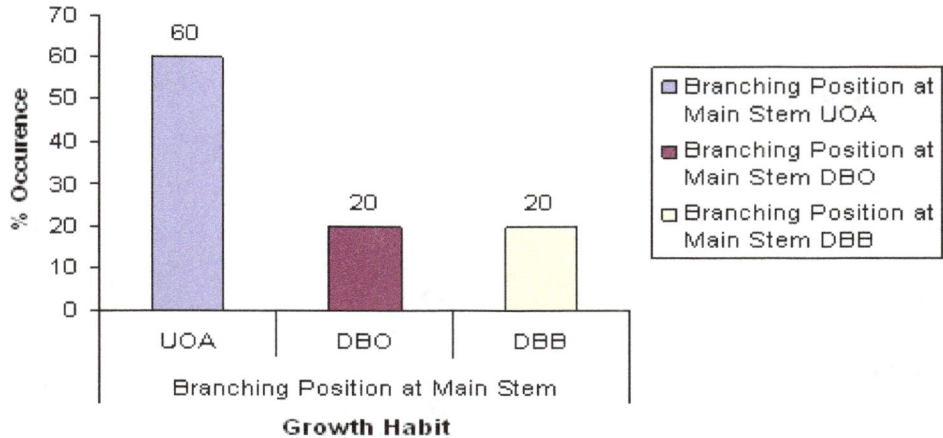

Figure 1. Variations in growth habit among okra accessions.

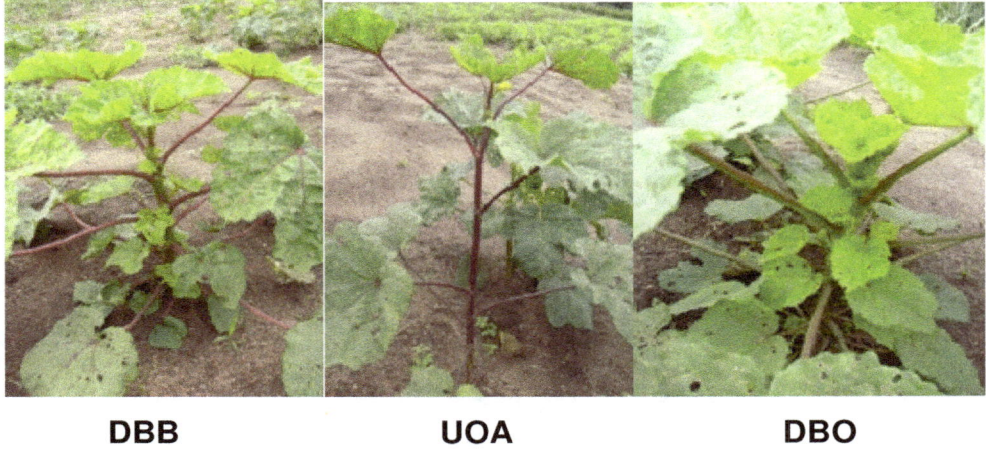

Figure 2. Variations in growth habit among okra accessions.

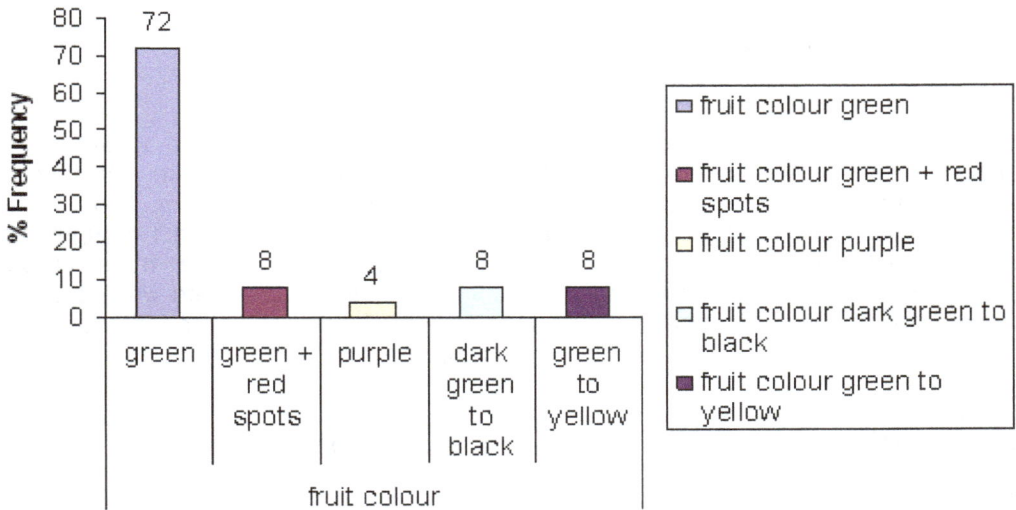

Figure 3. Variation in fruit colour among okra accessions.

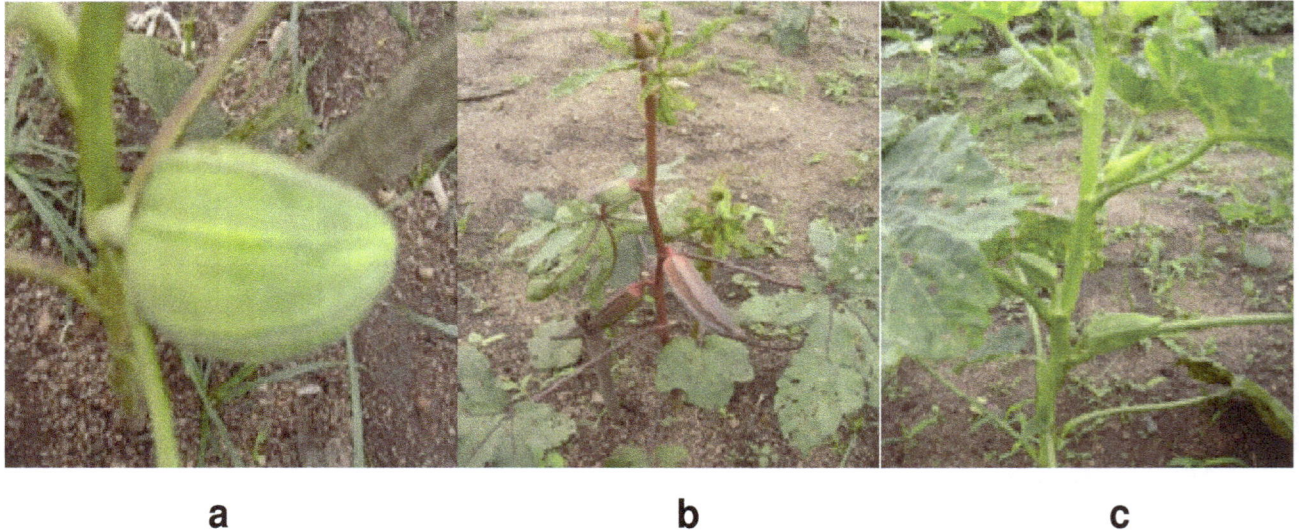

Figure 4. Variation in fruit colour among okra accessions. a, Yellowish-green okra fruits; b, Purple (GH3801 Pora) okra fruit; c, Okra with green fruits.

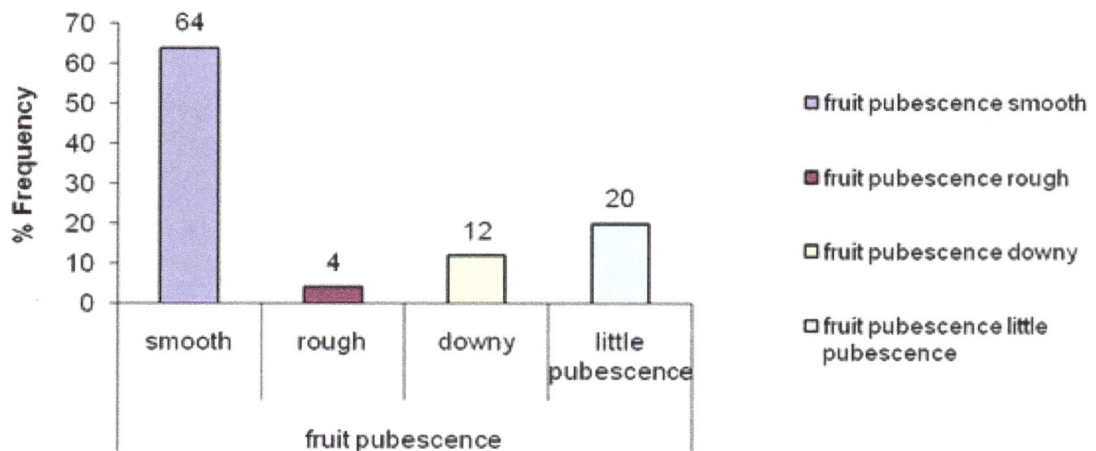

Figure 5. Variation in fruit pubescence among okra accessions.

okra. The number of ridges per fruit of the 25 entries ranged from 0-12 ridges. Accessions DA/08/004 Agbodro, GH 4964 Muomi, GH 4482 Muomi, GH 4487 Muomi recorded had no well-marked ridges on their fruits while Gbodro and Atuogya-tenten had five ridges on their fruits. The rest of the okra accessions, representing about 80% had very conspicuous ridges per fruit between 8 and 12 (Figure 8).

Fruit position on the stem

The position of fruits on the main stem of the accessions showed five distinct variations; erect intermediate, slightly falling, horizontal and drooping positions. Figure 9 showed that, 60% of the accessions had fruits that were intermediate on the stem, which means that they were half upright on the stem, 20% of the accessions bore fruits which were in erect (upright) position, 12% of the okra accessions bore fruits which were positioned horizontally on the stem while only 4% bore fruits which were slightly falling and drooping (fruits in an almost upside down position).

CLUSTERING

General clustering of okra accessions into groups

A cluster diagram obtained from the morphological

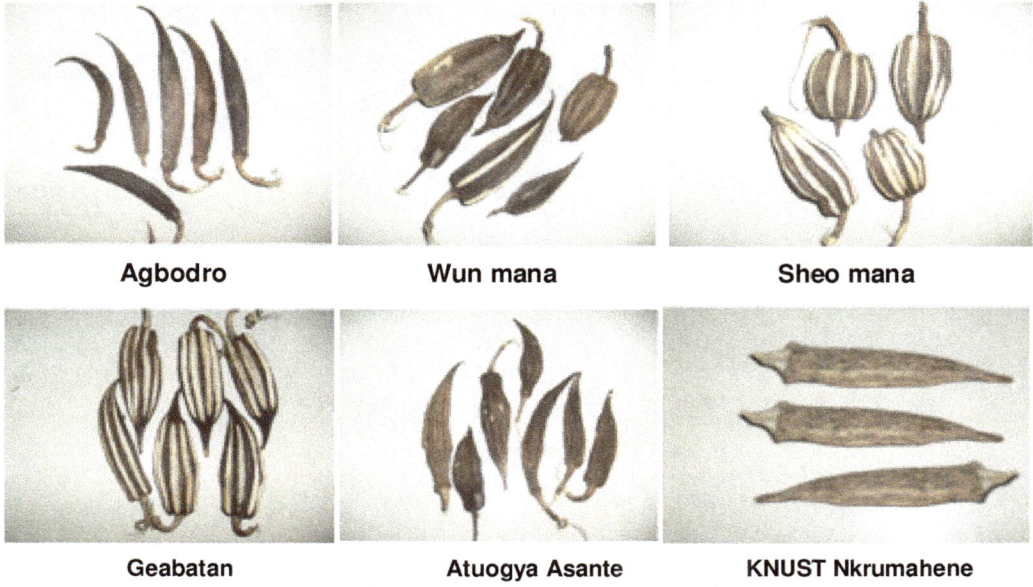

Figure 6. Variation in fruit shapes (forms) within accessions studied.

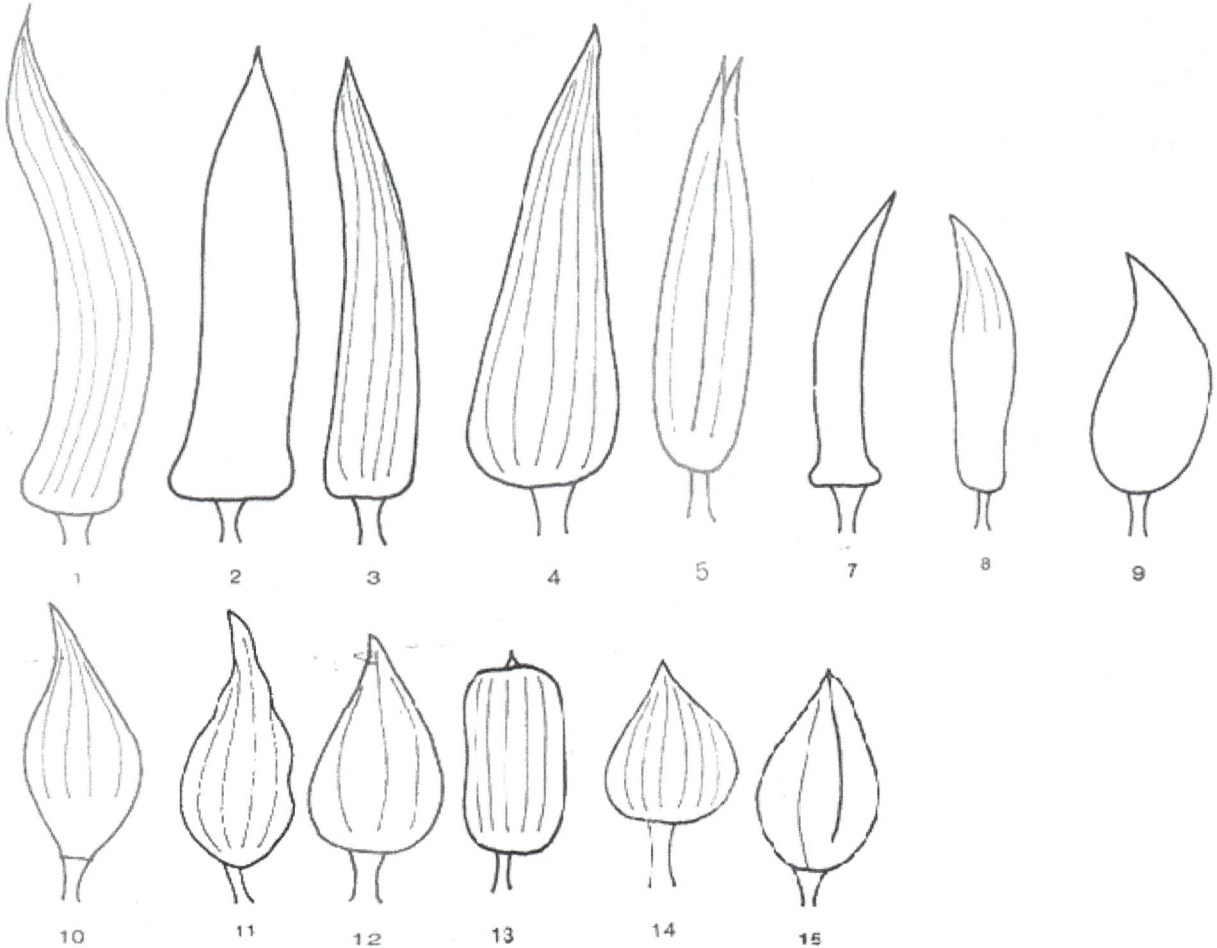

Plate 1. Variation in fruit shapes (IBPGR, 1991).

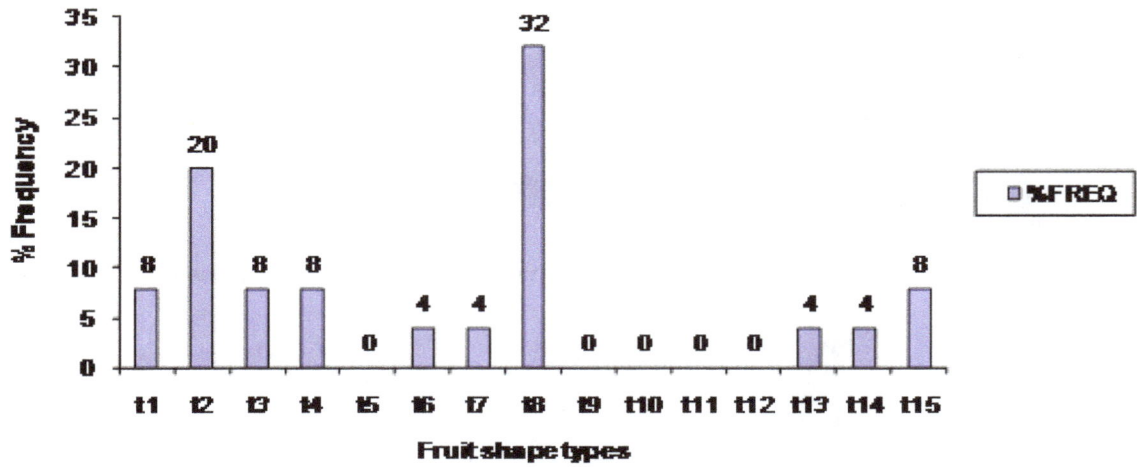

Figure 7. Variation in fruit shape among okra accessions.

Figure 8. Variation in fruit ridges among okra accessions.

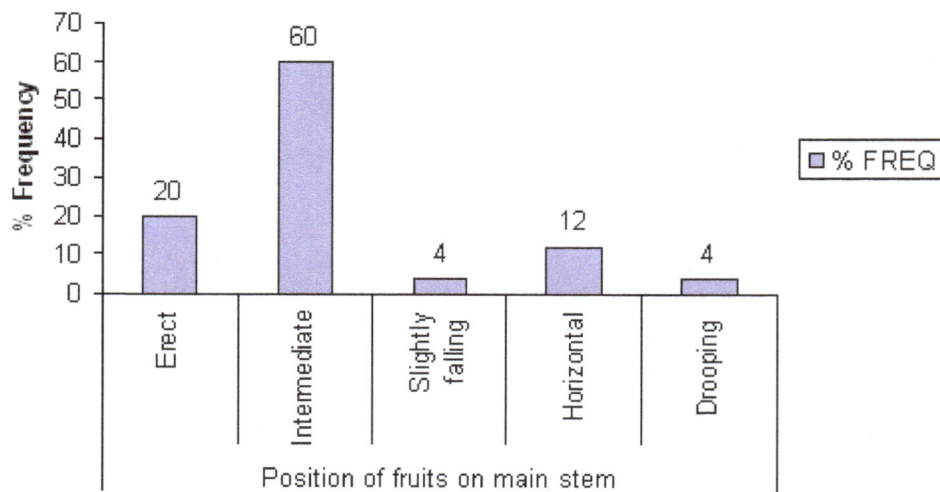

Figure 9. Variation in fruit position among okra accessions.

descriptors produced four main sub-cluster groups of okra accessions at a coefficient of 0.63. Accessions were put into cluster groups based on certain qualities unique to them. Cluster A, recorded the highest number of accessions (16) while cluster B consisted of only one accession.

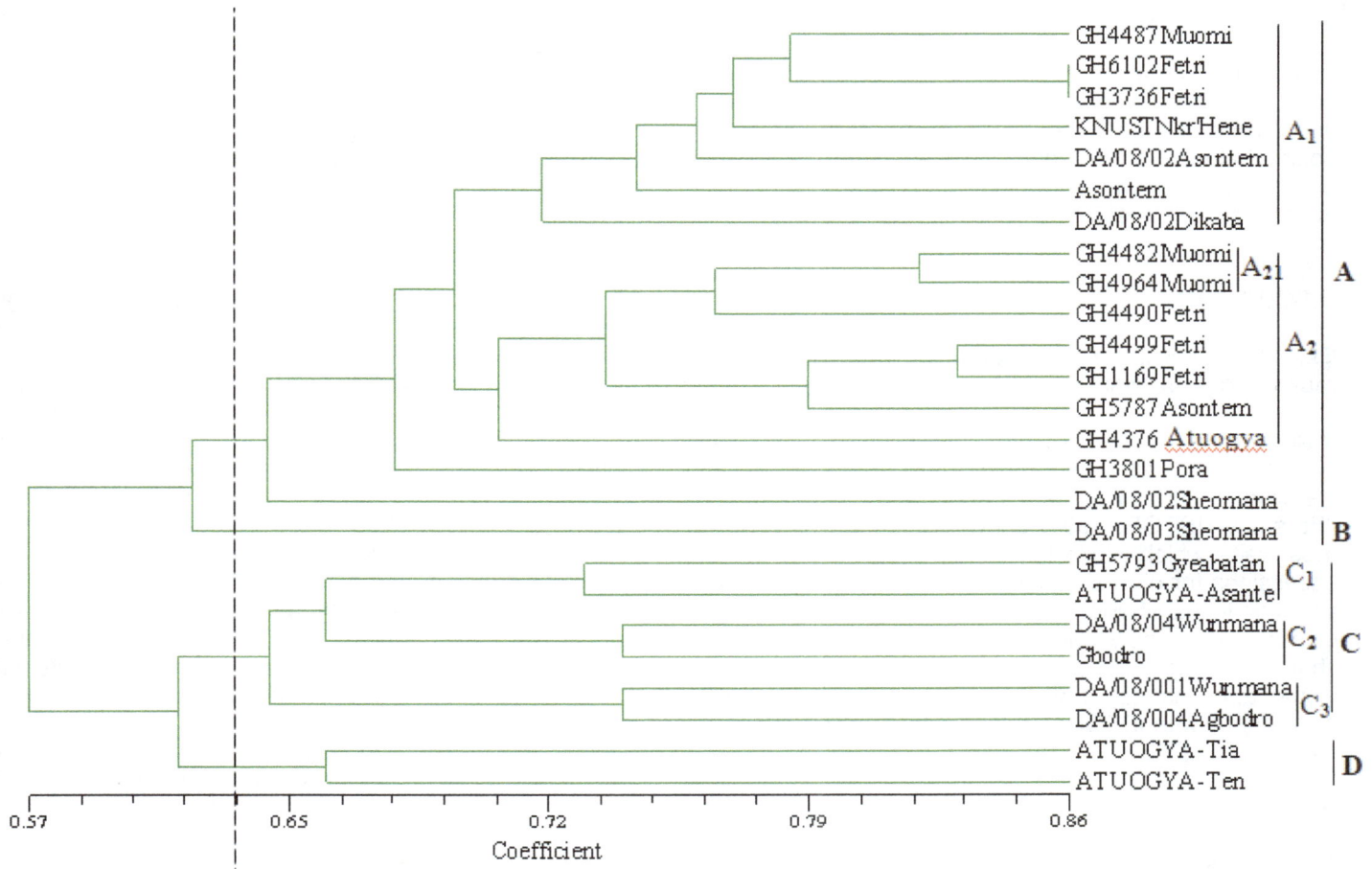

Figure 10. Dendrogram showing the relationship among 25 okra accessions revealed by UPGMA cluster analysis based on morphological characters.

It was observed that 23 out of 25 okra accessions under study were distinct accessions. A tie was recorded between accessions GH6102Fetri and GH3736Fetri. Similarity coefficient ranged from 45.8 to 86.5%. There was no unique relationship between the cluster groups and the regions of collection. Accessions with similar quantitative and qualitative morphological characters appeared well grouped in the same cluster. Okra accessions with common local names were also found in the same cluster (Figure 10).

Similarity per cluster group

Cluster A comprised of 16 accessions, differing from accession in the other clusters by having green fruit colour and golden yellow petal colour. Within cluster A, sub-cluster A_1 produced accessions with green stem colour as against A_2 with green-with-purple tinge stem colour. Within sub-cluster A_2, sub-sub-cluster A_{21}, consisting of accessions GH4482Muomi and GH4964Muomi, produced the same fruit shape (type 12)

and leaf shape (type 1).

Okra genotypes, GH3801 Pora and DA/08/02 Sheo mana were singled out of the two unique sub-cluster groups A_1 and A_2 due to the fact that, GH3801 Pora formed a unique fruit colour/pigmentation (purple), and leaf shape (type 7) while DA/08/02 Sheo mana had fruits with green colour and a leaf shape of type 3.

Cluster B comprises of only one okra accession, DA/08/03 Sheo mana, and this differed from accessions in the other clusters mainly in its fruit pubescence or the presence of hairs on its fruits (downy plus hairs), and fruit colour (green to yellow).

Cluster C, the second largest cluster with 6 okra accessions, composed of okra genotypes that were different from the others by their reduced fruit pubescence, yellow petal colour with stem colour having a combination of green with purple tinge. The orientation of fruits on main stem was drooping. Within cluster C, sub-cluster C_1 formed fruit with shape of type 4, and combined green with red-spotted fruits while sub-cluster C_2 showed fruits with a combination of dark green plus red and dark green-to-black as well as leaf shape of type

2. Sub-cluster C_3 included fruits with dark green-to-black colour only.

Cluster D consisted of 2 okra accessions, Atuogya-tiatia and Atuogya-tenten. These were different from accessions in the other clusters by their branching position-at-main-stem; being DBB and fruit pubescence being little rough with dark hairs. The orientation of fruits-at-main-stem of these okra varieties was horizontal.

DISCUSSION

Extent of variation in qualitative and quantitative morphological characteristics

Variation is an important attribute in breeding programmes (Hazra and Basu, 2000; Omonhinmin and Osawaru, 2005). The okra genotypes characterized in this study (IBPGR, 1991) showed a broad variation for most traits, which allows for the identification of promising accessions for okra breeding in Ghana and beyond. The variation in leaf shape, leaf rib colour, petal colour, petiole colour, stem colour, fruit colour and pubescence and fruit shape, among others, were easily recognizable with visual appraisal (by the use of a colour chart and in accordance with IBPGR, 1991 descriptor list for okra).

Vegetative characters

Growth habit

Different genotypes have different growth habits, as a result of selection or a natural adaptation mechanism. The commonest growth habit among all the landraces observed indeterminate growth habit with erect general growth appearance.

Erect plant type is advantageous to okra production, since it allows maximum and uniform exposure or distribution of all leaves and other vegetative parts for better interception of sunlight, and would also result in an increase in dry matter production and a subsequent increase in yield. This is similar to findings by Hanson (2005). Moreover, there is less chance of fruits touching the ground or soil thereby causing fruit rot.

The indeterminate nature of the okra landraces is a character which might have been selected for over the years by researchers and farmers because it allows for longer and continuous fruit harvest. This is an advantage when prices of the vegetable fluctuate. Farmers do not want these plants to produce long branches and would rather opt for more plants per area unit. Previous studies in Tomatoes by Hanson (2005) suggested this to be advantageous because it allows the combination of large numbers of fruit with many plants per unit space, which is an indicator of high yield.

Okra genotypes such as 'Wun mana' and 'Sheo mana' types produced dense branches with fruits closer to the soil, yet gave higher total fruit yields. This could be selected for and incorporated into other 'Asontem' varieties which had the typical UOA branching (erect) but moderate in plant height, in order to develop elite okra types that are less lodging with higher yields and can therefore pass for commercial production.

Fruit characters and production

Fruits displayed great diversity in size-shape and length. Earliness, expressed by the lower leaf axil in which flower buds appear is partly due to varietal characteristic.

Fruits with characteristics such as smooth, spineless, slender with green (light or dark) skin are very desirable in the Ghanaian local and export markets (Sinnadurai, 1992). Varieties such as KNUST/SL1/07Nkrumahene, DA/08/004 Agbodro and Asontem were among the landraces that displayed such traits. These can therefore be selected for breeding by crossing them with other local materials such as 'Atuogya', 'Sheo mana' or 'Wun mana' accessions which were comparatively high yielding with longer harvest duration and also highly resistant to environmental stresses such as diseases and pests, and drought but revealed undesirable fruit shape and colour.

Variation revealed in this study for fruit colour was highest for green and appreciably high for green-to-yellow as well as green-to-red spots. This is different from the results of Myanmar (1995) in which fruit colour was observed to be either green or yellow-to-green, though green fruit colour was found to be highest in the okra accessions studied. These results were expected, because Myanmar (1995) might have examined improved okra collections that were more uniform. The okra genotypes used in this study were landraces, and hence more variable or diverse.

Results obtained in this study show that the okra accessions exhibited varying degrees of fruit pubescence including smooth, rough, downy or little hairs on fruits but with the majority having smooth fruits. This result is in contrast to those of Bish et al. (1995) and Thomas (1991) who found downy type of fruit pubescence to be highest, followed by slightly rough while prickly fruits was the least in the okra accessions they studied. This shows that farmers in Ghana have selected the smooth fruit types as their preferred fruit and discarded the hairy types.

Variation in okra species has been investigated by several researchers (Bish et al., 1995; Akinyele and Oseikita, 2006; Duzyaman, 2005). They found that a large number of okra characters such as pigment colour and spines on the fruit surfaces are inherited in a simple fashion, suggesting that these characters are controlled by relatively few genes (monogenically inherited).

Ariyo (1993) indicated that the pattern of genetic variation observed in characters studied in West African okra suggests a lot of out crossing among the taxon. Wide morphological variation observed in okra characters studied could perhaps, be attributed to the preponderance of out crossing among different accessions of the okra studied.

Pattern of variation and description of the cluster groups

In cluster group A, the cluster analysis generally found accessions GH6102Fetri and GH3736Fetri (from Biakoye and Kpogadzi, respectively) in a tie. That is, the two accessions were placed at 100% similarity. These Fetri accession may therefore be identical. Perhaps, they may have been collected by the same farmer and misnamed due to the informal way of germplasm exchange from farmer to farmer, diverse languages or ethnic groups in the areas covered by the collection and marketing, which accounts for the differences in local names (Torkpo et al., 2006).

Within cluster group A other 'Fetri' accessions were also found (GH4490Fetri, GH4499Fetri and GH1169Fetri). These may again be the same accession but picked at different locations and time and named differently due to the informal means of germplasm collection, selection and dissemination by okra farmers. Similar reason could be ascribed to the three accessions of Muomi (GH4487, GH4482, and GH4964 from Bedoku, Prampram and Sutapong, respectively) also found in the same cluster group A.

Okra accessions, GH4376Atuogya and Atuogya-Asante were found in cluster groups A and C, respectively. One would expect these to be under one cluster group, however, they are not. Perhaps, the traits considered were inadequate or were not sufficiently discriminatory to permit their classification into one group. Similar results can be speculated of accessions such as DA/08/02 Sheo mana and DA/08/03 Sheo mana as well as DA/08/04 Wun mana and DA/08/001 Wun mana, which did not enter into one cluster group but were found in cluster B and sub-clusters C_2 and C_3, respectively.

The great difference in genetic relationship, particularly among the Atuogya, Wun mana, Sheo mana, Muomi and Fetri collections demand further classification of these collections by employing more discriminatory characters or by utilizing molecular markers.

It must be said that most of the okra genotypes found in the same cluster group, such as GH1169Fetri and GH4499Fetri (originally collected from Gabusa and Nyingutu respectively, both in the Northern region), Atuogya-tenten, Atuogya-tiatia, Atuogya-Asante, Asontem and KNUST/SL1/07 Nkrumahene (collected originally in the Ashanti region), DA/08/02 Sheo mana, DA/08/03 Sheo mana, and DA/08/04 Wun mana, DA/08/001 Wun mana (all from Sakogu in the Northern region), and scoring similar similarity indices may be eliminated from the germplasm collection. IBPGR (1991) reported that, repeated regional collections without proper documentation could account for duplication in germplasm collections.

From the analysis of the similarity matrix, similarity coefficient ranged from 45.8%, for the most distantly related accession, to 86.5% for those closely related. This is therefore indicative of a higher variability in the okra accessions studied.

From the similarity matrix, two pair of accessions, DA/08/03 Sheo mana and GH4482 Muomi, Atuogya-tiatia and GH4964 Muomi showed the widest variation in the characters measured, scoring 0.458 on the similarity matrix, meaning that these okra pairs are 45.8% similar and 54.2% dissimilar in the characters measured.

These variations are what plant breeders are very much interested in and therefore are suitable for further breeding purposes. This results is confirmed by Reid et al. (1998), who argued that, genetic diversity of crops in Africa have been naturally preserved for a longer time by virtue of the continent's relative traditional agriculture.

Four pairs of accessions; Atuogya-tiatia and GH4487 Muomi, DA/08/04 Wun mana and GH4487Muomi, DA/08/04 Wun mana and GH4490 Fetri and Atuogya-Asante and GH4499 Fetri showed a similarity matix of 0.49. This again implies that they were 49% similar. These okra accessions are therefore suitable for exploitation by breeders for further improvement of quality and yield. Mondal (2003) argues that genetic diversity is essential to meet the diversified goals of plant breeding which included increase in yield, diseases and pest toleance, wider adaptation and desirable consumer qualities.

Observation from the similarity matrix placed only three pairs of accessions; DA/08/03 Sheo mana and GH4376 Atuogya, DA/08/001 Wun mana and GH4376 Atuogya and Atuogya-Asante and GH 4490 Fetri at 50% similarity, meaning the two accession pairs matched at half the characters measured.

The similarity matrix again showed that of all the accessions measured, eight accessions gave a matrix score of 0.8 and above. This shows that the pairs are 80% or more similar, according to the similarity matrix. Studies by Irwin et al. (1998) affirmed that closely related accessions are normally located within 80-90% similarity. Crosses between accessions with similarity indices of 80-100% may, therefore, not be desirable; and that the potential for successful crossing of unrelated varieties may generate into an array of genotypes from which useful agronomic types may be selected, a similar observation was made by Gulsen et al. (2007). The large size of the accessions from a wide range of geographical areas is very essential for genetic distance estimation (Nei, 1978).

Findings Torkpo et al. (2006) indicate that the wide range of similarity indices, coupled with the clustering of accessions, suggested useful variability in the okra germplasm collection for genetic preservationists and plant breeders for improvement programmes.

Conclusion

Conclusively, the pubescence and pigmentation of

various plant parts as well as fruit characteristics, among qualitative traits of the okra landraces studied, proved to be most significant in the analysis of variability and contributed significantly to the total variation observed. The study has established that, varietal names are often descriptive; 'the early one', 'the late one', 'the dry season type', 'the wet season type'. The genetic affinity between the accessions from different regions observed in this study could however be attributed to the selection and exchange of okra germplasm between farmers from different regions and also between ethnic groups. Migrant farmers often carry seeds from their homes to their new locations, thus creating duplications in the germplasm. Nevertheless, there are distinct morphotypes in the Ghanaian okra germplasm, depicted by variation in petal colour, pubescence of the leaf and stem, fruit shape, anthocyanin pigmentation and number of days to 50% flowering. The clustering pattern indicated the presence of diverse forms in collections made from the same location, indicating tremendous opportunity to select the most desirable lines for that eco-geographical location.

REFERENCES

Akinyele BO, Oseikita OS (2006). Correlation and path coefficient analyses of seed yield attributes in okra (*Abelmoschus esculentus* (L.) Moench). Afr. J. Biotechnol., 14: 1330-1336.

Ariyo OJ (1993). Genetic diversity in West African Okra (*Abelmoschus caillei* (A. Chev.) Stevels): Multivariate analysis of morphological and agronomic characteristics. Genet. Resour. Crop Evol., 40: 25-32.

Bish IS, Mahajan RK, Rana RS (1995). Genetic diversity in South Asian okra (*Abelmoschus esculentus*) germplasm collection. Ann. Appl. Biol., 126:539-550.

Brown AHD, Marshall R, Frankel OH, Williams JT (1990). The use of plant genetic resources. Cambridge University Press, London.

De Vicente MC, Guzmán FA, Engels J, Ramaratha Rao V (2005). Genetic characterization and its use in decision making for the conservation of crop germplasm: The Role of Biotechnology, Villa Gualino, Turin, Italy – 5-7, 63 p.

Duzyaman E (2005). Phenotypic diversity within a collection of distinct okra (*Abelmoschus esculentus*) cultivars derived from Turkish landraces. Genet. Res. Crop Evol., 52: 1019-1030.

Grubben GJH, Tindall HD, Williams JT (1977). Tropical Vegetable and their Genetic Resources, International Board of Plant Genetics Resources, FAO, Rome, pp. 18-20.

Gulsen O, Karagul S, Abak K (2007). Diversity and relationships among Turkish germplasm by SRAP and Phenotypic marker polymorphism. Biologia, Bratislava, 62(1): 41-45.

Hanson P (2005). Lecture Notes on Tomato Breeding, Asian Vegetable Research and Development Center, Africa Regional Program Training, Arusha, Tanzania.

Hazra P, Basu D (2000). Genetic variability, correlation and path analysis in okra. Ann. Agric. Res., 21(3): 452-453.

IBPGR (1991). Report of an international workshop on okra genetic resources, held at the National Bureau for Plant Genetic Resources (NBPGR), New Delhi, India, 8–12 October, 1990. International Crop Network Series 5. International Board for Plant Genetic Resources (IBPGR), Rome, Italy, 133 p.

Irwin SV, Kaufusi P, Banks K, de la Peña R, Cho JJ (1998). Molecular characterisation of taro (*Colocasia esculenta*) using RAPD markers. Euphytica, 99: 183-189.

Joshi AB, Gadwal VR, Hardas MW (1974). Okra. *Abelmoschus esculentus* (Malvaceae), In Hutchinson J B (ed.). Evolutionary Studies in World Crops: Diversity and change in the Indian subcontinent. Cambridge, pp. 99-105.

Kumar S, Dagnoko S, Haougui A, Ratnadass A, Pasternak D, Kouame C (2010). Okra (*Abelmoschus* spp.) in West and Central Africa Potential and Progress on its Improvement. Afr. J. Agric. Res., 5(25) 3590-3598.

Lamont Jr. WJ (1999). Okra - A versatile vegetable crop. Hort. Technol. 9(2): 179-184.

Martin FW (1983). Natural outcrossing of okra in Puerto Rico. J. Agric Univ. Puerto., 67: 50-52.

Merriam-Webster (1991). Webster's ninth new collegiate dictionary Merriam-Webster Inc., publishers. Springfield, Massachusetts, USA.

Mondal AA (2003). Improvement of Potato (*Solanum tuberosum* L.) through Hybridization and *in-vitro* Culture Technique. PhD thesis Rajshahi University. Rajshahi, Bangladesh.

Myanmar AK (1995). Evaluation of Okra Germplasm. ARC-AVRDC Training Report, 12 p.

NARP (1993). National Agricultural Research Project, Horticultura crops. Vol. 3, July 1993. NARP, CSIR, Accra.

Nei M (1978). Estimation of average heterozygosity and genetic distance from a small number of individuals. Genetics, 89(3): 583-590.

Omonhinmin CA, Osawaru ME (2005). Morphological characterization of two species of Abelmoschus: *Abelmoschus esculentus* and *Abelmoschus caillei*. Genet. Resour. Newsl., 144: 51-55.

Reid R, Attere F, Toll J (1998). Germplasm Collection and Conservation in Africa: Role of IBPGR. Crop Genetic Resources of Africa, Ibadan Vol. 2.

Ren J, Mc Ferson J, Kresovich RLS, Lamboy WF (1995). Identities and Relationships among Chinese Vegetable Brassicas as Determined by Random Amplified Polymorphic DNA Markers, 120(3): 548-555.

Rohlf FJ (2005). Numerical Taxonomy and Multivariate Analysis System. *NTSYS*, Version 2.2, Applied Biostatistics Inc. Port Jefferson, New York.

Sinnadurai S (1973). Vegetable Production in Ghana. Acta Hort. (ISHS), 33: 25-28.

Sinnadurai S (1992). Vegetable Production in Ghana. Asempa Publishers Ltd., Accra, Ghana, 208 p.

Thomas TA (1991). Catalogue of Okra (*Abelmoschus esculentus* (L.) Moench) germplasm, Part II. NBPGR, New Delhi, 100 p.

Torkpo SK, Danquah EY, Offei SK, Blay ET (2006). Esterase, Total Protein and Seed Storage Protein Diversity in Okra (*Abelmoschus esculentus* (L). Moench). West Afr. J. Appl. Ecol., 9: 0855-4307.

26

DNA content of several bermudagrass accessions in Florida

Wenjing Pang[1]*, William T. Crow[1] and Kevin E. Kenworthy[2]

[1]Entomology and Nematology Department, University of Florida, Gainesville, FL 32611-0620, United States.
[2]Agronomy Department, University of Florida, Gainesville, FL 32611-0500, United States.

Bermudagrass (*Cynodon* spp.) is widely used on golf courses and sport fields in Florida. Most cultivars used on golf courses are triploid bermudagrass (*Cynodon dactylon* [L.] Pers. var. *dactylon* × *C. transvaalensis* Burtt-Davy) resulting from hybridizations between tetraploid common bermudagrass and diploid African bermudagrass. In order to breed and develop bermudagrass cultivars with superior characteristics, it is essential to know the ploidy levels of available bermudagrass accessions and select fertile ones. The objective of this study was to determine the DNA content and ploidy level of bermudagrass germplasm accessions in the University of Florida germplasm collection to aid in future cultivar breeding. Flow cytometry was used to determine the nuclear DNA contents of 48 bermudagrass accessions, and one diploid (2%), 19 triploid (40%), 24 tetraploid (50%), one pentaploid (2%) and three hexaploid (6%) accessions were identified. The range of the nuclear DNA contents was 1.17, 1.38 to 1.61, 1.94 to 2.24, 2.47 and 2.64 to 2.75 pg/2C nucleus^{-1} for the respective ploidy levels. As such, tetraploid and hexaploid accessions could be utilized for future breeding efforts. The triploid accessions could be the results of mutations that have occurred in existing commercial bermudagrass cultivars or from natural hybrids between diploid *C. transvaalensis* and tetraploid *C. dactylon*.

Key words: *Cynodon* spp., flow cytometry, nuclear DNA content, ploidy level.

INTRODUCTION

Bermudagrass (*Cynodon* spp.) is widely distributed in China, India, Africa, Australia, South America and the southern region of the United States (Abulaiti and Yang, 1998; Duble, 2010; Wu et al., 2006). In the United States, it is distributed throughout the warmer regions: from Florida northward to Maryland and New Jersey along the east coast, and westward along the southern border to California (Duble, 2010). In Florida, bermudagrass is one of the most common warm-season grasses. Its improved fine-textured cultivars produce a vigorous and dense turf that is widely used on golf courses, sports fields, lawns and parks (Duble, 2010; Trenholm et al., 2003). There are nine species in the genus, *Cynodon*, and the basic chromosome number is nine (Duble, 2010). Tetraploid *C. dactylon*, common bermudagrass (2n = 4x = 36), is the

most widespread species (de Silva and Snaydon, 1995; Duble, 2010), while *Cynodon transvaalensis* (2n = 2x = 18), African bermudagrass, is a diploid species (Forbes and Burton, 1963; Wu et al., 2006). Triploid *Cynodon* (2n = 3x = 27) from hybridizations of tetraploid *C. dactylon* and diploid *C. transvaalensis* produce fine-textured, dense bermudagrass cultivars that have become the standards for use on golf courses in Florida and other regions where warm season turfgrasses are utilized. Moreover, pentaploid (2n = 5x = 45) and hexaploid (2n = 6x = 54) plants have been previously reported (Burton et al., 1993; Hanna et al., 1990; Johnston, 1975; Kang, 2007; Wu et al., 2006). 'Tifton 10', released as a hexaploid cultivar, has been used on golf courses, athletic fields and home lawns (Hanna et al., 1990).

Flow cytometry (FCM) provides a rapid and accurate DNA content analysis and ploidy level determination for plant breeding programs (Arumuganathan and Earle, 1991; Dolezel et al., 1989; Schwartz et al., 2010). Genome sizes and ploidy levels of warm-season grass

*Corresponding author, E-mail: wpang@ufl.edu.

species, such as buffalograss [*Buchloe dactyloides* (Nutt.) Engelm.], *Paspalum* spp., *Zoysia* spp. and *Cynodon* spp. have been described by using flow cytometry (Jarret et al., 1995; Johnson et al., 1998; Schwartz et al., 2010; Taliaferro et al., 1997; Vaio et al., 2007). Conversely, Arumuganathan et al. (1999) reported the nuclear genome size of diploid, triploid and tetraploid bermudagrass genotypes. Triploid, tetraploid, pentaploid and hexaploid genotypes were identified in Chinese and Korean bermudagrass accessions, respectively, by Kang et al. (2007) and Wu et al. (2006).

However, information about the DNA content or ploidy level of bermudagrass accessions collected in Florida is limited. In order to develop hybrid cultivars with superior turf quality, it is essential to know the ploidy levels of the available bermudagrass genotypes in the University of Florida (UF) germplasm collection. Sterile accessions such as triploids or pentaploids could not be used for future cultivar breeding. The objective of this study was to determine the nuclear DNA content and ploidy level of selected superior UF bermudagrass germplasm accessions, and select fertile accessions for grass breeding.

MATERIALS AND METHODS

Plant materials

Forty-seven *Cynodon* spp. genotypes were selected for having superior turfgrass performance in Gainesville, FL, while three commercial cultivars, 'Tifway', 'Tifgreen' and 'Tifton 10', with known ploidy levels or nuclear DNA contents were included in this test (Table 1). A known diploid African bermudagrass accession, 'AB33', was also included as a reference. Each genotype was vegetatively propagated into 15 cm diam pots filled with 100% USGA specification green sand and was grown in a glasshouse at the University of Florida. The grass was maintained at a temperature range of 24 to 34°C under natural daylight in May, 2009. Grasses were watered for six minutes a day by an overhead automatic irrigation system and fertilized once every other week, using 24N-8P-16K at a rate of 0.5 kg N / 100 m^2 (1 lb N / 1000 ft^2) per growing month. However, grass leaves were clipped once a week except for aerial stolons.

Flow cytometry

Flow cytometry analyses were conducted in the forage evaluation support laboratory (FESL) at the University of Florida on a Partec PA, one-parameter flow cytometer (Partec GmbH, Otto-Hahn-Str. 32, D-48161 Munter, Germany) with a 100-watt HBO short arc lamp emitting UV light at 420 nm to excite fluorescence. The nuclear DNA content was measured by procedures modified from Arumuganathan and Earle (1991). When, at least, 10 aerial stolons were present in each pot (two months after planting the grass), flow cytometry analysis was started. A terminal node and tip from one stolon was removed from each accession and stored on ice before the flow cytometry analysis. A CyStain® PI Absolute P (05-5002, Partec North America, Inc., Mt. Laurel, NJ) nuclei extraction and DNA staining buffer kit was used to prepare samples. Triploid trout erythrocyte nuclei (BioSure® Inc., Grass Valley, CA) with a nuclear

DNA content of 7.2 pg/2C nucleus^{-1} (Hardie and Hebert, 2003; 2004) were used as an internal standard. About 50 mg of fresh nodal or stolon tip tissue from each accession was chopped with a razor blade into tiny pieces on a Petri dish, into which 400 µL of nuclear extraction solution was added. After incubation for one minute, the solution was transferred into a 5-mL test tube through a 50-µm Partec CellTrics® monofil nylon filter. After adding 1.6 mL DNA staining to the solution and incubating for another 10 min under room temperature, five drops of triploid trout erythrocyte nuclei were added into the test tube and were mixed well with the solution. The test tube with the solution was put into the flow cytometer and the DNA content of each plant sample was measured based on at least 10,000 scanned nuclei per sample. For each accession, three replications were measured on three different days. The sample's DNA content was calculated by the following formula:

Sample's nuclear DNA content = [(mean position of sample peak) / (mean position of the control peak)] × DNA content of the control (Arumuganathan et al., 1999).

The mean and standard deviation of the genome size were calculated for each genotype, and the ploidy levels of the genotypes were then determined by the genome size ranges of the respective ploidy levels reported previously (Arumuganathan et al., 1999; Kang et al., 2007; Taliaferro et al., 1997; Wu et al., 2006).

RESULTS AND DISCUSSION

Plant leaves were used for nuclear extraction and staining, but the results were not as consistent as those of terminal nodes or stolon tips. The same problems were observed in *Zoysia* spp. (Schwartz et al., 2010). Plant tissues under stress or with disease were also used; however, they yielded variable results. The peaks were not sharp or consistently clear for some samples. The reason could be that the DNA of the diseased plant tissue could be contaminated with that of the pathogen, and an interaction between them could counteract the peaks of the plant samples, which resulted in the unclear peaks. Therefore, healthy, non-stressed nodal plant tissue should be used for flow cytometry studies. Plants have been used as internal controls for grasses in previous FCM studies. Diploid barley (*Hordeum vulgare* L.), hexaploid wheat (*Triticum aestivum* L.) and tobacco (*Nicotiana tabacum* L) were used to test the nuclear DNA contents of 13 turfgrass species (Arumuganathan et al., 1999). Tetraploid bermudagrass, 'Savannah', was used as an internal standard for bermudagrass in FCM studies because its nuclear content was similar to other genotypes tested (Kang et al., 2007). In this study, we also used triploid bermudagrass cultivar, 'Tifway', with a known DNA nuclear content (Arumuganathan et al., 1999; Wu et al., 2006) as an internal control. However, clear and repeatable peaks were not obtained for all genotypes, especially for those with a nuclear DNA content similar to the control. As such, interactions occur between plant samples and the control, which counteract with the peaks of the samples. Plant control with a DNA content overlapping those of the samples was not a good internal standard in this study. Nonetheless, with a larger nuclear DNA content (7.2 pg/2C nucleus^{-1}), no interaction

Table 1. Nuclear DNA content and ploidy level of 48 bermudagrass accessions and three cultivars from the University of Florida.

Accession	DNA content mean ± SD pg/2C	Inferred ploidy (2n)	Accession	DNA content mean ± SD pg/2C	Inferred ploidy (2n)
102	1.61 ± 0.06	3x	355	2.24 ± 0.12	4x
131	2.02 ± 0.12	4x	445	2.05 ± 0.05	4x
132	2.20 ± 0.12	4x	481	1.94 ± 0.08	4x
157	2.00 ± 0.03	4x	489	1.38 ± 0.16	3x
171	2.09 ± 0.09	4x	490	1.50 ± 0.15	3x
173	1.45 ± 0.08	3x	525	1.54 ± 0.07	3x
227	2.20 ± 0.14	4x	528	2.08 ± 0.16	4x
282	1.49 ± 0.06	3x	PI 289922	2.47 ± 0.07	5x
283	1.54 ± 0.08	3x	PI 290868	2.08 ± 0.06	4x
285	1.48 ± 0.09	3x	PI 290872	2.00 ± 0.03	4x
286	1.49 ± 0.08	3x	PI 290895	1.51 ± 0.02	3x
291	1.47 ± 0.08	3x	PI 291590	1.94 ± 0.09	4x
293	1.52 ± 0.08	3x	UFC03	2.08 ± 0.15	4x
295	2.70 ± 0.01	6x	UFC06	2.19 ± 0.06	4x
296	1.48 ± 0.11	3x	UFC07	2.75 ± 0.09	6x
297	1.63 ± 0.03	3x	UFC11	2.16 ± 0.06	4x
299	1.98 ± 0.05	4x	UFC12	2.06 ± 0.01	4x
301	1.98 ± 0.10	4x	UFC25	1.59 ± 0.05	3x
304	2.16 ± 0.03	4x	UFC26	1.56 ± 0.03	3x
319	1.44 ± 0.02	3x	UFC29	2.00 ± 0.11	4x
320	1.60 ± 0.10	3x	UFC30	1.97 ± 0.06	4x
334	2.03 ± 0.02	4x	Tifway	1.53 ± 0.05	3x
343	2.10 ± 0.06	4x	Tifgreen	1.58 ± 0.05	3x
344	2.64 ± 0.06	6x	Tifton 10	3.06 ± 0.01	6x
347	2.06 ± 0.09	4x	AB33	1.17 ± 0.07	2x
352	1.50 ± 0.01	3x			

was found between trout erythrocyte nuclei and bermudagrass cells in that clear, consistent and repeatable peaks were obtained for all genotypes. Therefore, trout erythrocyte nuclei were a very good internal standard for bermudagrass FCM analysis. Other animal blood cells such as those of Channel catfish (*Ictalurus punctatus*) have also been reported as a good standard for bermudagrass nuclear DNA content measurement (Wu et al., 2006). Using this modified method, the CVs for the peaks of all genotypes were less than 6.0%. The standard deviations of nuclear DNA content ranged from 0.01 to 0.15 pg/2C nucleus[-1], which agreed with previous studies by Arumuganathan et al. (1999), Kang et al. (2007) and Wu et al. (2006) in that flow cytometry was a very precise method for bermudagrass genome size measurement.

Mean nuclear DNA contents and ploidy levels for the 48 bermudagrass accessions and three reference cultivars are presented in Table 1. Likewise, the histograms of the flow cytometry peaks are shown in Figure 1. The peaks of the trout erythrocyte nuclei were relatively smaller than

those of the plant samples because there were fewer cells in the triploid trout erythrocyte nuclei than the plant samples, but this did not affect the results. Clear and consistent peaks were obtained for all genotypes, and the cells in the G2 phase were also observed. However, the nuclear DNA content of the UF bermudagrass accessions varied from 1.38 to 2.75 pg/2C nucleus[-1] (Table 2). These values were lower than those of previously reported studies on Korean (Kang et al., 2007) and Chinese accessions (Wu et al., 2006). Using previously reported nuclear DNA content ranges (Arumuganathan et al., 1999; Kang et al., 2007; Taliaferro et al., 1997; Wu et al., 2006) to infer ploidy levels, UF accessions were classified as follows: one (2%) diploid accession with a genome size of 1.17 pg/2C nucleus[-1], 19 (40%) triploid accessions with a genome size of 1.38 to1.61 pg/2C nucleus[-1], 24 (50%) tetraploid accessions with a genome size of 1.94 to 2.24 pg/2C nucleus[-1], one (2%) pentaploid accession with a genome size of 2.47 pg/2C nucleus[-1] and three (6%) hexaploid accessions with a genome size of 2.64 to 2.75 pg/2C nucleus[-1] (Table 2). Moreover, the

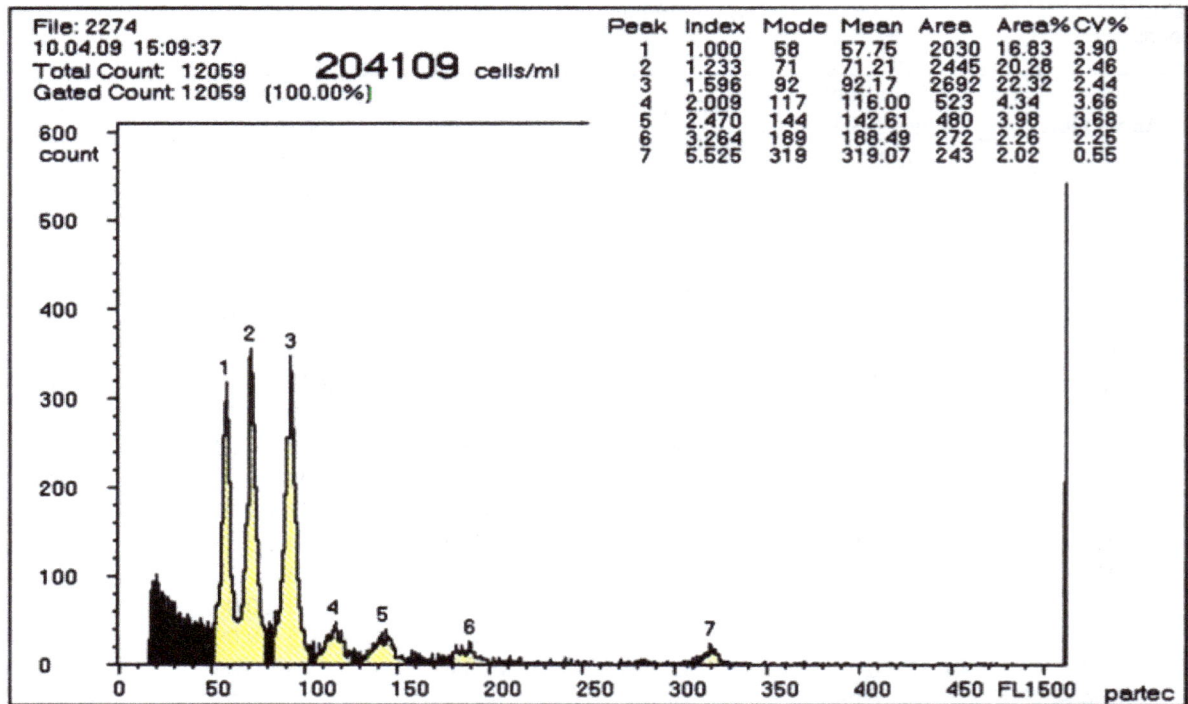

Figure 1. Flow cytometric histogram of diploid, triploid and tetraploid bermudagrass and trout erythrocyte nuclei.
Peak 1 = diploid accession; Peak 2 = triploid cultivar (Tifgreen) control; Peak 3 = tetraploid accession; Peak 4 = G2
phase of diploid accession; Peak 5 = G2 phase of triploid cultivar control; Peak 6 = G2 phase of tetraploid accession;
Peak 7 = trout erythrocyte nuclei.

Table 2. Nuclear DNA content ranges of 48 bermudagrass accessions and three cultivars from the University of Florida.

Ploidy level (2n)	DNA content (pg/2C)	Genotype
2x	1.17	AB33
3x	1.38 to 1.61	102, 282, 283, 285, 286, 291, 293, 296, 297, 319, 320, 352, 489, 490, 525, 290895, UFC25, UFC26, 'Tifway', 'Tifgreen', 'Tifton 10'
4x	1.94 to 2.24	131, 132, 157, 171, 173, 227, 299, 301, 304, 334, 343, 347, 355, 445, 481, 528, PI 290868, PI 290872, PI 291590, UFC03, UFC06, UFC11, UFC12, UFC29, UFC30
5x	2.47	PI 289922
6x	2.64-2.75	295, 344, UFC07

genome sizes of all accessions were previously reported
internally by the ploidy ranges.

The nuclear DNA contents of cultivars 'Tifway',
'Tifgreen' and 'Tifton 10' in this study were very close to
the values previously reported (Arumuganathan et al.,
1999; Wu et al., 2006), which verified the accuracy of this
assay. When compared to the Korean and Chinese
accessions (Kang et al., 2007; Wu et al., 2006), a lower
percentage of tetraploid genotypes were identified among

these superior UF accessions. Relatively, more triploid
accessions (40%) are identified, which are likely mutants
of commercial triploid cultivars. The 48 bermudagrass
accessions evaluated were selected based on their multi-
year performance, from a larger germplasm collection of
180 accessions. The 180 accessions were collected
primarily from managed turf sites in the state of Florida,
and many from golf courses that were likely planted with
a triploid bermudagrass cultivar. Due to the fact that

triploids are known for having superior turfgrass performance, it is very probable that inadvertently collected triploids would have been selected as part of the group of 48 accessions that represent those genotypes with the best overall multi-year performance. If the entire collection of 180 genotypes had been evaluated, it is the percentage of triploids that would have likely reduced and the percentage of tetraploids would have increased.

Conclusions

All tetraploid and hexaploid accessions could be used in bermudagrass breeding. Triploid accessions with good horticultural characters or disease resistance could be released in the future. This study provided valuable information for future bermudagrass cultivar development in Florida.

ACKNOWLEDGMENTS

The authors gratefully appreciate FTGA for funding the study and Mr. Richard P. Fethiere in Forage Evaluation Support Laboratory (FESL) at University of Florida for supporting this study technically. The authors would also like to acknowledge the discussion and help from Dr. Brian M. Schwartz at the University of Georgia and Dr. John E. Luc at the University of Florida.

REFERENCES

Abulaiti DSS, Yang G (1998). A preliminary survey of native Xinjiang bermudagrass. J. Xin j. Agric. Univ., 21: 124–127.

Arumuganathan K, Earle ED (1991). Estimation of nuclear DNA content of plants by flow cytometry. Plant Mol. Biol. Rep., 9: 229-233.

Arumuganathan K, Tallury SP, Fraser ML, Bruneau AH, Qu R (1999). Nuclear DNA content of thirteen turfgrass species by flow cytometry. Crop Sci., 39: 1518-1521.

Burton GW, Gates RN, Hill GM (1993). Registration of 'Tifton 85' bermudagrass. Crop Sci., 33:644–645.

de Silva PHAU, Snaydon RW (1995). Chromosome number in *Cynodon dactylon* in relation to ecological conditions. Ann. Bot. (London), 76: 535–537.

Dolezel J, Binarova P, Lucretti S (1989). Analysis of nuclear DNA content in plant cells by flow cytometry. Biol. Plant., 31: 113-120.

Duble RL (2010). Bermudagrass. The Sports Turf of the South. Available at http://plantanswers.tamu.edu/turf/publications/bermuda.html (verified 14 Sep. 2010). Texas Cooperative Extension, Department of Soil and Crop Sciences, Texas A&M University, College Station, TX.

Forbes I, Burton GW (1963). Chromosome numbers and meiosis in some *Cynodon* species and hybrids. Crop Sci., 3: 75–79.

Hanna WW, Burton GW, Johnson AW (1990). Registration of 'Tifton 10' turf bermudagrass. Crop Sci., 30: 1355–1356.

Hardie DC, Hebert PDN (2003). The nucleotypic effects of cellular DNA content in cartilaginous and ray-finned fishes. Genome, 46: 683-706.

Hardie DC, Hebert PDN (2004). Genome-size evolution in fishes. Can. J. Fish. Aquat. Sci., 61: 1636-1646.

Jarret RL, Ozias-Akins P, Phatak S, Nadimpalli R, Duncan R, Hiliard S (1995). DNA contents in *Paspalum* spp. determined by flow cytometry. Genet. Resour. Crop Evol., 42: 237-242.

Johnson PG, Riordan TP, Arumuganathan K (1998). Ploidy level determinations in buffalograss clones and populations. Crop Sci., 38: 478-482.

Johnston RA (1975). Cytogenetics of some hexaploid × tetraploid hybrids in *Cynodon*. M.S. thesis. Oklahoma State Univ., Stillwater.

Kang SY, Lee GJ, Lim KB, Lee HJ, Park IS, Chung SJ, Kim JB, Kim DS, Rhee HK (2007). Genetic Diversity among Korean Bermudagrass (*Cynodon* spp.) Ecotypes Characterized by Morphological, Cytological and Molecular Approaches. Mol. Cells, 25(2): 163-171.

Schwartz BM, Kenworthy KE, Engelkec MC, Genovesic AD, Odomb RM, Quesenberryb KH (2010). Variation in 2c nuclear DNA content of *Zoysia* spp. as determined by flow cytometry. Crop Sci., 50: 1519-1525.

Taliaferro CM, Hopkins AA, Henthorn JC, Murphy CD, Edwards RM (1997). Use of flow cytometry to estimate ploidy level in *Cynodon* species. Int. Turfgrass Soc. R. J., 8: 385–392.

Trenholm LE, Cisar JL, Unruh JB (2003). Bermudagrass for Florida lawns. Florida Lawn Handbook. Environmental Horticulture Department, Florida Cooperative Extension Service, Institute of Food and Agricultural Sciences, University of Florida, Gainesville, FL.

Vaio M, Mazzella C, Porro V, Speranza P, Lopez-Carro B, Estramil E, Folle GA (2007). Nuclear DNA content in allopolyploid species and synthetic hybrids in the grass genus *Paspalum*. Plant Syst. Evol., 265: 109-121.

Wu YQ, Taliaferro CM, Bai GH, Martin DL, Anderson JA, Anderson MP, Edwards RM (2006). Genetic analyses of Chinese *Cynodon* accessions by flow cytometry and AFLP markers. Crop Sci., 46: 917-926.

27

Development and use of microsatellites markers for genetic variation analysis, in the Namibian germplasm, both within and between populations of marama bean (*Tylosema esculentum*)

M. Takundwa[1], E. Nepolo[1], P. M. Chimwamurombe[1]*, A. C. Cullis[2], M. A. Kandawa-Schulz[4] and K. Kunert[3]

[1]Department of Biological Sciences, University of Namibia, P. Bag 13301, Windhoek, Namibia.
[2]Department of Biology, Case Western Reserve University, Cleveland, Ohio, USA.
[3]Department of Plant Science, University of Pretoria, 0001 South Africa.
[4]Department of Chemistry and Biochemistry, University of Namibia, Windhoek, Namibia.

Tylosema esculentum (marama) has long been identified as a candidate crop for arid and semi-arid environments due to its success in these environments and the high nutritional value of the seed. Molecular markers are essential for the assessment of the levels of genetic variation present within and between populations of marama as well for future marker-assisted breeding efforts. Microsatellites were isolated using a modified FIASCO enrichment technique. Eighty pairs of primers were designed to amplify across a selected set of perfect microsatellite repeats with greater than 5 repeat units. Of the 80 primer pairs screened, 76% were able to detect polymorphism and 21% gave monomorphic bands while the other 3% gave inconsistent results. Four of the polymorphic SSR's were used for genetic variation analysis and have proved to be useful and informative markers for assessing intra-specific and inter-specific variability of marama bean. Heterozygosity (H) within and between populations of marama bean in the Namibian germplasm ranged from 0.30 to 0.74. Some of the populations had low genetic variation while others had high genetic variation.

Key words: *Tylosema esculentum*, microsatellites, FIASCO, polymorphism, genetic variation, heterozygosity.

INTRODUCTION

The marama bean, *Tylosema esculentum* (Burchell.) A. Schreiber, belong to the Fabaceae (Legumionosae) family, sub family Caesalpinioideae (Dubois et al., 1995). It is presently known only in the wild state and is not cultivated. Marama bean is found in South Africa in the western and Northwestern Gauteng and in the Northern part of the Cape Province, Botswana and North-east of Namibia (Bower et al., 1988; Naomab, 2004). Its primary agronomic potential is based upon the high nutritional value of the seeds. The seeds and tubers are edible after roasting and cooking respectively. There is an increasing

interest in its cultivation, due to its potential as a cash crop and food source especially in the face of climate change since it occurs in a very harsh environment.

Plant microsatellites, simple sequence repeats (SSRs), have been developed for germplasm conservation, cultivar identification and for assessing genetic diversity not only in crops such as tomato (He et al., 2003), sweet orange (Novelli et al., 2006), soybean (Rongwen et al., 1995; Akkaya et al., 1992), peanut (Fergurson et al., 2004) and common bean (Gaitán-Solis et al., 2002), but also in perennial plants such as the *Melaleuca* (the tea tree) (Rossetto et al., 1999). Until recently developing SSR markers for new species was a laborious and costly exercise thus limiting their potential application to novel plant species. The development of SSR enrichment techniques, in which selective genomic libraries con-

*Corresponding author. E-mail: pchimwa@unam.na.

Table 1. Location, number of accessions of marama individuals collected and region where the locations are within the Namibian germplasm used for genetic variation analysis.

Location	Number of accession	Region
Omitara	19	Khomas
Otjovanatje	20	Omaheke
Sanveld	21	Omaheke
Harnas	25	Omaheke
Ozondema	26	Otjozondjupa
Epukiro/Post 3	30	Omaheke
Omipanda	31	Omaheke
Osire	40	Otjozondjupa
Ombujondjou	40	Otjozondjupa
Otjiwarongo	40	Otjozondjupa
Okomumbonde	40	Omaheke

taining pre-screened inserts are prepared, has increased the efficiency of SSR characterization in new species (Zane et al., 2002). The availability of such technology opens new opportunities for large scale SSR characterization of species with no previous knowledge such as marama bean. The fast isolation by AFLP of sequences containing repeats (FIASCO) technique is one of such method and a variation of it was used in this study.

This paper describes the isolation of an enriched microsatellite fraction from marama, the sequencing of this enriched fraction using 454 sequencing, the identification of SSR's and the use of Polymerase Chain Reaction (PCR) with primers designed in the flanking sequences to characterize the level of polymorphism within the marama collection assembled (Santana et al., 2009). In this study four of these markers were used to characterize the 332 individual marama plants in the germplasm.

MATERIALS AND METHODS

Distribution of marama bean and sample collection

A total of 332 *T. esculentum* individuals representing 11 populations described in Nepolo et al. (2009) were sampled in the Namibian germplasm (Table 1). The 11 populations were from several localities in the Omaheke, Otjozondjupa and Khomas regions of Namibia. Leaf materials were collected from randomly selected marama plants at each location. Leaf material was stored at -4 °C in the laboratory following field collections.

DNA extraction

DNA was extracted from leaves of each of the plant samples collected from the 11 sampling sites using the manufacturer's protocol for the DNeasy (Qiagen, Valencia, CA, USA) mini-kit for purification of genomic DNA from plant tissue. The DNA concentration was determined on a 1% agarose gel stained with ethidium bromide using known molecular weight standards and stored in clearly labeled microcentrifuge tubes at -20 °C. DNA with a

concentration of 25 – 250 µg/µl. DNA samples were then diluted to 10 ng/µl for PCR amplification.

Microsatellite isolation using a modified FIASCO technique

Marama DNA was enriched for microsatellites by a modified Fast Isolation by AFLP of Sequences Containing microsatellite (FIASCO) technique (Zane et al., 2002) as described below. Nine marama bean microsatellite libraries enriched for different SSR motifs using biotinylated $(AAG)_7$, $(GTT)_7$, $(AGG)_7$, $(GAG)_7$, $(CA)_{10}$, $(CT)_{10}$, $(TCC)_7$, $(CA)_{15}$ and $(CAC)_7$ were created. The steps in the protocol followed are described below:

Restriction enzyme digests and purification

One microgram of DNA was digested with the restriction enzymes, *Msp* 1, H1, *Csp* 6I and *Sau* 3A, as per the supplier's instructions. The digest was then cleaned using the Qiagen PCR purification kit.

Adaptor ligation and amplicon preparation

Adaptors that were used were those designed for representational difference analysis (Oh et al., 2007) so that a wide range of restriction enzymes could be used to fragment the genome. The ligation was performed with 500 ng of the restriction-digested DNA, 1 µl of 12 mer adapter, 1 µl 24 mer adapter and 3 µl ligation buffer in a final volume of 28 µl. The reaction mixture was heated to 72 °C for 3 min then cooled by 1 °/min to 4 °C. 2 µl ligase was added and the reaction was incubated at 4 °C for 16 h. The 12 bp adaptors were removed by heating to 72 °C for 5 min to melt off the 12 mer followed by purification using the QIAquick PCR purification kit. Next, the ligated DNA was amplified with PCR by combining 5 µl PCR buffer, 5 µl (20 mM) MgCl2, 4 µl (10 mM) dNTPs, 2 µl adapter (100 µM), 34.75 µl water, and 1 µl ligated DNA. The reaction was heated at 72 °C for 5 min, 5 units of *Taq* polymerase enzyme was added and incubated for 5 min at 72 °C. The DNA was amplified for 20 cycles of 95 °C for 30 s and 72 °C for 90 s with a final hold at 72 °C for 5 min. Following the PCR, another QIAquick PCR cleanup was performed.

SSR enrichment

The SSR enrichments were repeated four times (with each enzyme of the preparation steps) with $(AAG)_7$, $(GTT)_7$, $(AGG)_7$, $(GAG)_7$, $(CA)_{10}$, $(CT)_{10}$, $(TCC)_7$, $(CA)_{15}$ and $(CAC)_7$ biotinylated primers. First the amplified digest was denatured and annealed to the biotinylated primer by combining 20 µl PCR product (200 µg) and 1 µl primer (10 µM) and heating at 95 °C for 5 min, followed by incubation at room temperature for 30 min. Before combining the primed DNA with streptavidin beads, 10 µl of unrelated DNA (sheared herring sperm at 1 mg/ml) was added to minimize non-specific binding. The annealed DNA mixture was then added to 1 mg of magnetic beads and incubated for 30 min at room temperature, allowing the streptavidin beads to bind to the biotinlyated primers. Five washes with TEN100 (Tris/EDTA/NaCl) followed by 5 washes with 0.2X SSC, 0.1% SDS were performed to remove non-specific DNA binding. Next, two denaturation steps were performed to separate DNA containing SSRs from the beads. The first denaturation was done by adding 50 µl of TE (Tris-HCl 10 mM, EDTA 1 mM) and heating at 95 °C for 5 min. The remaining solution was separated magnetically and stored. The second denaturation used 12 µl 0.5 N NaOH, which was neutralized with 12 µl 0.5 N HCl and separated magnetically. Each denaturation product (2 per enzyme) was amplified separately with PCR by adding 5 µl PCR buffer, 3 µl $MgCl_2$ (25 mM), 4 µl dNTPs (each 2.5 mM), 2 µl adapter (10 µM, 34.75 µl water, 0.25 µl r*Taq* (2 units) and 1 µl DNA into a PCR tube. The mixture was cycled 20 times from 95 to 72 °C. Each of the amplified denaturation products were separated on a 1.5% agarose gel.

Sequencing

All the microsatellite-enriched genomic DNA pools were combined and a total of 5 µg of the extracted microsatellites were sequenced and analyzed with the Roche 454 GS-FLX platform at Inqaba Biotech (Pretoria, Gauteng, South Africa). Sample preparation and analytical processing such as base calling, were performed at Inqaba Biotech using the manufacturer's protocol as described previously (Santana et al., 2009). The sequence information gathered was sorted into contig and single read files that were used for SSR identification in the sequences and subsequently in primer design.

SSR identification and primer design

The Simple Sequence Repeat Identification Tool (SSRIT) software SSRIT, (http://www.gramene.org/db/markers/ssrtool) was used to identify the microsatellites in the contig files obtained from 454 sequence data. The search parameters were for SSRs from dimers to pentamers with a minimum of 5 exact repeats required for each type of microsatellite. Primers were then developed in the flanking region around the SSR sites using Primer 3 (http://frodo.wi.mit.edu/primer3) and the primers were synthesized by Inqaba Biotech. Eighty microsatellite primers pairs were then used for amplification of marama DNA from the 11 different Namibian sites as well as 1 location in South Africa (Pretoria). Each microsatellite primer marker was given a name consisting of the prefix "MARA" followed by a number (001 - 080).

SSR primer screening

Each of the primer sets was initially used to screen individual marama DNAs from 19 individuals from the Omitara population. PCR amplifications were performed in 25 µl reaction volumes, with a 2X PCR master mix from Fermentas. Each PCR reaction contained 1 µl template genomic DNA, 1 µl (1 µM) of SSR forward primer, 1 µl of SSR reverse primer, 12.5 µl of the 2X PCR master mix and 9.5 µl nuclease free water. The PCR reaction profile used

involved an initial denaturation step of 95 °C for 4 min, followed by 35 cycles of denaturation at 95 °C for 30 s, an annealing at between 55 and 65 °C (primer sequence dependent) for 60 s and an extension at 72 °C for 2 min, a final extension at 72 °C for 5 min and then held at 4 °C. Agarose gel (2%) visualization of PCR products was then used to determine if a primer pair was polymorphic or monomorphic based on its separation of amplification products and banding patterns generated on the agarose gels in the different DNA templates. A total of 80 microsatellite primers were screened and described as polymorphic, monomorphic or unable to amplify. Four of the 80 microsatellites were then used for diversity analysis of the 332 individual marama bean plants from the Namibian germplasm.

Genetic diversity analysis

Polymorphic primers MARA 001, MARA 065, MARA 068 and MARA 077 were used to determine genetic variation in each of the 12 ecotypes using the same reaction profile and conditions described in the screening for polymorphic microsatellites. Agarose gels (2.5%) were used to visualize amplification products. The DNA bands obtained were scored with 1 denoting presence of a band and 0 absence of a band, generating a binary data matrix and analysed using Primer 5 (Version 5.2.0) software.

RESULTS

Characterization of the 80 SSR primers

Microsatellite loci were isolated from the marama bean germplasm using the modified FIASCO enrichment technique and used to design 80 microsatellite primers based on perfect microsatellites. The primers (Table 2) were screened using the Omitara sub-population. Microsatellite loci were described (Table 3) as: Group1-Monomorphic (for example *MARA 004)*; Group 2-Polymorphic in typical SSR type pattern (for example *MARA 001)*; Group 3-Polymorphic but not SSR type pattern (for example *MARA 045)*; Group 4-Not amplifying some individuals or poor quality amplification products (for example *MARA 067)*.

Of the 80 primers screened, 76% were able to detect polymorphism (Group 2 and Group 3) yet 21% of them gave monomorphic bands (Group 1). The remaining 2 primers out of the 80 primers did not give clearly scorable amplifications or they gave no product at all. Four primer pairs (MARA 001, MARA 065, MARA 068 and MARA 077) out of the 80 primer pairs screened that gave reproducible polymorphic patterns were used for analysis of genetic variation of marama bean.

Characterization of individuals from 11 populations using 4 primer pairs

Primer pairs: MARA 001, MARA 065, MARA 068 and MARA 077 were used to amplify and detect polymorphism in all 11 populations of the Namibian germplasm. The properties of the four primer pairs used are given in Table 4. Genetic variation in each of the populations was

Table 2. The 80 microsatellite loci and primers designed to screen for polymorphic microsatellite loci in this study.

	Marama microsatellites designed and screened for polymorphism (80)					
Primer	Left primer (L)	Right primer (R)	SSR	Repeats	Annealing (L/R)	Expected product (bp)
MARA001	GCACAACCAATTTCCTGCTT	TCCCTCACTGGCCTATATCC	gag	5	60.12/58.96	137
MARA002	CTCCCTCCTCCTCCTCGTAG	CGGGAGCAAATAGACCCTTT	acc	8	60.34/60.44	106
MARA003	TCTCACCGACCGGGTCTC	CCTCTATCCGCTCCCTATC	ctc	5	62.30/60.02	160
MARA004	TGCAGGCTTCACCAGAGTAA	TCTAAGACTGCGCACACAGC	ga	5	59.59/60.36	170
MARA005	GCTATCCGAGGGAGGATCA	GTGTCTATGTGTGCGCGTGT	ag	6	60.13/60.82	128
MARA006	GCTATCCGAGGGAGGATCA	TCCCATTCAGCCATTTAGG	tg	7	60.13/59.89	171
MARA007	TATCCGAGGGAGGGATCATGT	TCACATCCTAAGACTCGAACTTCA	ac	6	59.32/60.29	150
MARA008	GCTGGTCCATGGCTTCAT	TTTGTAATCGGTTGACACTTTGA	tg	5	59.59/59.54	185
MARA009	GGGAGGATCAACCTCAACAA	TGTACAAAAGCAGGCTCCA	gaa	5	59.90/59.46	216
MARA010	TGTGCTATCCGAGGGAGGAT	ACGTCGCGATTAAACAAACC	aag	7	61.92/60.00	152
MARA011	TGTCAACGCTTACGTTGGTC	TCATTTGAAACCCTTGTACTGC	tc	8	59.76/59.13	169
MARA012	ATATGGTGGCTCGTCGATGT	GCACATAATTCGAACAGAACACA	ag	5	60.37/60.05	163
MARA013	GCTTCTCGTACATGGGCTTT	GCATTTATCGCGAATACAGCA	tc	5	59.34/61.10	154
MARA014	GGTGGTGGTGTAGGGAGGAGA	GACTTGAGTGCATGCCATTT	agg	5	59.96/58.73	167
MARA015	ACTCATCCCGCTCCTAAGGT	AAACAGGCTCGTATTTTATCTTCG	tg	5	60.10/60.05	204
MARA016	TTCAATTTCTTCACCACAAACTC	ACAGGAAGGTCTTCCACAGC	ca	7	59.56/59.30	102
MARA017	ACCCTTGAATTGTGGTAGGG	ACTGTGCTATCCGAGGGAGA	ct	6	58.76/59.83	105
MARA018	ATTTTGGCTTTACCGCACAC	AGCACTCTCCAGCCTCTCAC	cttga	3	60.00/59.74	158
MARA019	CCGGAACAGGAGAAGCTATG	TCAACTTTGCAATGAACGAA	ctt	4	59.83/59.33	161
MARA020	TGTCTTCCCCTCCTTCCT	TTGACACTTTGGGACTGCTG	cag	4	60.19/59.87	175
MARA021	GAGGGAGGATCACCACTCAG	TGGCCATCAATCATGTTACG	tgt	4	59.64/60.34	166
MARA022	CCCCTGTACCCAAGACTCTG	TCCATGAAGTCAGGAGAAGGA	tagc	3	59.57/59.79	171
MARA023	ATGGGGATACTCCCGAAACT	AATGGGAGCAAGAATTTCCA	aaag	3	59.65/59.50	250
MARA024	CCAAGAGTGGGGGATGAAAGA	TTGGAATAGTTCCCCCTTCC	aga	4	60.04/60.12	227
MARA025	CACGTGGGTTGTACTTATCTGC	TAATGTGTTGAGCGCCGTAG	tcttc	3	59.56/59.90	160
MARA026	GCTGTTGGGAACCGTAGAAA	CCTATTGTCAGTGTAAGCAACCA	tc	4	60.11/59.21	208
MARA027	TTGTTCCAAACCACCAGGTCA	TGGCCATCTCCCAATTTTAC	ca	4	59.98/59.76	194
MARA028	CTCCGCATCTGACTTCAAAA	CCTCCTCCCTGATTTCC	aga	3	59.00/60.01	150
MARA 029	CCGAGGGAGTAGTGCTTCAT	CGCCACTTAGCATTTGCTTT	tg	4	59.31/60.40	155
MARA030	GAGCCAAAGCCATGATCCTA	CCCATGTTGTATATTCGTGGAA	caa	3	60.18/59.59	178
MARA031	CTCAGCACTCTCCAGCCTCT	CCGAGGGAGGATCATTAACA	gga	7	59.88/59.89	126
MARA032	AGACGCACTCCCTCTCACC	GCTATCCGAGGGAGGGATCA	aca	5	60.41/60.13	151
MARA033	GCACTCAGGCAACTGTGCTA	AGCACTCTCCAGCCTCTCAC	aac	7	60.21/59.74	132
MARA034	CTCAGCACTCTCCAGCCTCT	AGGGAGGAGGATCACCTCCAAAC	gag	8	59.88/60.31	168

Table 2. Contd.

MARA035	GACGCACTCAGCAACTCTCC	TCCAGCCTCTCACCGATTAC	acc	8	61.17/60.22	152
MARA036	GACGCACTCAGGCAACTGT	CCGAGGGAGGATCAAAGAAT	gga	11	60.05/60.40	180
MARA037	GGGAGGATCAATCTTCACCA	TCCGAGAGAAGAGGAGGAAA	gagt	5	59.86/59.09	165
MARA038	TGTTGATGAACTAGTGCTAGTGGT	AGCACTCTCCAGCCTCTCA	tgg	7	58.02/58.79	126
MARA039	TCATTAAAGGGCTCCATTGC	ATGCCCAAAATCACCAACAT	aga	7	60.04/60.06	176
MARA040	GACGCACTCAGGCAACTGT	CTGGCCTATATCCCCTCCTC	gga	8	60.05/59.88	192
MARA041	AGACGCACTCAGCACTCTCC	GCTATCCGAGGGAGGATCAC	gga	7	60.77/60.96	150
MARA042	CAAATAGCCAAAGCCCGTTA	ACTCTCAAACCGTGGCACAT	agg	7	60.09/60.58	184
MARA043	TGTTGATGAACTAGTGCTAGTGGT	AGCACTCTCCAGCCTCTCA	tgg	7	58.02/58.79	126
MARA044	AGACGCACTCAGCATTCTCC	GGTCTCGTCTTCCCCTTCAT	gag	8	60.56/60.46	156
MARA045	GACGCACTCAGGCAACTGT	CTGGCCTATATCCCCTCCTC	gga	8	60.05/59.88	192
MARA046	GCACTCAGGCAACTGTGCTA	TGGCTGGCACTCTGATTAAG	cta	10	60.21/59.02	169
MARA047	GCACTCAGGCAACTGTGCTA	TGACTAGTCCCGTGATGGT	caa	7	60.21/60.39	188
MARA048	AGACGCACTCCACCACTGTA	TGCTGAAACCGTGAGAGAGA	ct	12	59.34/59.70	212
MARA049	GCACTCAGGCAACTGTGCTA	GGCGAACTAGTGCTATCGAG	ct	13	60.21/57.75	183
MARA050	AGACGCACTCAGCACTCTCC	TGTGCTATCCGAGGGAGGAT	cac	7	60.77/61.92	116
MARA051	GCACTCAGGCAACTGTGCTA	AGCCTCTCACCGATTACTGC	ca	15	60.21/59.46	139
MARA052	GCACTCAGGCAACTGTGCTA	CACGCCTCTCACAAGAAACA	ct	14	60.21/60.02	249
MARA053	CTCAGCACTCTCCAGCCTCT	CCCTCATCTCCCTTTCCTTC	gga	9	59.88/60.01	238
MARA054	GCACTCAGGCAACTGTGCTA	AGCACTCTCCAGCCTCTCAC	gtt	8	60.21/59.74	159
MARA055	GACGCACTCAGCAACTCTCC	TCCAGCCTCTCACCGATTAC	acc	8	61.17/60.22	150
MARA056	AGACGCACTCAGGCAACTGT	ATCGGAGGAGGATCATTAAA	gga	8	61.07/59.75	150
MARA057	GACGCACTCAGGCAACTGT	TGAAGATCCTCCCTCGGATA	gga	11	60.05/59.58	171
MARA058	GCACTCAGGCAACTGTGCTA	ACGACGAACGTAGTCGTCTC	ca	20	60.21/57.96	246
MARA059	GACGCACTCCTGTGTATCC	ACGTCGCGATTAAACAAACC	aag	7	60.83/60.00	162
MARA060	CTCAGCACTCTCCAGCCTCT	CCTTCGTGTTTTACAGTTGTCG	gtg	11	59.88/59.71	179
MARA061	GAGGGAGGATCAAGGACAC	AGCACTCTCCAGCCTCTCAC	ctc	7	60.86/59.74	183
MARA062	GTGCACTCCTGTGTATCC	TGCGCAAGGACAATGATTAC	aag	7	60.83/59.69	202
MARA063	GTGCAAGACCCGTTTAGGAA	AGGACGAACACGTGCGTATC	ct	9	60.11/61.13	185
MARA064	GGAGGAGGAGGAGGAGTTTG	GAGGATCCACTCCCTCACTG	gag	5	60.19/59.64	192
MARA065	TGGTGGTAGGGTGGTGGTAT	CCACTTTTCACAGGCAAACA	ttc	6	59.97/59.73	191
MARA066	GCACTCAGGCAACTGTGCTA	GCTATCCGAGGGAGGAGAGA	cct	5	60.21/60.83	161
MARA067	AGACGCACTCAGCCTCTCAC	CCTCTATCCCGGTCCCTATC	ctc	5	60.77/60.02	174
MARA068	GGAGGAGGAGGAGGAGTTTG	GAGGATCCACTCCCTCACTG	gag	5	60.19/59.64	192
MARA069	GGGAGGGATCAACCTCAACAA	TGTACAAAAGCAGGCTCCA	gaa	5	59.90/59.46	216
MARA070	GACGCACTCAGGCAACTGT	GGGAGGGATCACTTCCACTCTC	gga	5	60.05/60.07	168

Table 2. Contd.

MARA071	CTCAGCACTCTCCAGCCTCT	CCTGTTGGGGGAGTTGTTGTT	cac	6	59.88/59.86	202
MARA072	CTCAGCACTCTCCAGCCTCT	GATTGCTGTTGTTGGCAGTG	caa	6	59.88/60.31	225
MARA073	AGCACTCTCCAGCCTCTCAC	ATGTTGAGGCAGAGGAGGAA	cat	5	59.74/59.80	161
MARA074	CCAGCCTCTCAACCGATTAC	AGGCACAGCCCTAGACTCCT	aag	6	59.69/60.41	250
MARA075	GGTGGTGGTGTAGGGAGGAGA	TCCAGCCTCTCACCGATTAC	agg	5	59.96/60.22	202
MARA076	ATTTTGCATCAGCAACAGC	ATCCGAGGGAGGATCCTATG	aca	5	58.92/60.25	193
MARA077	CTCAGCACTCTCCAGCCTCT	GGGTTGGTTGAAGAGGGAGT	aag	5	59.88/60.35	197
MARA078	GCACTCAGGCAACTGTGCTA	GTTCACCATCCCCTCTCTGA	gct	5	60.21/60.05	223
MARA079	GACGCACTCCAACTGTGCTA	TCATTTGAAACCCTTGTACTGC	tc	8	60.06/59.13	238
MARA080	ACCGACCGAGAGAATGAAGA	GAGTCCTCAACAGGGAGCTG	atc	6	59.80/59.99	153

calculated as heterozygosity for the 2 micro-satellite loci MARA 001 and MARA 068 as these 2 loci gave the characteristic diploid type profile enabling calculation of heterozygosity (Figures 1 and 2). Out of 20 individuals from Otjovanatje amplified with primer pair MARA001, ten (50%) were heterozygous (Figure 1). Comparatively, eleven individuals (55%) out of 20 individuals from Otjovanatje amplified using primer pair MARA068 were heterozygous (Figure 2). These two primer pairs gave between one and two bands (alleles) with dominant upper bands in all 11 populations. The calculation of heterozygosity for the two microsatellite loci is tabulated in Table 5.

Analysis of SSR amplicons of primer pair (MARA065)

Primer pair "MARA065" gave many separation. This profile pattern was observed in Harnas as well as the other 10 populations amplified using the same primer pair. Individual 1,2,10 and 19 from Harnas gave three distinctive bands, while individual 20, 22 and bands with minimum 25, two bands each. Other individuals such as 4, 6 and 12 bands each.

gave bands that cannot be clearly distinguished (Figure 3).

Analysis of SSR amplicons of primer pair "MARA077"

Amplification of all 11 populations using primer pair "MARA077" gave polymorphic bands (alleles) which ranged from one to four bands among individuals of marama bean. The amplicons obtained with primer pair MARA 077 shows smears together with bands for certain individuals. For instance, in individuals 1, 4 6, 9, 15 and 21 smears together with associated bands were observed (Figure 4).

DISCUSSION AND CONCLUSION

The isolation and use of microsatellite markers for genetic variation analysis in *T. esculentum* in this study was successful. Polymorphisms for both length band intensities were observed. From the 80 microsatellites primers developed, 5% of the SSR's were used for genetic variation analysis

and these SSR primers have proved that microsatellites are useful and informative markers for assessing intra-specific and inter-specific variability of marama bean populations as in other legumes like soybean (Rongwen et al., 1995). The information reported here is the first for the use of perfect microsatellites whose primers were designed for genetic variation analysis.

Microsatellite markers have been used to assess genetic diversity in large numbers of cultivars in legumes like soybean (Rongwen et al., 1995) and the number of alleles amplified was 11 to 26 (Akkaya et al., 1992). In the present study one to six alleles per primer pair were amplified from the 332 marama bean individuals at the four polymorphic loci. Heterozygosity was used as an estimate of variability within and between populations of marama bean in the Namibian germplasm and ranged from 0.30 to 0.74. It is a good measure of variation because it estimates the probability that two alleles taken at random from a population are different (Mason et al., 2005). Some of the populations had low genetic variation while others had high genetic variation. This suggests genetic variation within as well as between populations of marama bean is high.

Table 3. Characterization of 80 microsatellite primers screened.

Number of primers screened	Number of Group 1 primers	Number of Group 2 primers
80	17	17

Number of Group 1 primers

Number of Group 2 primers

Number of group 3 primers

44

Number of group 4 primers

2

In the group 2 type of loci, usually 2 bands for any SSR were observed suggesting a single copy region with 2 alternative forms (Figure 1). The distribution of alleles within populations was not as expected in random outcrossing populations that started out with equal numbers of the 2 forms of the alleles at each locus. The founding population in such cases could have been skewed towards the presence of a particular allele that led

Table 4. List of selected primer pairs used to detect polymorphism and for the analysis of genetic variation in 11 marama populations of the Namibian germplasm.

Primer	Sequence 5'→ 3'	Repeat	Expected PCR product size (bp)
MARA001	L - GCACAACCAATTTCCTGCTT R – TCCCTCACTGGCCTATATCC	$(gag)_5$	137
MARA065	L - TGGTGGTAGGGTGGTGGTAT R – CCACTTTTCACAGGCAAACA	$(ttc)_6$	191
MARA068	L - GGAGGAGGAGGAGGATTTG R – GAGGATCCACTCCCTCACTG	$(gag)_5$	192
MARA077	L- CTCAGCACTCTCCAGCCTCT R - GGGTTGGTTGAAGAGGGAGT	$(aag)_5$	197

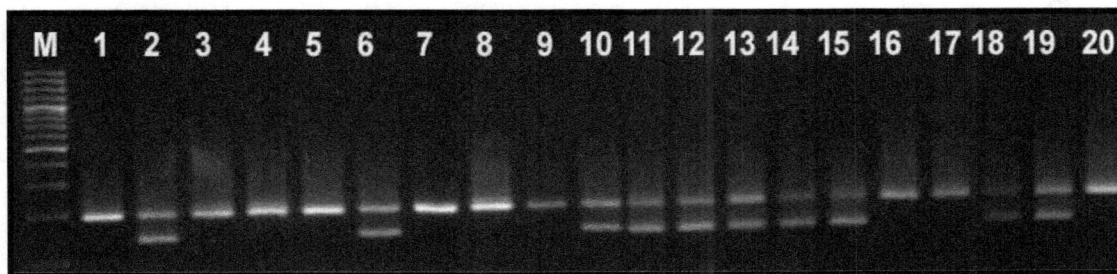

Figure 1. A 2.5 % agarose gel electrophoresis of amplification products 20 marama bean individuals from Otjovanatje using primer pair MARA001, M indicates the DNA molecular size marker (O'gene ruler 100 bp).

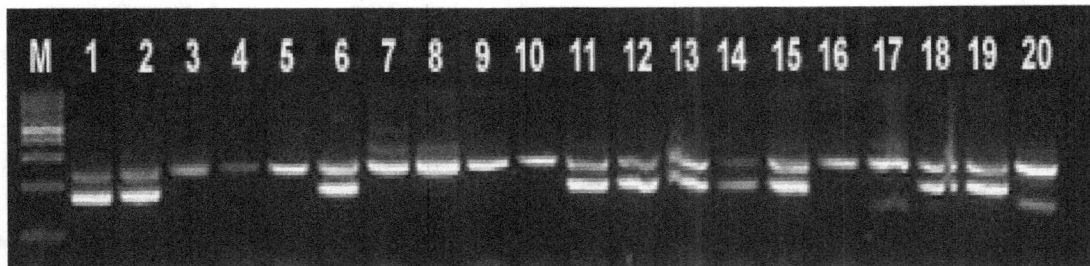

Figure 2. A 2.5% agarose gel electrophoresis of amplification products 20 marama bean individuals from Otjovanatje using primer pair MARA068 , M indicates the DNA molecular size marker (O'gene ruler 100 bp).

to a founder effect or these observed dominant type bands could be linked to a trait that is advantageous in the environment it is occurring in so the allele type has an increased frequency. Furthermore, in some cases observed bands were found to be either more than 30 bp above or below the expected SSR band size. It is not yet known whether such observed patterns were caused by SSRs or by insertions that have nothing to do with SSRs. Experiments to investigate this question through sequencing amplicons need to be initiated.

Microsatellite primers are now available for use as markers in the breeding of marama bean and in conservation efforts. It is desirable to isolate and characterize more DNA markers in marama bean for, more productive genetic studies such as genetic mapping, molecular marker assisted selection and gene discovery. Therefore the development of microsatellite markers for marama bean holds a promise for such studies. The primers can also be tested and used across species amplification of closely related legume species.

ACKNOWLEDGEMENTS

The work described in this study was supported by

Table 5. Heterozygosity and average heterozygosity at two loci (*MARA 001* and *MARA 068*) in populations of the Namibian marama bean germplasm as determined by electrophoresis.

Population	Number of Individuals			Heterozygosity		Average heterozygosity at two loci (MARA 001 and MARA 068)
	Heterozygotes (MARA 001)	Heterozygotes (MARA 068)	Total	Locus MARA 001	Locus MARA 068	
Omitara	8	11	19	0.42	0.58	0.50
Otjovanatje	10	11	20	0.50	0.55	0.53
Sandveld	9	7	21	0.43	0.33	0.38
Harnas	10	12	25	0.40	0.48	0.44
Ozondema	12	12	26	0.46	0.46	0.46
Epukiro/Post 3	13	16	30	0.43	0.53	0.48
Omipanda	19	27	31	0.61	0.87	0.74
Osire	14	20	40	0.35	0.50	0.43
Ombujondjou	17	37	40	0.43	0.93	0.68
Otjiwarongo	16	37	40	0.40	0.93	0.67
Okomumbonde	10	14	40	0.25	0.35	0.30
Average				0.43	0.51	0.51

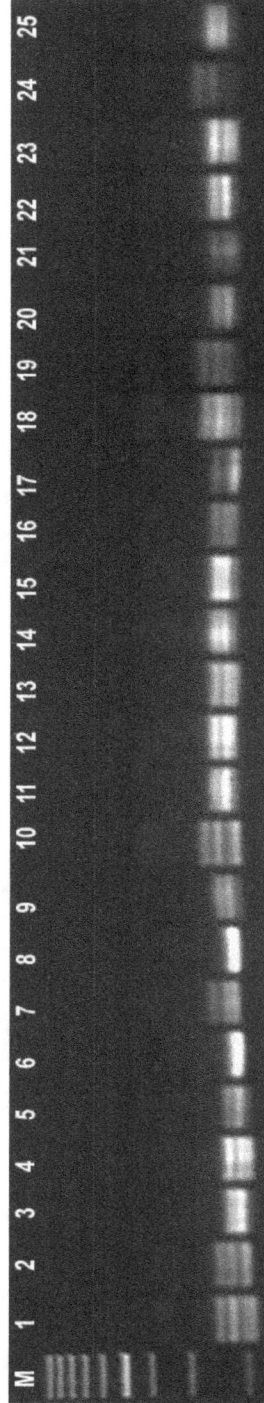

Figure 3. Amplification of individuals from Harnas obtained by the primer pair (MARA065) and visualized on a 2% agarose gel, M indicates the DNA molecular size marker (O'gene ruler 100 bp).

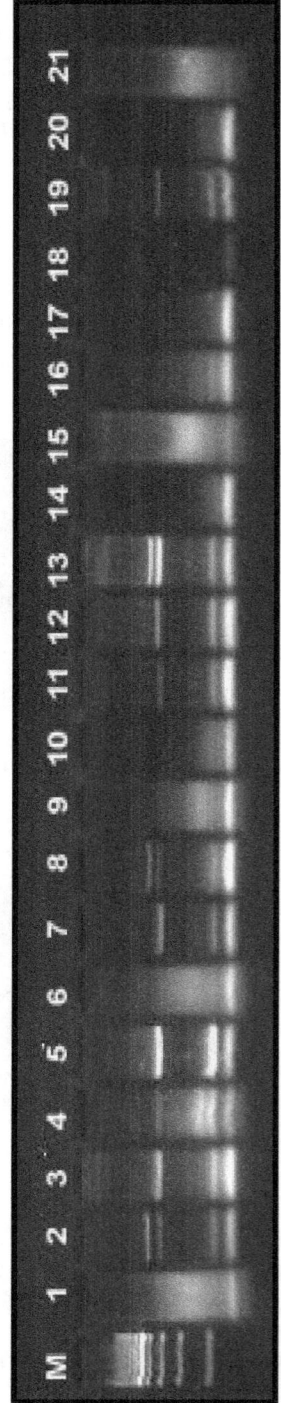

Figure 4. Amplification of individuals from sandveld obtained by the primer pair (MARA077) and visualized on a 2% agarose gel, M indicates the DNA molecular size marker (O'gene ruler 100 bp).

funding from the Kirkhouse Trust, United Kingdom (to PC) and by a World-Wide Learning Experience grant, a McGregor Fund initiative in the College of Arts and Sciences (to CAC). We would also like to acknowledge Mr. J.D. Uzabakiriho for his never ending support in the field and laboratory throughout this study.

REFERENCES

Akkaya MS, Bhagwat AA, Cregan PB (1992). Length polymorphisms of simple sequence repeat DNA in soybean. Genetics. 132: 1131-1139.

Bower N, Hertel K, Oh J, Storey R (1988). Nutritional evaluation of Marama bean (*Tylosema esculentum*, Fabaceae): analysis of seed. Econ. Bot., 42: 533-540.

Dubois M, Lonay G, Baudart E (1995). Chemical characterization of *Tylosema fassoglensis* (Kotschy) Torre & Hillc Oilseed. J. Food Sci. Agric., 67: 163-167.

Ferguson ME, Burow MD, Schulze SR (2004). Microsatellite identification and characterization in peanut (*A. Hypogea* L.). Theor. Appl. Genet., 108: 1064-1070.

Gaitán-Solis E, Duquqe MC, Edwards KJ, Tohme J (2002). Microsatellite repeats in Common Bean (*Phaseolus vulgaris*): Isolation, characterization and cross-species amplification in *Phaseolus* ssp. Crop Sci., 42: 2128-2136.

He C, Poysa V, Yu K (2003). Development and characterization of simple sequence repeat (SSR) markers and their use in determining relationships among *Lycopersicon esculentum* cultivars. Theor. Appl. Genet., 106: 363-373.

Mason SL, Stevens MR, Jellen EN, Bonifacio A, Fairbanks DJ, Coleman CE, McCarty RR, Rasmussen AG, Maughan PJ (2005). Development and use of microsatellite markers for germplasm characterization in Quinoa (*Chenopodium quinoa* Willd.). Crop Sci., 45: 1618-1630.

Naomab E (2004). Assessment of genetic variation in natural populations of Marama bean *(Tylosema esculentum)* using molecular markers. Unpublished MSc. Thesis, University of Namibia, Windhoek, Namibia.

Nepolo E, Takundwa M, Chimwamurombe P, Cullis CA, Kunert K (2009). A review of the geographical distribution of marama bean [*Tylosema esculentum* (Burchell) Schreiber] and genetic diversity in the Namibian germplasm. Afr. J. Biotechnol., 8: 2088-2093.

Novelli VM, Cristofani M, Souza AA, Machado MA (2006). Development and characterization of polymorphic microsatellite markers for the sweet orange (*Citrus sinensis* L. Osbeck). Genet. Mole. Biol., 29: 90-96.

Oh TJ, Cullis MA, Kunert K, Engelborghs I, Swennen R, Cullis CA (2007). Genomic changes associated with somaclonal variation in banana (*Musa* spp.). Physiol. Plant, 129: 766-774.

PRIMER 5 (Plymouth Routines in Multivariate Ecological Research) for windows (2001) Version 5.2.0. Primer-E ltd, UK.

Rongwen J, Akkaya MS, Bhagwat AA, Lavi U, Cregan PB (1995). The use of microsatellite DNA markers for soybean genotype identification. Theor. Appl. Genet., 90: 43-48.

Rosseto M, Melauchan A, Harris FCL (1999). Abundance and polymorphism of microsatellite markers in the tea tree (*Melaleuca alternifolia*, Myrtaceae). Theor. Appl. Genet., 98: 1091-1098.

Santana QC, Coetzee MPA, Steenkamp ET (2009). Microsatellite discovery by deep sequencing of enriched genomic libraries. Biotechniques, 46: 217-223.

Zane L, Bargelloni L, Pattarnello T (2002). Strategies for microsatellite isolation: A review. Mol. Ecol., 11: 1-16.

Permissions

All chapters in this book were first published in JPBCS, by Academic Journals; hereby published with permission under the Creative Commons Attribution License or equivalent. Every chapter published in this book has been scrutinized by our experts. Their significance has been extensively debated. The topics covered herein carry significant findings which will fuel the growth of the discipline. They may even be implemented as practical applications or may be referred to as a beginning point for another development.

The contributors of this book come from diverse backgrounds, making this book a truly international effort. This book will bring forth new frontiers with its revolutionizing research information and detailed analysis of the nascent developments around the world.

We would like to thank all the contributing authors for lending their expertise to make the book truly unique. They have played a crucial role in the development of this book. Without their invaluable contributions this book wouldn't have been possible. They have made vital efforts to compile up to date information on the varied aspects of this subject to make this book a valuable addition to the collection of many professionals and students.

This book was conceptualized with the vision of imparting up-to-date information and advanced data in this field. To ensure the same, a matchless editorial board was set up. Every individual on the board went through rigorous rounds of assessment to prove their worth. After which they invested a large part of their time researching and compiling the most relevant data for our readers.

The editorial board has been involved in producing this book since its inception. They have spent rigorous hours researching and exploring the diverse topics which have resulted in the successful publishing of this book. They have passed on their knowledge of decades through this book. To expedite this challenging task, the publisher supported the team at every step. A small team of assistant editors was also appointed to further simplify the editing procedure and attain best results for the readers.

Apart from the editorial board, the designing team has also invested a significant amount of their time in understanding the subject and creating the most relevant covers. They scrutinized every image to scout for the most suitable representation of the subject and create an appropriate cover for the book.

The publishing team has been an ardent support to the editorial, designing and production team. Their endless efforts to recruit the best for this project, has resulted in the accomplishment of this book. They are a veteran in the field of academics and their pool of knowledge is as vast as their experience in printing. Their expertise and guidance has proved useful at every step. Their uncompromising quality standards have made this book an exceptional effort. Their encouragement from time to time has been an inspiration for everyone.

The publisher and the editorial board hope that this book will prove to be a valuable piece of knowledge for researchers, students, practitioners and scholars across the globe.

List of Contributors

Bnejdi Fethi
Laboratoire de Génétique et Biométrie, Département de Biologie, Faculté des Sciences de Tunis, Université Tunis, El Manar, Tunis 2092, Tunisia

El Gazzah Mohamed
Laboratoire de Génétique et Biométrie, Département de Biologie, Faculté des Sciences de Tunis, Université Tunis, El Manar, Tunis 2092, Tunisia

Leonardo Anabalón Rodríguez
Laboratorio de Mejoramiento Vegetal, Escuela de Agronomía. Universidad Católica de Temuco. Temuco. Chile

Max Thomet Isla
CET Sur. Temuco. Chile

P. A. Sofi
Directorate of Research, SKUAST-K, Shalimar, 191121, J and K, India

Shafiq A. Wani
Directorate of Research, SKUAST-K, Shalimar, 191121, J and K, India

A. G. Rather
Rice Research Station, SKUAST-K, Khudwani, India

Shabir H. Wani
Department of Plant Breeding, Genetics and Biotechnology, PAU, Ludhiana, India

Tan FeiQuan
State Key Laboratory of Plant Breeding and Genetics, Sichuan Agriculture University, Ya'an, Sichuan 625014, China

Fu ShuLan
State Key Laboratory of Plant Breeding and Genetics, Sichuan Agriculture University, Ya'an, Sichuan 625014, China

Tang ZongXiang
State Key Laboratory of Plant Breeding and Genetics, Sichuan Agriculture University, Ya'an, Sichuan 625014, China

Ren ZhengLong
State Key Laboratory of Plant Breeding and Genetics, Sichuan Agriculture University, Ya'an, Sichuan 625014, China

School of Life Science and Technology, University of Electronic Science and Technology of China, Chengdu 610054, China

Zhang HuaiQiong
State Key Laboratory of Plant Breeding and Genetics, Sichuan Agriculture University, Ya'an, Sichuan 625014, China

M. H. Pahlavani
Department of Plant Breeding and Biotechnology, Gorgan University of Agricultural Sciences and Natural Resources, P.O. Box 386, Gorgan, Iran

S. E. Razavi
Department of Plant Protection, Gorgan University of Agricultural Sciences and Natural Resources, P.O. Box 386, Gorgan, Iran

F. Kavusi
BS students of Plant Breeding; Gorgan University of Agricultural Sciences and Natural Resources, P.O. Box 386, Gorgan, Iran

M. Hasanpoor
BS students of Plant Breeding; Gorgan University of Agricultural Sciences and Natural Resources, P.O. Box 386, Gorgan, Iran

P. N. Njau
Kenya Agricultural Research Institute (KARI) Njoro, Kenya

R. Wanyera
Kenya Agricultural Research Institute (KARI) Njoro, Kenya

G. K. Macharia
Kenya Agricultural Research Institute (KARI) Njoro, Kenya

J. Macharia
Egerton University Njoro. P. O. Box 536 Njoro, Kenya

R. Singh
CIMMYT Mexico. Apdo - Postal 6-641, 06000 Mexico, DF

B. Keller
Institute of Plant Biology, University of Zurich, Switzerland

Salah B. Mohamed Ahmed
Department of Crop Science, Faculty of Agriculture and
Environmental Sciences, University of Gedaref, P.O. Box
449, Gedaref, Sudan

Abdel Wahab H. Abdella
Department of Agronomy, Faculty of Agriculture,
University of Khartoum, P. O. Box 13314, Shambat, Sudan

M. I. B. Efombagn
Institute of Agricultural Research for Development
(IRAD), P. O. Box 2067 or 2123, Yaoundé, Cameroon

O. Sounigo
CIRAD, UPR31, 34398 Montpellier Cedex 5, France

S. Nyassé
Institute of Agricultural Research for Development
(IRAD), P. O. Box 2067 or 2123, Yaoundé, Cameroon

M. Manzanares-Dauleux
Agrocampus de Rennes, 65 Rue de St Brieuc, 35042,
Rennes, France

A. B. Eskes
CIRAD, UPR31, 34398 Montpellier Cedex 5, France

M. J. Mahasi
Kenya Agricultural Research Institute (KARI), P. O.
Private Bag, Njoro, Kenya

F. N. Wachira
Egerton University, P. O. Box 536, Njoro, Kenya

R. S. Pathak
Egerton University, P. O. Box 536, Njoro, Kenya

T. C. Riungu
KARI – Muguga South P.O. Box 30148 Nairobi - Kenya

Peng-fang Zhu
Forestry Faculty, Shenyang Agricultural University,
Shenyang, China, 110161, China

Yu-tang Wei
Horticulture Faculty, Shenyang Agricultural University,
Shenyang, China, 110161, China

L. N. Gitonga
Kenya Agricultural Research Institute, National
Horticultural Research Center, P. O. Box 01000 - 220,
Thika

A. W. T. Muigai
Department of Botany, Jomo Kenyatta University of
Agriculture and Technology, P. O. Box 00200-62000,
Nairobi

E. M. Kahangi
Department of Horticulture, Jomo Kenyatta University
of Agriculture and Technology, P. O. Box 00200-62000,
Nairobi

K. Ngamau
Department of Horticulture, Jomo Kenyatta University
of Agriculture and Technology, P. O. Box 00200-62000,
Nairobi

S. T. Gichuki
Kenya Agricultural Research Institute, Biotechnology
Center, P. O. Box 00200-57811, Nairobi

M. Cherif
Institut National Agronomique de Tunisie. 43 Avenue
Charles Nicolle. 1082 Cité Mahrajène, Tunis, Tunisia

S. Rezgui
Institut National Agronomique de Tunisie. 43 Avenue
Charles Nicolle. 1082 Cité Mahrajène, Tunis, Tunisia

P. Devaux
Laboratoire de Biotechnologie, Florimond Desprez, PB
41, 59242 Capelle en Pévèle, France

M. Harrabi
Institut National Agronomique de Tunisie. 43 Avenue
Charles Nicolle. 1082 Cité Mahrajène, Tunis, Tunisia

Anim-Kwapong Esther
Cocoa Research Institute of Ghana, P. O. Box 8, New
Tafo-Akim, Ghana

Boamah Adomako
Cocoa Research Institute of Ghana, P. O. Box 8, New
Tafo-Akim, Ghana

Sunday E. Aladele
National Centre for Genetic Resources and Biotechnology
(NACGRAB), P. M. B. 5382, Moor Plantation, Ibadan,
Nigeria

Firdissa Eticha
National Wheat Research Project, Kulumsa Agricultural
Research Center, P. O. Box 489, Asella, Ethiopia

Heinrich Grausgruber
Department of Applied Plant Sciences and Plant
Biotechnology, BOKU–University of Natural Resources
and Applied Life Sciences, Gregor-Mendel-Str. 33, 1180
Vienna, Austria

Emmerich Berghoffer
Department of Food Science and Biotechnology, BOKU–
University of Natural Resources and Applied Life
Sciences, Mutthgasse, Vienna, Austria

L. Chaudhary
Department of Biotechnology Engineering, Ambala College of Engineering and Applied Research, Mithapur, Ambala Cantt., Haryana, India- 133 101

A. Sindhu
Department of Botany and Plant Physiology, CCS Haryana Agricultural University, Hisar-125 004, India

M. Kumar
Department of Botany and Plant Physiology, CCS Haryana Agricultural University, Hisar-125 004, India

R. Kumar
Department of Botany and Plant Physiology, CCS Haryana Agricultural University, Hisar-125 004, India

M. Saini
Department of Plant, Soil and Agricultural System, Southern Illinois University, Carbondale-62901, USA

Yongle Li
Plant Breeding, Technische Universität München/Centre of Life and Food Sciences Weihenstephan, 85350 Freising, Germany

Sankalp Bhosale
Institute for Plant Breeding, Seed Science, and Population Genetics, University of Hohenheim, 70593 Stuttgart, Germany

Bettina I. G. Haussmann
International Crops Research Institute for the Semi-Arid Tropics (ICRISAT), BP 12404, Niamey, Niger

Benjamin Stich
Max Planck Institute for Plant Breeding Research, 50829 Cologne, Germany

Albrecht E. Melchinger
Institute for Plant Breeding, Seed Science, and Population Genetics, University of Hohenheim, 70593 Stuttgart, Germany

Heiko K. Parzies
Institute for Plant Breeding, Seed Science, and Population Genetics, University of Hohenheim, 70593 Stuttgart, Germany

Sunday Clement Olubunmi Makinde
Department of Botany, Faculty of Science, Lagos State University, Ojo Campus, P.O Box 001, LASU Post Office Ojo, Lagos State, Nigeria

Omolayo Johnson Ariyo
Department of Plant Breeding and Seed Technology, College of Plant Science University of Agriculture Abeokuta, P. M. B. 2240 Abeokuta, Ogun State, Nigeria

C. Bermejo
CONICET, Zavalla, Argentina

V. P. Cravero
CONICET, Zavalla, Argentina

F. S. López Anido
Department of Plant Breeding, Rosario National University, (UNR), CC 14, Zavalla S2125ZAA, Argentine

E. L. Cointry
Department of Plant Breeding, Rosario National University, (UNR), CC 14, Zavalla S2125ZAA, Argentine

Abrar B. Yasin
Department of Genetics and Plant Breeding, College of Agriculture, Allahabad Agricultural Institute, Deemed University, Naini Allahabad- 211007 (U. P.), India

Shubhra Singh
Department of Genetics and Plant Breeding, College of Agriculture, Allahabad Agricultural Institute, Deemed University, Naini Allahabad- 211007 (U. P.), India

Venkateswarlu Yadavalli
Department of Biochemistry, School of Life Sciences, University of Hyderabad. Hyderabad (AP), India-500 046, India
Centre for Plant Molecular Biology and Biotechnology, Tamil Nadu Agricultural University, Coimbatore, Tamilnadu, India-641003, India

Gajendra P. Narwane
Centre for Plant Molecular Biology and Biotechnology, Tamil Nadu Agricultural University, Coimbatore, Tamilnadu, India-641003, India

P. Nagarajan
Centre for Plant Molecular Biology and Biotechnology, Tamil Nadu Agricultural University, Coimbatore, Tamilnadu, India-641003, India

M. Bharathi
Department of Entomology, Tamil Nadu Agricultural University, Coimbatore, Tamilnadu, India-641003, India

Asif M. Iqbal
Division of Plant Breeding and Genetics, Sher-e-Kashmir University of Agricultural Sciences and Technology of Kashmir, Shalimar, 191121, J and K, India

F. A. Nehvi
Division of Plant Breeding and Genetics, Sher-e-Kashmir University of Agricultural Sciences and Technology of Kashmir, Shalimar, 191121, J and K, India

Shafiq A. Wani
Division of Plant Breeding and Genetics, Sher-e-Kashmir University of Agricultural Sciences and Technology of Kashmir, Shalimar, 191121, J and K, India

H. Qadri
Division of Plant Breeding and Genetics, Sher-e-Kashmir University of Agricultural Sciences and Technology of Kashmir, Shalimar, 191121, J and K, India

Z. A. Dar
Division of Plant Breeding and Genetics, Sher-e-Kashmir University of Agricultural Sciences and Technology of Kashmir, Shalimar, 191121, J and K, India

Aijaz A. Lone
Division of Plant Breeding and Genetics, Sher-e-Kashmir University of Agricultural Sciences and Technology of Kashmir, Shalimar, 191121, J and K, India

Joginder Singh
Department of Economics and Sociology, Punjab Agricultural University, Ludhiana, India

Asghar Ebadi segherloo
Department of Plant Breeding, Faculty of Agriculture, Tarbiat Modarres University, Tehran, Iran

Sayyed Hossain Sabaghpour
Dry land Agricultural Research Institute, Kermanshah, Iran

Hamid Dehghani
Department of Plant Breeding, Faculty of Agriculture, Tarbiat Modarres University, Tehran, Iran

Morteza kamrani
Department of Agronomy and Plant Breeding, Faculty of Agriculture, University of Tabriz, Tabriz, Iran

D. Oppong-Sekyere
Department of Crop and Soil Sciences, KNUST-Kwame Nkrumah University of Science and Technology,Kumasi, Ghana

R. Akromah
Department of Crop and Soil Sciences, KNUST-Kwame Nkrumah University of Science and Technology,Kumasi, Ghana

E. Y. Nyamah
Department of Horticulture, KNUST-Kwame Nkrumah University of Science and Technology, Kumasi, Ghana

E. Brenya
CRIG-Cocoa Research Institute of Ghana, P. O. Box 8, New Tafo-Akim, Ghana

S. Yeboah
Department of Crop and Soil Sciences, KNUST-Kwame Nkrumah University of Science and Technology,Kumasi, Ghana

Wenjing Pang
Entomology and Nematology Department, University of Florida, Gainesville, FL 32611-0620, United States

William T. Crow
Entomology and Nematology Department, University of Florida, Gainesville, FL 32611-0620, United States

Kevin E. Kenworthy
Agronomy Department, University of Florida, Gainesville, FL 32611-0500, United States

M. Takundwa
Department of Biological Sciences, University of Namibia, P. Bag 13301, Windhoek, Namibia

E. Nepolo
Department of Biological Sciences, University of Namibia, P. Bag 13301, Windhoek, Namibia

P. M. Chimwamurombe
Department of Biological Sciences, University of Namibia, P. Bag 13301, Windhoek, Namibia

A. C. Cullis
Department of Biology, Case Western Reserve University, Cleveland, Ohio, USA

M. A. Kandawa-Schulz
Department of Chemistry and Biochemistry, University of Namibia,Windhoek, Namibia

K. Kunert
Department of Plant Science, University of Pretoria, 0001 South Africa